T0227820

Lightwave Engineering

OPTICAL SCIENCE AND ENGINEERING

Founding Editor
Brian J. Thompson
University of Rochester
Rochester, New York

RECENTLY PUBLISHED

Lightwave Engineering, *Yasuo Kokubun*

Handbook of Optical and Laser Scanning, Second Edition, *Gerald F. Marshall and Glenn E. Stutz*

Computational Methods for Electromagnetic and Optical Systems, Second Edition, *John M. Jarem and Partha P. Banerjee*

Optical Methods of Measurement: Wholefield Techniques, Second Edition, *Rajpal S. Sirohi*

Optoelectronics: Infrared-Visible-Ultraviolet Devices and Applications, Second Edition, *edited by Dave Birtalan and William Nunley*

Photoacoustic Imaging and Spectroscopy, *edited by Lihong V. Wang*

Polarimetric Radar Imaging: From Basics to Applications, *Jong-Sen Lee and Eric Pottier*

Near-Earth Laser Communications, *edited by Hamid Hemmati*

Laser Safety: Tools and Training, *edited by Ken Barat*

Slow Light: Science and Applications, *edited by Jacob B. Khurgin and Rodney S. Tucker*

Dynamic Laser Speckle and Applications, *edited by Hector J. Rabal and Roberto A. Braga Jr.*

Biochemical Applications of Nonlinear Optical Spectroscopy, *edited by Vladislav Yakovlev*

Tunable Laser Applications, Second Edition, *edited by F. J. Duarte*

Optical and Photonic MEMS Devices: Design, Fabrication and Control, *edited by Ai-Qun Liu*

The Nature of Light: What Is a Photon?, *edited by Chandrasekhar Roychoudhuri, A. F. Kracklauer, and Katherine Creath*

Introduction to Nonimaging Optics, *Julio Chaves*

Introduction to Organic Electronic and Optoelectronic Materials and Devices, *edited by Sam-Shajing Sun and Larry R. Dalton*

Fiber Optic Sensors, Second Edition, *edited by Shizhuo Yin, Paul B. Ruffin, and Francis T. S. Yu*

Terahertz Spectroscopy: Principles and Applications, *edited by Susan L. Dexheimer*

Photonic Signal Processing: Techniques and Applications, *Le Nguyen Binh*

Smart CMOS Image Sensors and Applications, *Jun Ohta*

Organic Field-Effect Transistors, *Zhenan Bao and Jason Locklin*

Coarse Wavelength Division Multiplexing: Technologies and Applications, *edited by Hans Joerg Thiele and Marcus Nebeling*

Microlithography: Science and Technology, Second Edition, *edited by Kazuaki Suzuki and Bruce W. Smith*

Physical Properties and Data of Optical Materials, *Moriaki Wakaki, Keiei Kudo, and Takehisa Shibuya*

*Please visit our website **www.crcpress.com** for a full list of titles*

Lightwave Engineering

Yasuo Kokubun

CRC Press
Taylor & Francis Group
Boca Raton   London   New York

CRC Press is an imprint of the
Taylor & Francis Group, an **informa** business

CRC Press
Taylor & Francis Group
6000 Broken Sound Parkway NW, Suite 300
Boca Raton, FL 33487-2742

First issued in paperback 2017

© 2013 by Taylor & Francis Group, LLC
CRC Press is an imprint of Taylor & Francis Group, an Informa business

No claim to original U.S. Government works

Version Date: 20120709

ISBN 13: 978-1-4200-4648-9 (hbk)
ISBN 13: 978-1-138-07203-9 (pbk)

This book contains information obtained from authentic and highly regarded sources. Reason-
able efforts have been made to publish reliable data and information, but the author and publisher
cannot assume responsibility for the validity of all materials or the consequences of their use. The
authors and publishers have attempted to trace the copyright holders of all material reproduced in
this publication and apologize to copyright holders if permission to publish in this form has not
been obtained. If any copyright material has not been acknowledged please write and let us know so
we may rectify in any future reprint.

Except as permitted under U.S. Copyright Law, no part of this book may be reprinted, reproduced,
transmitted, or utilized in any form by any electronic, mechanical, or other means, now known or
hereafter invented, including photocopying, microfilming, and recording, or in any information
storage or retrieval system, without written permission from the publishers.

For permission to photocopy or use material electronically from this work, please access www.
copyright.com (http://www.copyright.com/) or contact the Copyright Clearance Center, Inc.
(CCC), 222 Rosewood Drive, Danvers, MA 01923, 978-750-8400. CCC is a not-for-profit organiza-
tion that provides licenses and registration for a variety of users. For organizations that have been
granted a photocopy license by the CCC, a separate system of payment has been arranged.

Trademark Notice: Product or corporate names may be trademarks or registered trademarks, and
are used only for identification and explanation without intent to infringe.

Visit the Taylor & Francis Web site at
http://www.taylorandfrancis.com

and the CRC Press Web site at
http://www.crcpress.com

Contents

Part I   Introduction

Part II   Description of Light Propagation through Electromagnetism

List of Figures

List of Tables

Preface

Since the invention of lasers and optical fibers, optoelectronics have developed rapidly and have been used in a very broad range of fields such as optical communication, optical disk storage, optical information processing, optical sensors, and laser processing.

I decided to entitle this book *Lightwave Engineering* as a basic academic discipline that deals with the behavior of electromagnetic waves and the propagation of light that forms the basis of optoelectronics.

"Lightwave engineering" is a term that may be unfamiliar to readers, but this book was published as one in a series of books called the "Opto-Electronics Series" in Japan. It played a role in the series by explaining the basic theories on the behavior of electromagnetic waves of light, taking into consideration basics such as the propagation and guided waves of laser light and their spatial and temporal behaviors across a wide range of application areas in optoelectronics.

Clues to understand optical propagation are not given by merely explaining optical propagation using mathematical equations, but rather the author also has thought of ways wherein the state of the propagation can be drawn as an image in the mind. However, there is a saying that "natural phenomenon can be explained by the language of mathematics," so if we cannot first derive the meaning of such phenomena using mathematical equations, we will not be able to accurately understand them. Consequently, this book describes the full mathematical derivations as much as possible and takes care to discuss the physical meaning of such mathematical equations. Therefore, this book is full of mathematical equations, and readers who pick up the book may think at first that it is difficult. However, the mathematical equations were carefully derived and are not really that difficult; so if readers continue to read on, they will be able to understand them.

However, by only conceptually explaining the basic theories, we will not be able to understand the operating principles and key points of instruments that actually make use of laser light. So, I took care in relating the explanations with problems in optical communication, optical disk storage, and other laser light applications. Consequently, after explaining the concept, the author has included several sample problems tailored to illustrate the concept to make it easier to understand quantitatively.

These sample problems also have another significance. Since merely explaining the basic theory is boring, the author added the sample problems in such a way that the readers will be able to think of them as explaining the next concept that is derived from a previous concept.

In regular textbooks, there are sections that appear to explain certain theorems, but in this book, theorems are sometimes explained in the sample problems.

It is the author's hope that readers will solve these sample problems by themselves. The solutions to the sample problems are written immediately after the problem, but it is important for readers to first solve the problems by themselves to be able to understand them.

The objective of this book is to provide a wide range of depth of contents in levels from undergraduate to graduate level, and in addition, the book contains material that would serve as a review of the basic theories for professional engineers. The reason for this is related to the contents of the lectures being given by the author at his university.

Keeping in mind the continuous learning from undergraduate to graduate school, the author decided that this book will explain the same content in a spiral manner, like that of climbing a spiral staircase. In other words, the same concept will first be explained using a simple example so that it will spring to mind as an image, and then, the same content will be drilled down further, and the theory will be expanded using a more detailed mathematical equation.

The embodiment of this approach is in the division of the discussion into Part I and Part II with the first half corresponding to undergraduate lectures and the latter half corresponding to graduate lectures. Moreover, since the discussion in the first half starts from the explanation of the concept of waves, there are some parts that describe content from physics at the high school level. Thus, those readers who thought at first that the book contained a lot of equations may next question why the book mentions such easy concepts. However, explanations have to be laid out in a successive manner to some extent, and even content that is very easy from the start may help readers move on to more difficult concepts.

In Part II of the book, I added more difficult concepts and assumed that professional engineers would use that part of the book as a handbook, and so I wrote many formulas on optical propagation.

In the practice exercises, hints are given as much as possible and supplementary explanations are also included. Nevertheless, due to space limitations, I had to omit content that I would have wanted to include such as the propagation in anisotropic medium, electro-optical effects, magneto-optical effects, and the operating principle of simple optical devices. For these, I want the reader to refer to the respective technical books.

As for the units, unless they are confused with other symbols, the units of the physical quantities in the text of this book are shown without any brackets. When there is a need to write the units inside the equation, they will be displayed in brackets [ ]. The units of the physical quantities use the SI unit system.

Moreover, in describing the propagation of optical waves, the sinusoidal wave is expressed as a complex number. However, in optics, which was developed as a field of physics, the imaginary unit is expressed as i, and the phase change is expressed as $e^{i(\beta z - \omega t)}$. At the outset of the invention of lasers during a stage in the development of optoelectronics, in the fashion by which electronic engineers dealt with light, the imaginary unit was expressed as j (since in electronics engineering, current is expressed as i), and the phase change was usually denoted as $e^{j(\omega t - \beta z)}$. This book follows the latter style (with the imaginary unit as j and the phase change as $e^{j(\omega t - \beta z)}$), so please take note of this when comparing with equations in other books such as those on quantum mechanics.

Author Biography

Yasuo Kokubun received his BE degree from Yokohama National University, Yokohama, Japan, in 1975, and ME and Dr. Eng. degrees from the Tokyo Institute of Technology, Tokyo, Japan, in 1977 and 1980, respectively. After working with the Research Laboratory of Precision Machinery and Electronics, Tokyo Institute of Technology, as a research associate from 1980 to 1983, he joined the Yokohama National University as an associate professor in 1983 and is now a professor in the Department of Electrical and Computer Engineering. From 2006 to 2009, he served as dean of the Faculty of Engineering and is now the vice-president of Yokohama National University. His current research is in integrated photonics, including waveguide-type functional devices and three-dimensional integrated photonics, and also in optical fibers, including multi-core fibers. From 1984 to 1985, he worked with AT&T Bell Laboratories, Holmdel, New Jersey, as a visiting researcher and was engaged in the study of a novel waveguide on a semiconductor substrate (ARROW) for integrated optics. From 1996 to 1999, he served as project leader of the three-dimensional microphotonics project at the Kanagawa Academy of Science and Technology.

Professor Kokubun is a fellow of the Institute of Electrical and Electronics Engineers, a fellow of the Japan Society of Applied Physics, a fellow of the Institute of Electronics, Information and Communication Engineers, and a member of the Optical Society of America.

Part I

Introduction

1 Fundamentals of Optical Propagation

1.1 PARAMETERS AND UNITS USED TO DESCRIBE LIGHT

Light is an ultra high frequency electromagnetic wave. Light used for general illumination and imaging, which has a wide spectrum (includes wavelengths of many components) is called incoherent light, wherein interference is less likely, and focusing the light on a small spot is difficult. On the other hand, laser light, which is used as a light source for optical communication devices and optical disk storage, is a single wavelength spectrum, wherein interference can be strong, and focusing light at a wavelength-sized small spot is possible. This light posses different optical properties and is actually a collection of *photons*; the individual photons of specific frequencies (wavelengths) propagate at a speed of $c = 2.9979 \times 10^8$ m/s in vacuum. The nature of the photons can be roughly described by its energy and momentum, wherein the energy E of the photon with angular frequency ω (the frequency or oscillation frequency is $v = \frac{\omega}{2\pi}$) is

$$E = hv = \hbar\omega = \frac{hc}{\lambda} \quad \text{[J]}, \tag{1.1}$$

where $h = 6.626 \times 10^{-34}$ J·s and in Planck's constant $\hbar = \frac{h}{2\pi}$, and momentum p is

$$p = \hbar k_0 = \frac{h}{\lambda} \quad \text{[N·s]}, \tag{1.2}$$

where $k_0 = \frac{2\pi}{\lambda}$ is the propagation constant of plane waves in vacuum.

These photons gather to configure the "light" we usually get to see; however, though we are aware of the intensity of the light, we do not think of how many photons have gathered. This is because light that we can normally perceive contains a large number of photons and is not of a level where we can count only one or two photons.* Now, let us estimate the number of photons contained in a light of intensity that we can normally perceive. A 40-W straight tube fluorescent lamp with an estimated luminous efficiency of 20% will emit 8 W of power. As a lighting device, fluorescent lamps emit a spectrum that spans a wide range of visible light, but for simplicity, let us assume that it has a

---

* A value that can be counted using a natural number such as $1, 2, \ldots$ is called a discrete number, whereas a value that can be written with many digits after a decimal point like that of real numbers is called a continuous number.

wavelength of 0.5 μm. Supposing the length of the fluorescent lamp is 1 m and the emitted light diffuses cylindrically, then about $2 \times 10^{-4}\%$ of the emitted light will enter the human eye, with a pupil diameter of about 4 mm, which is at a distance of 1 m from the fluorescent lamp, and signifies that the energy of the light is 16 μW. This light is quite weak. Because the energy of one photon is 4×10^{-19} J according to Equation (1.1), 4×10^{13} photons enter the human eye per second. This number is so large that it cannot be treated as a discrete number (treating it as a discrete number would make no sense) and is usually indicated as a continuous number.

Now, what unit can be used for the optical intensity indicated by this continuous number? As a fluorescent lamp is a lighting device, the unit used is lux, but in optoelectronics watt [W] (alternatively, every time it becomes smaller by 10^{-3} times, then μW, nW, pW, etc.) is used.* Moreover, to indicate a relative value, for example, in Figure 1.1, wherein P_{in} is the optical power of the light incident on an optical circuit (transmission lines such as optical fibers are also fine) and P_{out} is the optical power of the light emitted from an optical circuit, the gain or loss of the optical circuit is represented by

$$\alpha = 10 \log_{10} \frac{P_{out}}{P_{in}} \quad \text{[dB]}. \tag{1.3}$$

This unit is read, "decibel."† If the value for Equation (1.3) is positive, then the optical intensity is amplified, whereas a negative value implies that it is attenuated. However, if the phenomenon is known beforehand to be attenuated, then the sign is omitted, and the value is often expressed as loss of 3 dB. This unit is convenient for calculating the total gain or loss in optical circuits that are connected in multiple cascading steps. Why is this so? If an optical circuit with a 50% loss is connected in two steps of cascading, the loss will be $0.5 \times 0.5 = 0.25$ or 25%, which means that repeated multiplication has to be done. On the contrary, if expressed in dB, 50% loss is -3 dB ($\log_{10} 2 = 0.301$), which after simple addition will yield $-3 - 3 = -6$ dB.

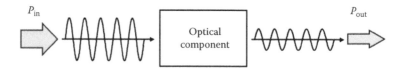

FIGURE 1.1   Input and output power of the optical circuit.

---

* 10^{-3} is milli (m), 10^{-6} is micro (μ), 10^{-9} is nano (n), 10^{-12} is pico (p), 10^{-15} is femto (f), and these are written before various units.
† "bel" comes from Alexander Graham Bell, the inventor of the telephone.

To apply this convention in calculating the absolute value of optical power, 1 mW will be used as the basis of dB value, and then the unit shown below can be used

$$10 \log_{10} \frac{P[\text{mW}]}{1[\text{mW}]} \quad [\text{dBm}]. \tag{1.4}$$

That is, 10 mW is $+10$ dBm, 0.1 mW is -10 dBm, and 1 nW is -60 dBm. If this unit is used when the power of the incident ray is P_{in} [dBm] and the loss of the optical circuit is α [dB], the optical power of the outgoing beam P_{out} [dBm] can be determined by

$$P_{\text{out}}[\text{dBm}] = P_{\text{in}}[\text{dBm}] - \alpha[\text{dB}]. \tag{1.5}$$

To convert the unit of P_{out} [dBm] to [mW], back calculate from Equation (1.4), and then use the formula $10^{\frac{P_{\text{out}}[\text{dBm}]}{10}}$ [mW].

In addition, log base 10 is not used to express relative value, but rather what is used is e (the base of the natural logarithm, approximately 2.718) and the equation

$$\alpha = \log_e \frac{P_{\text{out}}}{P_{\text{in}}} \quad [\text{Neper}]. \tag{1.6}$$

This unit is read "ney-per." Because attenuation and amplification that occur in nature is given a function of the form e^{α}, it is convenient to assign the above unit when using differential equations to describe attenuation and amplification. Many textbooks write [cm^{-1}] as the unit for gain g in an active layer of a semiconductor laser; however, this unit should be written correctly as [Neper/cm] or [Neper/mm] if SI units are used. The conversion factor, 1 Neper = 4.34 dB, can be used to convert dB to Neper and vice versa ($\log_{10} x = \log_{10} e \cdot \log_e x = 0.43429 \log_e x$).

Next, let us think about the wavelength and frequency of light. The speed of light in vacuum is $c = 2.9979 \times 10^8$ m/s, and the frequency ν [Hz] of light with wavelength λ [μm] is $\nu = \frac{c}{\lambda}$. According to this equation, the frequency of an infrared light with a wavelength of 1.550 μm is 193.4 THz.* In comparison, this frequency is higher by five orders of magnitude than that of radio waves in cellular phones. The generation and detection of such high-frequency waves are not done using antennas, but rather with semiconductor lasers and semiconductor photodetectors, wherein interaction takes place between photons and materials. The wavelength range of light that we can normally perceive is

---

* THz is 10^{12} Hz and is read "terahertz." Every time the value becomes bigger by 10^3 times, the unit will be assigned a specific prefix such that 10^3 kilo (k), 10^6 is mega (M), 10^9 is giga (G), 10^{12} is tera (T), and 10^{15} is peta (P).

much shorter, as this is a light with a wavelength of about 0.375–0.75 μm. This wavelength range is called the visible wavelength range. The wavelength of light source for optical disk memories such as CDs and DVDs should be as short as possible, which is advantageous as the light-focusing spots can be made smaller. Hence, the visible wavelength range (currently, the red laser with a wavelength of approximately 0.6 μm) is used. On the other hand, the low loss wavelength range of optical fiber is the long infrared wavelength range, with a wavelength that is longer than the visible wavelength range as shown in Figure 1.2.

PROBLEM 1-1

As shown in Figure 1.2, optical fiber loss is about 0.16 dB/km for infrared light in the vinicity of a wavelength of range 1.55 μm. When laser light with an output power of 2 mW and a wavelength of 1.55 μm is incident on an optical fiber 100 km in length, how much light is emitted at the output end?

SOLUTION 1-1

The total loss α_T of an optical fiber 100 km in length is $\alpha_T = 0.16 \times 100 = 16$ dB. On the other hand, as the power of the incident light is 3 dBm, the

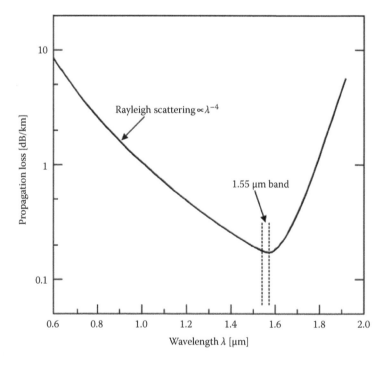

FIGURE 1.2   Loss spectrum of optical fiber.

power of the emitted light will be −13 dBm. Consequently, 0.05 mW (attenuation is $\frac{1}{10}$ for −10 dB, and because half is attenuated at −3 dB, the total will be $\frac{1}{20}$ of 1 mW) will be emitted.   **[End of Solution]**

PROBLEM 1-2

What is the output power when the length of the optical fiber is 1100 km?

SOLUTION 1-2

In this case, the emitted light of −163 dBm is very weak. Weak light can be expressed mathematically by any number, but 5×10^{-20} W is an extremely weak light, and the energy of 1 photon with a wavelength of 1.55 μm is 1.28×10^{-19} J, only 1 photon arrives in about 2.56 sec. This exceeds the limit, wherein light can be expressed as a continuous number, and we have entered a world ruled by quantum and statistical mechanics, which is called photon counting.   **[End of Solution]**

1.2 OPTICAL COHERENCE

Even if the frequency of a single photon is the same, a large number of photons gather, and the differences in the nature of the electromagnetic waves will surface. For example, monochromatic light sources such as sodium lamps, which are used in an atomic emission spectrum, have a spectrum with very sharp peaks, so even if the light is focused using lenses, the focused spot cannot be so small compared with wavelength. Moreover, even if the light is emitted from a double heterojunction* semiconductor, such as AlGaAs/GaAs or GaInAsP/InP, laser oscillation is not induced, and it will have the spectrum spreading of a light-emitting diode. Thus, light can neither be effectively focused into a single-mode fiber even with the use of a lens nor be used as a light source for CDs and DVDs. This light, generated from the transition of atoms between different energy levels and the recombination of electrons and holes in semiconductors, is called a *spontaneous emission* as individual photons will come out arbitrarily. On the other hand, laser lights such as semiconductor laser and He–Ne laser can be focused using a lens to a spot as small as a wavelength. What is the source of this difference?

First, the process of generating a spontaneous emission involves the transition of the atom from an excited state (excited level) to the ground state (ground level) (or the recombination of electrons and holes in semiconductors),

---

* For semiconductor light sources, the optical waveguide structure discussed in Section 1.3 is fabricated using a semiconductor heterojunction structure of different band gap energies and is used for the confinement of both the light and the carrier. For more details, please read the other textbooks mentioned at the end of the book.

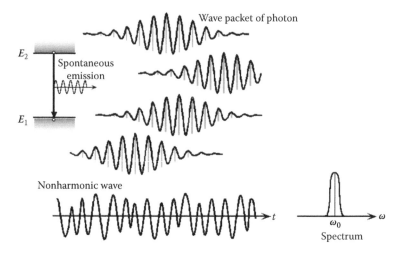

FIGURE 1.3 Temporal coherence of spontaneous emission.

with the photon being generated by the energy difference between states (almost equivalent to the band gap energy in semiconductors) at an angular frequency determined by Equation (1.1). As shown in Figure 1.3, the electric and magnetic fields of each photon will grow up and be attenuated during the period of transition between states (normally in the order of picoseconds). This may not be a continuous wave, but because it is an electromagnetic wave, phase is added as a basic parameter in addition to the frequency.* During a spontaneous emission, individual photons will be emitted arbitrarily regardless of their phases, so the average frequency of light, which is composed of a collection of these photons, will be almost equivalent to the frequency determined by Equation (1.1); however, phase will greatly change in an irregular manner, and the wave will not be sinusoidal, but rather it will be a so-called "irregular" wave. This kind of light is called *incoherent light*.

On the other hand, when the transition from the excited state to the ground state of the atom is not natural but is rather triggered by an incident photon with energy equal to the energy difference between states, such electromagnetic radiation is called a *stimulated emission*. During a stimulated emission, the new phase of the photon generated during the transition is the same as the phase of the photon that became the basis to induce transition. Therefore, the number of photons having the same energy is amplified, and they are all of the same phase. Furthermore, as shown in Figure 1.4, a nearly sinusoidal, electromagnetic wave of the same phase can be obtained on laser oscillation when feedback is applied by a resonator. This kind of light is called a *coherent light*.

---

* Frequency is related to the duration time of the photon and the uncertainty.

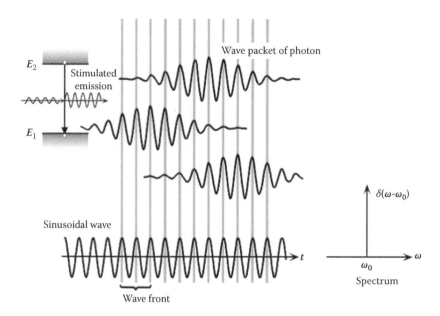

FIGURE 1.4 Temporal coherence of laser light.

Now, the concept of coherence mentioned earlier is the one viewed from the change of amplitude of light as a one-dimensional function of time, and thus it is called the temporal coherence. However, light is an electromagnetic wave, and the amplitude is also a function of spatial coordinates, thus there is also the concept of spatial coherence. Well, how should we think of spatially coherent and incoherent light? We can understand it simply by the difference in the focused light resulting from the spatial propagation of light through a lens. First, let us think of the case when radiation from light bulbs is focused at a focal point using a lens as shown in Figure 1.5.

Because radiation emitted from the filament of the light bulb is from a heated object, similar to black-body radiation, light of various wavelengths is emitted. To simplify, let us assume that a monochromatic light (as in a sodium lamp) is generated. Nevertheless, in sources of incoherent light, the photon emitted from the atom at the end of the filament and the other photons emitted right beside this atom or from the opposite end of the filament will be emitted arbitrarily without synchronizing their phase relationship with each other. Consequently, the photon emitted from the point at the end of the filament, S_1, as shown in Figure 1.5, will propagate independently as a single photon (as a spherical wave, if the atom's transition between excited and ground states is a transition between S orbitals). A portion of this wave front will undergo transformation with the lens (will be refracted when talking about light rays)

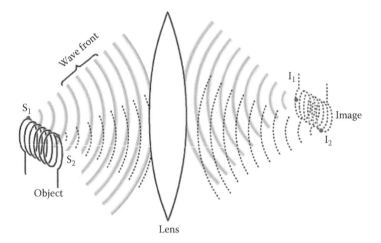

FIGURE 1.5   Spatial coherence of spontaneous emission.

and will be focused at point I_1, which is near the focal point of the lens. However, the photon that was emitted at point S_2, which is at the opposite side of the filament, also has an erratic phase relationship with another photon, and so will propagate independently and be focused at point I_2. Then, each of the photons that are emitted independently from various places in the filament are focused at separate locations near the focal point of the lens according to the difference in the original radiation position. Once these focal point positions are connected, the image of the filament can be created as shown in Figure 1.5. Consequently, even if an incoherent light is focused using a lens, the image near the focal point cannot be focused to a small spot that is the size of a wavelength.

In contrast, light of the same phase is emitted by stimulated emission for coherent light (laser light); thus, a spatially constant phase relationship can be formed. That is, a wave front (equiphase plane) is formed even with the numerous photons. If such a wave front becomes a parallel beam as shown in Figure 1.6, it will become a single spherical wave after wave front transformation occurs with a lens and can then be focused to a small spot that is the size of a wavelength. Consequently, when the light from the source is focused on a single-mode fiber core (the diameter is about the size of the wavelength) or when the light beam is focused on the pit (the width is about the size of a wavelength) of an optical disk, a coherent laser light can be used as a light source.

Based on the explanation given earlier, incoherent light and coherent light are classified from the standpoint of the focusing property of the lens; however, interference of incoherent light is still possible with an interferometer. This concept of coherence does not mean that the presence or absence of

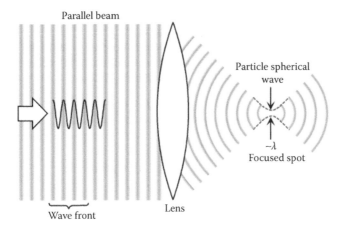

FIGURE 1.6 Spatial coherence of laser light.

coherence can only be expressed by the numbers 0 or 1, but rather it can be expressed as a continuous number depending on the coherence length (correlation length). The explanation for this is left to other textbooks on wave optics.

Unless otherwise noted, coherent light will be dealt within the following sections of this book.

1.3 FUNDAMENTAL EQUATIONS OF THE ELECTROMAGNETIC FIELDS AND PLANE WAVES

1.3.1 ELECTROMAGNETIC WAVE EQUATIONS

The electromagnetic field distribution of electromagnetic waves is derived from Maxwell's equations. Maxwell's equations consist of the following four equations, which relate the electric field E, magnetic field H, electric flux density D, the magnetic flux density B, current density J, and charge density ρ.[*]

$$\nabla \times E = -\frac{\partial B}{\partial t}, \tag{1.7}$$

$$\nabla \times H = J + \frac{\partial D}{\partial t}, \tag{1.8}$$

$$\nabla \cdot D = \rho, \tag{1.9}$$

$$\nabla \cdot B = 0. \tag{1.10}$$

---

[*] Maxwell's equations have three additional auxiliary equations ($D = \varepsilon E, B = \mu H, J = \sigma E$) for a total of seven equations. However, if an electromagnetic field changes with time like that of electromagnetic waves, two of the equations $\nabla \cdot D = 0$ and $\nabla \cdot B = 0$ can be derived from Equations (1.7) and (1.8). (Refer to the footnote on p. 134 of Section 5.1.1 in Chapter 5.)

Here, if the propagation is under vacuum or inside a dielectric material (insulator), the conductivity $\sigma = 0$, and hence the current density $J = 0$. Moreover, as the medium is a dielectric material, there is no charge, and since it is not magnetic, the permeability is approximately equal to that in vacuum. If these conditions are written as equations, they will be as follows:

$$\sigma = 0 \quad \rightarrow \quad J = 0, \tag{1.11}$$

$$\rho = 0, \tag{1.12}$$

$$\mu = \mu_0. \tag{1.13}$$

With these conditions, Equations (1.7) to (1.10) will become

$$\nabla \times E = -\mu_0 \frac{\partial H}{\partial t}, \tag{1.14}$$

$$\nabla \times H = \varepsilon \frac{\partial E}{\partial t}, \tag{1.15}$$

$$\nabla \cdot E = 0, \tag{1.16}$$

$$\nabla \cdot H = 0. \tag{1.17}$$

Here, if the curl on both sides of Equation (1.14) is taken off

$$\nabla \times \nabla \times E = -\mu_0 \frac{\partial}{\partial t} (\nabla \times H)$$

$$\uparrow$$

Substitute with Equation (1.15)

$$= -\varepsilon \mu_0 \frac{\partial^2 E}{\partial t^2} \tag{1.18}$$

is derived. On the other hand, from the vector formula

$$\nabla \times \nabla \times E = \nabla \cdot (\nabla \cdot E) - \nabla^2 E$$

$$\uparrow$$

0 from Equation (1.16)

$$= -\nabla^2 E \tag{1.19}$$

is produced, and from Equations (1.18) and (1.19) the following *wave equation* is derived:

$$\nabla^2 E - \varepsilon \mu_0 \frac{\partial^2 E}{\partial t^2} = 0. \tag{1.20}$$

Thus, the next wave equation can be obtained exactly in the same way for the magnetic field.

$$\nabla^2 H - \varepsilon \mu_0 \frac{\partial^2 H}{\partial t^2} = 0. \tag{1.21}$$

By solving these equations and finding a general solution under appropriate boundary conditions, various forms of wave propagation are obtained.

1.3.2 PLANE WAVE PROPAGATION CONSTANT

PROBLEM 1-3

In the spatial distribution of electromagnetic waves, the simplest form is the *plane wave*. Consider a plane wave propagating at any direction with an angular frequency ω. Show that the propagation direction, the electric field vector, and the magnetic field vector are orthogonal to each other using Cartesian coordinates (x, y, z coordinates).

SOLUTION 1-3

A plane wave propagating at any direction in vacuum with a dielectric constant ε is expressed by

$$E = Ae_p \exp[j(\omega t - k \cdot r)], \tag{1.22}$$

where A is the amplitude, which is constant regardless of position, e_p is a unit vector showing the polarization direction (direction of the oscillation of the electric field), k is the *propagation vector* ($= k_x e_x + k_y e_y + k_z e_z$) with length equal to the propagation constant facing the direction of propagation, and r is the position vector ($= xe_x + ye_y + ze_z$).* First, substitute this equation in $\nabla \times E$ to obtain

$$\nabla \times E = -j(k \times E). \tag{1.23}$$

(Even without using the vector formula, this can be derived immediately using $\frac{\partial E}{\partial x} = -jk_x E$.) On the other hand, if the electric field has time dependency $e^{j\omega t}$, the magnetic field will also possess the same time dependency, and Equation (1.14) will become

$$\nabla \times E = -j\omega\mu_0 H. \tag{1.24}$$

The following equation can be obtained from Equations (1.23) and (1.24):

$$H = \frac{1}{\omega\mu_0}(k \times E). \tag{1.25}$$

That is to say that the electric field E and the magnetic field H are orthogonal. At the same time, this also shows that the magnetic field H is also orthogonal to the propagation vector k.

---

* Later in this book, the unit vectors for the x-axis direction, y-axis direction, and z-axis direction are expressed as e_x, e_y, and e_z, respectively.

Next, let us show that E is orthogonal to the propagation vector k. When Equation (1.22) is substituted to div E, the equation becomes

$$\nabla \cdot E = -jA(k \cdot e_p)\exp[j(\omega t - k \cdot r)]. \tag{1.26}$$

However, the right side of Equation (1.26) must be 0 according to Equation (1.16), thus

$$k \cdot e_p = 0 \tag{1.27}$$

is derived. This equation shows that the electric field and the direction of propagation are orthogonal. In other words, a plane wave is a transverse wave.* **[End of Solution]**

PROBLEM 1-4

Show the plane wave propagation constant k using the angular frequency ω, permittivity ε, and permeability μ.

SOLUTION 1-4

When Equation (1.22) is substituted to the first term $\nabla^2 E$ on the left side of the wave equation (1.20)

$$\nabla^2 E = -(k_x^2 + k_y^2 + k_z^2)E$$
$$= -k^2 E \tag{1.28}$$

is obtained.

On the other hand, when Equation (1.22) is substituted to the second term (the second-order derivative for time) on the left side of the wave equation (1.20), the equation becomes

$$-\varepsilon\mu_0\frac{\partial^2 E}{\partial t^2} = \omega^2\varepsilon\mu_0 E, \tag{1.29}$$

and from Equations (1.28) and (1.29) and the wave equation (1.20)

$$k = \omega\sqrt{\varepsilon\mu_0} \tag{1.30}$$

is obtained. This is the propagation constant of the plane wave propagating in a uniform dielectric medium with a dielectic constant ε. In vacuum, the dielectric constant is ε_0, so the propagation constant k_0 of a plane wave propagating in vacuum is

$$k_0 = \omega\sqrt{\varepsilon_0\mu_0}. \tag{1.31}$$

[End of Solution]

---

* Another solution would be to start with the wave equation (1.21) to obtain $k \times H = -\omega\varepsilon E$, and then using Equation (1.25), we will know that E, H, and k are orthogonal to each other.

Now, the two exercises above assume that direction of the propagation and the direction of the electric field are arbitrary. However, as the electric field E, the magnetic field H, and the propagation vector k are shown to be orthogonal to each other in Exercise 1-3, no generality is lost by taking the electric field to be polarized in the x-axis direction and the propagation direction for a plane wave propagating inside a uniform dielectric medium to be the z-axis, as shown in Figure 1.7. The direction of the electric field is called the polarization direction,* and in the case of Figure 1.7, it is polarized in the x-axis direction, so it is called x-polarized light (or x-polarized wave).[†] Because a plane wave is defined as a wave where the amplitude is uniform in a cross-section perpendicular to the propagation direction, $\frac{\partial A}{\partial x} = 0$ and $\frac{\partial A}{\partial y} = 0$ hold true. Consequently, the wave equation (1.20) with only z and t can be written as

$$\frac{\partial^2 E_x}{\partial z^2} - \varepsilon \mu_0 \frac{\partial^2 E_x}{\partial t^2} = 0. \qquad (1.32)$$

The general solution to this equation, with arbitrary constants A and B, gives

$$E_x = A \exp[j(\omega t - kz)] + B \exp[j(\omega t + kz)]. \qquad (1.33)$$

However, k is the propagation constant given in Equation (1.30). The first term at the right side of Equation (1.33) represents a forward traveling wave, and to maintain the same phase with an increase in time t, the propagation distance z also has to increase. (That is, the wave moves toward the positive z-axis direction.) On the other hand, the second term at the right side represents the backward traveling wave, and this wave moves toward the negative z-axis direction (corresponding to the reflected wave).

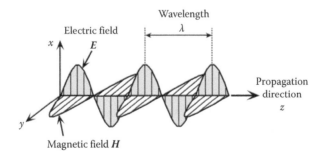

FIGURE 1.7 Electromagnetic field distribution of the plane wave.

---

* In textbooks, there are cases in which the direction of the magnetic field is defined as the polarization direction, so care must be taken.

[†] If the plane of vibration of the electric field is in one direction, then it is called a linear polarization.

Next, because the electric field is an x component and its direction of propagation is the z-axis direction, the magnetic field will be in the y-direction. Then, when Equation (1.15) is used

$$-\frac{\partial H_y}{\partial z} = \varepsilon \frac{\partial E_x}{\partial t} \tag{1.34}$$

can represent H_y with E_x, and when substituted with Equation (1.33) to calculate H_y, the next equation is obtained

$$H_y = \sqrt{\frac{\varepsilon}{\mu_0}}\{A \exp[j(\omega t - kz)] - B \exp[j(\omega t + kz)]\} \tag{1.35}$$

Here, if only the forward traveling wave or the backward traveling wave is considered, the amplitude of the magnetic field will have a constant proportional relationship, in particular, if the medium is vacuum and the dielectric constant $\varepsilon_0 = 8.85419 \times 10^{-12}$ [F/m] and the permeability constant $\mu_0 = 4\pi \times 10^{-7}$ [H/m] are used, a ratio of

$$Z_0 = \frac{E_x}{H_y} = \sqrt{\frac{\mu_0}{\varepsilon_0}} = 376.73 \quad [\Omega] \tag{1.36}$$

can be calculated. This is called the *vacuum impedance* (or the impedance of space). Moreover, it is important to be aware that the sign of the backward traveling wave is minus in Equation (1.35). This is because the direction of the Poynting vector (described later) is in the negative direction.

To make things simpler, let us consider only the forward traveling wave shown in Figure 1.7. Then, the electric field and the magnetic field are expressed by the following equations:

$$\begin{aligned} \boldsymbol{E} &= E_x \cdot \boldsymbol{e_x} \\ &= Ae^{j(\omega t - kz)} \cdot \boldsymbol{e_x}, \end{aligned} \tag{1.37}$$

$$\begin{aligned} \boldsymbol{H} &= H_y \cdot \boldsymbol{e_y} \\ &= \sqrt{\frac{\varepsilon}{\mu_0}} Ae^{j(\omega t - kz)} \cdot \boldsymbol{e_y}. \end{aligned} \tag{1.38}$$

Here, a wave front of a certain phase at time t_0 will be at position z_0 of the z-axis, as shown in Figure 1.8. At that same time, the point at which the phase has progressed by only 2π will become position $z_0 + \lambda_g$ with λ_g as the wavelength.

That is, since

$$\omega t_0 - kz_0 + 2\pi = \omega t_0 - k(z_0 + \lambda_g) \tag{1.39}$$

holds true, then

$$k = \frac{2\pi}{\lambda_g} \tag{1.40}$$

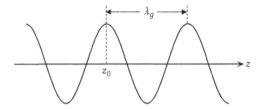

FIGURE 1.8   Phase change due to plane wave propagation.

Here λ_g is called the *cavity wavelength.** In a dielectric medium with the dielectric constant ε, λ_g is different from the wavelength λ in vacuum. In vacuum

$$k_0 = \omega\sqrt{\varepsilon_0\mu_0} = \frac{2\pi}{\lambda} \tag{1.41}$$

is obtained. λ is the wavelength in vacuum.

1.3.3   PROPAGATION VELOCITY AND POWER FLOW DENSITY OF A PLANE WAVE

PROBLEM 1-5

Find the propagation velocity of a plane wave.

SOLUTION 1-5

Let us assume the case in which the wave front with phase ϕ of a plane wave located at position z_0 at time t_0 moved to position $z_0 + \delta z$ at time $t_0 + \delta t$. This represents a wave front of the same phase resulting in

$$\phi = \omega t_0 - k z_0 = \omega(t_0 + \delta t) - k(z_0 + \delta z). \tag{1.42}$$

Because the propagation velocity is the ratio of the propagation distance δz and the time δt, it can be expressed as

$$v_p = \frac{\delta z}{\delta t} = \frac{\omega}{k}. \tag{1.43}$$

This velocity is called the *phase velocity.* Moreover, when Equation (1.43) is substituted to the k of Equation (1.30)

$$v_p = \frac{1}{\sqrt{\varepsilon\mu_0}} \tag{1.44}$$

---

* Cavity means a cavity resonator and refers to a microwave waveguide resonator. However, an optical resonator is also called a cavity. Also, the wavelength of light that propagates in a medium is called a cavity wavelength.

is obtained. In vacuum, ε is replaced with ε_0 and it follows that

$$c = \frac{1}{\sqrt{\varepsilon_0 \mu_0}}. \tag{1.45}$$

This is generally called the velocity of light, and using the electric constant $\varepsilon_0 = 8.85419 \times 10^{-12}$ [F/m] and the permeability of $\mu_0 = 4\pi \times 10^{-7}$ [H/m] in vacuum, 2.997925×10^8 [m/s] is obtained.*

[End of Solution]

PROBLEM 1-6

With the plane wave as an example, find the power flow density per unit area.

SOLUTION 1-6

In an electromagnetic wave, there is an energy flow in the direction of propagation, and the energy flow that flows per unit time through a unit cross-sectional area perpendicular to the direction of propagation is called the power flow density. As the power flow density is expressed by the Poynting vector, first let us derive that definition. Subtraction of the scalar product of E and Equation (1.8) from the scalar product of H and Equation (1.7) gives

$$H \cdot (\nabla \times E) - E \cdot (\nabla \times H) = -\left(E \cdot \frac{\partial D}{\partial t} + H \cdot \frac{\partial B}{\partial t} \right) - E \cdot J. \tag{1.46}$$

When the vector formula (see Appendix D, Equation (D.9)) is used here, the equation can be transformed to the following:

$$H \cdot (\nabla \times E) - E \cdot (\nabla \times H) = \nabla \cdot (E \times H). \tag{1.47}$$

By substituting Equation (1.46) to the left side of Equation (1.47) using $D = \varepsilon E$, $B = \mu H$, and $J = \sigma E$, Equation (1.46) transforms to the following:

$$\nabla \cdot (E \times H) - \sigma E^2 = -\frac{\partial}{\partial t} \left(\frac{1}{2} \varepsilon E^2 + \frac{1}{2} \mu H^2 \right). \tag{1.48}$$

The right side of Equation (1.48) represents the electric field energy and the magnetic field energy accumulated per unit volume.[†] Here, let us consider a three-dimensional space V enclosed by a closed surface S. Equation (1.48)

---

* Actually, the phase velocity of the plane wave in vacuum is defined according to the dielectric constant $\varepsilon_0 = \frac{10^7}{4\pi c^2}$.

[†] $E^2 = |E|^2 = E \cdot E$ and the time dependency $e^{j\omega t}$ still remains because this was not taken as an absolute value of a complex number.

is integrated over this three-dimensional area, and the following equation is obtained when the volume integral of divergence on the left side of the equation is rewritten by the surface integral on surface S according to Gauss' formula.*

$$\iint_S (E \times H) da + \iiint_V \sigma E^2 dr^3$$
$$= -\frac{\partial}{\partial t} \iiint_V \left(\frac{1}{2} \varepsilon E^2 + \frac{1}{2} \mu H^2 \right) dr^3. \tag{1.49}$$

The second term on the left side of this equation represents the power consumed as heat when there is a conductor inside this volume, whereas the right side represents the time rate of change of the sum of the electric field energy and the magnetic field energy stored in this three-dimensional area. The problem is with the first term on the left side, which represents the inflow of power (the sign in this case is positive but will be negative in the case of an outflow of power) through surface S. That is, Equation (1.49) means that the sum of the power that transforms into Joule heat inside a closed area and the power flowing as an electromagnetic radiation from the surface is equivalent to the rate of change of the electric and magnetic energies accumulated inside a closed area (because this will decrease, a negative sign is added to the right side of Equation (1.49)).

As the first terms of Equation (1.49) represents the electromagnetic power that is emitted from the surface of the closed area, the inner product of the vector S given by Equation (1.50) and the unit normal vector n of surface S is represented by the power emitted per unit area.

$$S = E \times H. \tag{1.50}$$

This vector is called the *Poynting vector.*

Because E and H included in Equation (1.50) have a temporal dependency[†] $e^{j\omega t}$, the real parts of E and H are used to determine the time-averaged power flow density to obtain

$$\frac{1}{T} \int_0^T S dt = \frac{1}{T} \int_0^T \left[\frac{1}{2}(E + E^*) \times \frac{1}{2}(H + H^*) \right] dt$$
$$= \frac{1}{4}[E \times H^* + E^* \times H] = Re\left[\frac{1}{2}(E \times H^*) \right] \tag{1.51}$$

---

* $\iiint_V \nabla \cdot A dr^3 \equiv \iint_S A da$ holds true for any vector A. Here, $\iint_S da$ represents the surface integral, whereas $\iiint_V dr^3$ represents the volume integral, where da is the infinitesimal surface element vector facing an outward direction perpendicular to the surface S.

† The Poynting vector may be thought to vibrate temporally with time dependency $e^{j2\omega t}$; however, this is a mistake as the time average is 0 here. For the time dependency of E and H, which is represented by $e^{j\omega t}$, the real part of the number is presumed to correspond to a physical quantity. To be exact, as the electric and the magnetic fields have a $\cos \omega t$ time dependency, the Poynting vector will have a $\cos^2 \omega t$ time dependency. As a consequence, power flows in a certain direction even when time averaged.

(the real part of any complex number z can be calculated using $Re[z] = \frac{1}{2}[z + z^*]$.) Here $T = \frac{2\pi}{\omega}$ is the period, whereas $*$ represents the complex conjugate. Because the terms $E \cdot H$ and $E^* \cdot H^*$ include $e^{j2\omega t}$ and $e^{-j2\omega t}$, it will become 0 when the time average is taken. However, the complex time dependency of the terms E^* and H^* involved in $E^* \cdot H$ and $E \cdot H^*$ become $e^{-j\omega t}$, thus the original time dependency $e^{j\omega t}$ is offset and becomes 1, and so when the time average is taken, it will still remain as 1. The inside of the real part symbol of Equation (1.51)

$$\tilde{S} = \frac{1}{2}(E \times H^*) \qquad (1.52)$$

is called the *complex Poynting vector*.

The $\frac{1}{2}$ placed on the right side of Equation (1.52) can be interpreted as taking the time average of $\cos^2 \omega t$.

Finally, from Equations (1.37) and (1.38), the complex Poynting vector in a plane wave is expressed by

$$\tilde{S} = \frac{1}{2}\sqrt{\frac{\varepsilon}{\mu_0}}A^2(e_x \times e_y) = \frac{1}{2}\sqrt{\frac{\varepsilon}{\mu_0}}A^2 e_z = \frac{n}{2Z_0}A^2 e_z, \qquad (1.53)$$

which means that a power of $\frac{1}{2}\sqrt{\frac{\varepsilon}{\mu_0}}A^2$ [W/m^2] per unit area is flowing in the z-direction (direction of propagation). This power flow contains an unknown coefficient A, but conversely, the amplitude constant can be normalized (so that it can become 1 [mW/cm^2]) by the power flow density using Equation (1.53). An unknown coefficient is also included in beam waves as well as in plane waves, and such beam waves can be normalized depending on the power that is carried by the beam.

[End of Solution]

1.4 REFLECTION AND REFRACTION OF PLANE WAVES

In this section, let us consider the reflection and refraction on the boundary surface of different media using a plane wave, which has the simplest spatial distribution pattern.

1.4.1 REFRACTIVE INDEX AND SNELL'S LAW

Let us suppose the case of a plane wave incident on a medium with a dielectric constant ε from vacuum (or air) as shown in Figure 1.9. When a straight line is drawn from the equiphase plane* of the plane wave to the extended direction of the Poynting vector, such a straight line represents the direction of propagation

---

* This is also called a wave front.

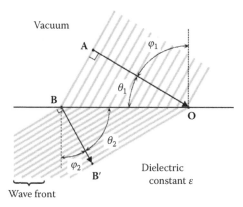

FIGURE 1.9 Refraction of a plane wave.

of that light and is called a ray.* The boundary of the vacuum and the medium is a plane, and the angle between the normal line and incident ray is φ_1, and the angle of the refracted ray and the normal line is φ_2.

During the period of time when a ray of light passing through point A on an equiphase plane reaches point O on the boundary surface, a ray of light that took off from point B on the boundary surface of the same coordinate phase plane as point A will reach point B'. That is, using the phase velocity of Equations (1.45) and (1.44), the relationship in the following equation is obtained:

$$\frac{\overline{AO}}{c} = \frac{\overline{BB'}}{v_p}. \tag{1.54}$$

Here, $\overline{AO}$ represents the distance between point A and point O (this notation is applied elsewhere with the same meaning). On the other hand, we can obtain

$$\overline{AO} = \overline{BO} \sin \varphi_1, \tag{1.55}$$
$$\overline{BB'} = \overline{BO} \sin \varphi_2. \tag{1.56}$$

using simple geometry. When these equations are substituted to Equation (1.54) and then rearranged

$$\frac{\sin \varphi_1}{\sin \varphi_2} = \frac{c}{v_p} = \sqrt{\frac{\varepsilon \mu_0}{\varepsilon_0 \mu_0}} = \sqrt{\frac{\varepsilon}{\varepsilon_0}} \tag{1.57}$$

---

* If the dielectric constant distributed in a medium changes along with the propagation of the wave front, the direction of the ray of light will also change.

is obtained. Because the *refractive index* n is defined as the ratio of the phase velocity c in vacuum to the phase velocity v_p in another medium, then

$$n \triangleq \frac{c}{v_p} = \sqrt{\frac{\varepsilon \mu_0}{\varepsilon_0 \mu_0}} = \sqrt{\frac{\varepsilon}{\varepsilon_0}} = \sqrt{\varepsilon_r}. \tag{1.58}$$

Then, Equation (1.57) expresses *Snell's law* (or the law of refraction) because the right-hand side of Equation (1.57) becomes exactly equal to the refractive index. In the equation, ε_r is the relative permittivity.* Moreover, when the angles $\theta_1 (= \frac{\pi}{2} - \varphi_1)$ and $\theta_2 (= \frac{\pi}{2} - \varphi_2)$ measured tangent from the boundary surface are used instead of angles φ_1 and φ_2 between the normal line to the boundary surface and the ray of light, and when the refractive index of the medium of the incoming light is expressed by n_1 and the refractive index of the medium after the light rays were refracted expressed by n_2, Snell's law is rewritten:

$$n_1 \cos \theta_1 = n_2 \cos \theta_2. \tag{1.59}$$

Equation (1.59) shows that the refraction angle can be calculated if the refractive index of the medium of the incoming side and the refractive index of the exiting side are known, even if the medium consists of several layers.

As the refractive index has been defined by Equation (1.58), the propagation vector of the plane wave in the medium with refractive index n can be expressed by $k = k_0 n$. Consequently, the term for the time and propagation distance dependency in Equation (1.22) can be expressed by

$$\exp[j(\omega t - n k_0 \cdot r)]. \tag{1.60}$$

Here, when a plane wave propagates only for a distance L from point P to Q in a medium with a refractive index n, the phase difference $\Delta \Phi_{PQ}$ between point P and Q, if the time is fixed, is given by

$$\Delta \Phi_{PQ} = \arg \left[\frac{E_{out}}{E_{in}} \right] = -j k_0 n L = -j k_0 S_{PQ}. \tag{1.61}$$

In Equation (1.61),

$$S_{PQ} = nL \tag{1.62}$$

is called the optical path length, and if k_0 is multiplied to this, the phase difference between the two points is obtained. This is the length that can actually be felt when the light wave is propagating.

---

* Here, it is necessary to note that the dielectric constant is 1 at the frequency of light. The dielectric constant published in other textbooks on electromagnetism to calculate the capacitance of a capacitor is given for much lower frequencies than that of light. Because the dielectric constant at the frequency of light is small, the refractive index cannot be calculated using this value.

1.4.2 AMPLITUDE REFLECTANCE AND POWER REFLECTIVITY

Snell's law is applicable only for the direction of the refraction, and information on the percentage of incident power transmitted (transmittance) and the
percentage reflected (reflectance) cannot be obtained from it. It is important to
use conditional expressions (boundary conditions) for a boundary surface that
an electromagnetic field must satisfy to determine the reflectance and transmittance.* Let us consider flat boundary surfaces for medium 1 with a dielectric
constant ε_1 and medium 2 with a dielectric constant ε_2 (the refractive index for
each is $n_1 = \sqrt{\frac{\varepsilon_1}{\varepsilon_0}}$ and $n_2 = \sqrt{\frac{\varepsilon_2}{\varepsilon_0}}$, respectively) as shown in Figure 1.10.

The electric and magnetic fields in medium 1 are expressed as E_1 and H_1,
whereas those in medium 2 are expressed as E_2 and H_2. The unit normal
vector directed from the boundary surface to medium 1 is defined as n. If there
is no surface charge or surface current at the boundary, the electromagnetic
fields E_1, H_1, E_2, and H_2 should satisfy the following boundary conditions:[†]

$$E_1 \times n = E_2 \times n, \tag{1.63}$$

$$H_1 \times n = H_2 \times n, \tag{1.64}$$

$$D_1 \cdot n = D_2 \cdot n, \tag{1.65}$$

$$B_1 \cdot n = B_2 \cdot n. \tag{1.66}$$

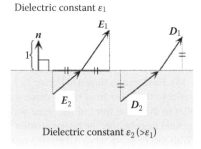

Dielectric constant ε_1

Dielectric constant $\varepsilon_2 (>\varepsilon_1)$

FIGURE 1.10 Boundary conditions at the boundary of two media with different dielectric constants.

---

* Generally, many of the laws of physics can be expressed by differential equations; however, their
general solutions may contain unknown coefficients. To describe a specific physical phenomenon,
the unknown coefficient should be determined, and if it is a differential equation that includes
time, the initial conditions should be given in addition to boundary conditions for the space.
† If a surface charge density ω [C/m^2] and a surface current density K [A/m] exist, Equations (1.64)
and (1.65) will become $(H_1 - H_2) \times n = K$ and $(D_1 - D_2) \cdot n = \omega$.

These equations show that the component (tangential component) parallel to the boundary surface of each of the electric field E and the magnetic field H in Equations (1.63) and (1.64) is equal at both sides of the boundary, whereas the component (normal component) perpendicular to the boundary surface of each of the electric flux density D and the magnetic flux density B in Equations (1.65) and (1.66) is equal at both sides of the boundary. The boundary condition Equations (1.63) and (1.64) of the tangential component, containing the time derivatives, are derived from Maxwell's equations (1.7) and (1.8), whereas the boundary condition Equations (1.65) and (1.66) of the normal component, containing the divergence operator, are derived from Maxwell's equations (1.9) and (1.10), respectively. However, Maxwell's equations (1.9) and (1.10) can be derived from Equations (1.7) and (1.8) containing the time derivatives, and as long as Equations (1.63) and (1.64) of the tangential component are satisfied in electromagnetic fields that change with time like electromagnetic waves, Equations (1.65) and (1.66) of the normal component will be automatically satisfied. Thus, there is no need to think about them.*

PROBLEM 1-7

Show that the reflection and transmission of a plane wave at the dielectric boundary can be resolved into two states, namely, the direction of the electric field as parallel to the boundary surface and the direction of the magnetic field as parallel to the boundary surface.

SOLUTION 1-7

A plane wave is incident on the boundary of the medium with a different refractive index at an angle φ_1 (angle measured from the normal line) as shown in Figure 1.11. The electromagnetic field of the plane wave as expressed in Equations (1.22) and (1.25) should satisfy Maxwell's equations (1.14) and (1.15). However, as shown in Figure 1.11, the y-axis direction is always parallel to the wave front of the plane wave, and because the direction parallel to the wave front of the plane wave does not have an electromagnetic field distribution, the differentiation with respect to y is zero ($\frac{\partial}{\partial y} = 0$). With this in mind, when Maxwell's equations (1.14) and (1.15) are resolved and written using Cartesian coordinates (x, y, z coordinate system), Table 1.1 will be obtained. Here, n_i is n_1 for medium 1 and n_2 for medium 2.

Six of the equations contain three (E_x, E_y, E_z) electric field components and three magnetic field components (H_x, H_y, H_z) for a total of six unknown values, which look like six linear simultaneous equations. Actually, Equations (1.67), (1.69), and (1.71) contain only E_y, H_x, and H_z (the components enclosed in ☐) of the six electromagnetic field

---

* It is necessary to think of all the boundary conditions in electrostatic fields and static magnetic fields, which do not vary with time.

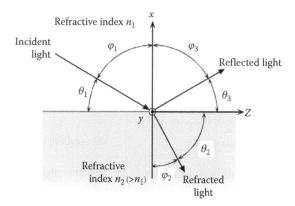

FIGURE 1.11 Reflection and refraction of a plane wave at the boundaries of two media with different refractive indices.

TABLE 1.1

$x\,y\,z$ Components of Maxwell's Equation for Plane Waves

	Equation (1.14)		Equation (1.15)	
x	$j\beta\boxed{E_y} = -j\omega\mu_0\boxed{H_x}$	(1.67)	$j\beta\left(H_y\right) = j\omega\varepsilon_0 n_i^2\left(E_x\right)$	(1.70)
y	$-j\beta\left(E_x\right) - \dfrac{\partial\left(E_z\right)}{\partial x} = -j\omega\mu_0\left(H_y\right)$	(1.68)	$-j\beta\boxed{H_x} - \dfrac{\partial\boxed{H_z}}{\partial x} = j\omega\varepsilon_0 n_i^2\boxed{E_y}$	(1.71)
z	$\dfrac{\partial\boxed{E_y}}{\partial x} = -j\omega\mu_0\boxed{H_z}$	(1.69)	$\dfrac{\partial\left(H_y\right)}{\partial x} = j\omega\varepsilon_0 n_i^2\left(E_z\right)$	(1.72)

components and the remaining Equations (1.68), (1.70), and (1.72) contain the H_y, E_x, and E_z components (the components enclosed in $\bigcirc$), and each of these are three independent linear simultaneous equation systems. In this system of equations, the plane wave containing the components E_y, H_x, and H_z enclosed in $\square$ is called the *TE-polarized light* or *TE wave* (transverse electric wave* and is an *s* wave[†] in physics), whereas the plane wave containing the components H_y, E_x, and E_z enclosed in $\bigcirc$ is called

---

* Originally, this was the term for the optical waveguide, wherein light was repeatedly reflected inside the waveguide traveling in the *z*-direction and was named as such because it only had the horizontal component E_y, which was perpendicular to the propagation direction of an electromagnetic wave.

[†] An abbreviation for senkretcht, which is German for perpendicular.

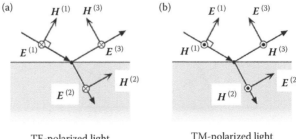

TE-polarized light   TM-polarized light

FIGURE 1.12 Two intrinsic polarizations in the reflection and refraction of plane waves: (a) TE-polarized light and (b) TM-polarized light.

the *TM wave* (abbreviation for transverse magnetic wave, and is a *p* wave* in physics) or TM-polarized light. Light with an electric field vibrating at a specific direction is called a polarized light,[†] and a plane wave with arbitrary polarization direction can be resolved into TE-polarized light and TM-polarized light.[‡] Accordingly, when the electric field and the magnetic field of a polarized light are written, Figure 1.12a and b are obtained.

[End of Solution]

PROBLEM 1-8

Find the formula giving the electric field reflection coefficient (amplitude reflectance) and the electric field permeability coefficient (amplitude transmittance) of a plane wave on a dielectric boundary surface.

SOLUTION 1-8

First, let us find the electric field reflection coefficient and the electric field permeability coefficient of the TE-polarized light (*s* wave). The procedure to find these coefficients involves the sum of the incident wave and reflected wave for the electromagnetic field in medium 1 and only that of the refracted wave in medium 2. Therefore, equations should be formulated so that the tangential components (E_y and H_z for the TE-polarized light)

---

[*] This is an abbreviation for parallel, which is German for parallel, although the same spelling is used in English.

[†] Normally, the direction of the electric field vector is defined as the polarization direction; however, sometimes there are cases when the direction of the magnetic field vector is defined as the polarization direction. Thus, care must be taken when reading references.

[‡] The polarized component of a specific direction can be produced from an incoherent light using a polarizing filter; however, the phases between orthogonal polarized components are not correlated to each other. Thus, the incoherent light is not polarized.

for the boundary surface of such electromagnetic fields become equivalent at the boundary surface to set up the amplitude coefficients.

When the electric fields $E^{(1)}$ and $E^{(3)}$ of the incident wave and the reflected wave in medium 1 are written using the format of Equation (1.22), the following is obtained. (However, the time dependency $e^{j\omega t}$ is common to all equations, so it will be omitted here.)

$$\begin{aligned}
E^{(1)} &= A e_y \exp[-j k_1 \cdot r] \\
&= A e_y \exp[-j k_0 n_1 (-x \cos \varphi_1 + z \sin \varphi_1)], \quad (1.73) \\
E^{(3)} &= B e_y \exp[-j k_0 n_1 (x \cos \varphi_3 + z \sin \varphi_3)]. \quad (1.74)
\end{aligned}$$

That is, the electric field of the TE-polarized light only has the y component E_y. On the other hand, the magnetic field can be calculated from the electric field using Equations (1.25), (1.58), (1.31), and (1.36) as follows:

$$H^{(i)} = \frac{n_1}{Z_0} E^{(i)} \left(\frac{k_i}{k_i} \times e_y \right) \quad (i = 1 \text{ or } 3). \quad (1.75)$$

This magnetic field has an x component and a z component, but for the boundary condition, only the tangential component H_z is necessary. When $H_z^{(1)}$ and $H_z^{(3)}$ are determined from Equation (1.75)

$$H_z^{(1)} = -\frac{n_1}{Z_0} A \cos \varphi_1 \exp[-j k_0 n_1 (-x \cos \varphi_1 + z \sin \varphi_1)] \quad (1.76)$$

$$H_z^{(3)} = \frac{n_1}{Z_0} B \cos \varphi_3 \exp[-j k_0 n_1 (x \cos \varphi_3 + z \sin \varphi_3)] \quad (1.77)$$

are obtained.

On the other hand, E_y and H_z in medium 2 are expressed using the following equations:

$$E_y^{(2)} = C \exp[-j k_0 n_2 (-x \cos \varphi_2 + z \sin \varphi_2)], \quad (1.78)$$

$$H_z^{(2)} = -\frac{n_2}{Z_0} C \cos \varphi_2 \exp[-j k_0 n_2 (-x \cos \varphi_2 + z \sin \varphi_2)]. \quad (1.79)$$

Here, as the boundary condition is that the tangential components of the electric field and the magnetic field on both sides of the boundary surface become equal at $x = 0$, we obtain

$$[E_y^{(1)}(x, z) + E_y^{(3)}(x, z)]_{x=0} = [E_y^{(2)}(x, z)]_{x=0}, \quad (1.80)$$

$$[H_z^{(1)}(x, z) + H_z^{(3)}(x, z)]_{x=0} = [H_z^{(2)}(x, z)]_{x=0}. \quad (1.81)$$

When Equations (1.73) to (1.79) are substituted to these equations, the following equations are obtained:

$$A \exp(-jk_0 n_1 z \sin \varphi_1) + B \exp(-jk_0 n_1 z \sin \varphi_3)$$
$$= C \exp(-jk_0 n_2 z \sin \varphi_2), \tag{1.82}$$

$$\frac{n_1}{Z_0}[-A \cos \varphi_1 \exp(-jk_0 n_1 z \sin \varphi_1)$$
$$+ B \cos \varphi_3 \exp(-jk_0 n_1 z \sin \varphi_3)]$$
$$= -\frac{n_2}{Z_0} C \cos \varphi_2 \exp(-jk_0 n_2 z \sin \varphi_2), \tag{1.83}$$

For these equations to hold independent of z

$$n_1 \sin \varphi_1 = n_1 \sin \varphi_3 = n_2 \sin \varphi_2 \tag{1.84}$$

must hold up. From this equation, the following well-known formula is derived:

$$\varphi_1 = \varphi_3 \qquad \text{(law of reflection)}, \tag{1.85}$$
$$n_1 \sin \varphi_1 = n_2 \sin \varphi_2 \qquad \text{(Snell's law)}. \tag{1.86}$$

Moreover, when these equations are used in Equations (1.82) and (1.83), the following two equations are obtained:

$$A + B = C, \tag{1.87}$$
$$n_1 \cos \varphi_1(-A + B) = -n_2 C \cos \varphi_2. \tag{1.88}$$

Here, when the electric field reflection coefficient r_i (amplitude reflectance) and the electric field permeability coefficient t_i (amplitude transmittance) are defined as

$$r_i = \frac{B}{A}, \quad t_i = \frac{C}{A} \quad (i = s \text{ or } p), \tag{1.89}$$

then,

$$r_s = \frac{n_1 \cos \varphi_1 - n_2 \cos \varphi_2}{n_1 \cos \varphi_1 + n_2 \cos \varphi_2}$$
$$= -\frac{\sin(\varphi_1 - \varphi_2)}{\sin(\varphi_1 + \varphi_2)}, \tag{1.90}$$
$$t_s = \frac{2n_1 \cos \varphi_1}{n_1 \cos \varphi_1 + n_2 \cos \varphi_2}$$
$$= \frac{2 \cos \varphi_1 \sin \varphi_2}{\sin(\varphi_1 + \varphi_2)}, \tag{1.91}$$

can be obtained from Equations (1.87) and (1.88).

When a similar calculation is done for the TM-polarized wave (p wave), the following two equations are obtained:

$$r_p = \frac{n_2 \cos \varphi_1 - n_1 \cos \varphi_2}{n_2 \cos \varphi_1 + n_1 \cos \varphi_2}$$
$$= \frac{\tan(\varphi_1 - \varphi_2)}{\tan(\varphi_1 + \varphi_2)}, \tag{1.92}$$

$$t_p = \frac{2n_1 \cos \varphi_1}{n_2 \cos \varphi_1 + n_1 \cos \varphi_2}$$
$$= \frac{2 \cos \varphi_1 \sin \varphi_2}{\sin(\varphi_1 + \varphi_2) \cos(\varphi_1 + \varphi_2)}. \tag{1.93}$$

[End of Solution]

When light is incident perpendicular to the boundary, the amplitude reflectance and the amplitude transmittance by putting $\varphi_1 = 0$ in Equations (1.90) to (1.93) (this would of course result in $\varphi_2 = \varphi_3 = 0$) are as follows:

$$r_s = \frac{n_1 - n_2}{n_1 + n_2}, \tag{1.94}$$

$$r_p = \frac{n_2 - n_1}{n_1 + n_2}, \tag{1.95}$$

$$t_s = t_p = \frac{2n_1}{n_1 + n_2}. \tag{1.96}$$

For vertical incidence, although there is no difference between the reflectance of the TE-polarized light and the TM-polarized light, the amplitude reflectance of the TM-polarized light is different from the amplitude reflectance of the TE-polarized light by a phase of π as seen in Equations (1.94) and (1.95). This is because the positive direction of the electric field after the reflection was defined based on the normal component as shown in Figure 1.12b, which resulted in the positive coordinate axis direction after the reflection reversing from that before the reflection. Consequently, there will be no confusion if the equation for the s wave is used for the vertical incidence.

We have already considered the reflection and refraction of a light that was beamed from one side to another side of a medium (refractive index n_1 to n_2). Now, let us consider the amplitude reflectance r_i' and the amplitude transmittance t_i' ($i = s$ or p) when a light follows a path that is exactly opposite to that of the optical path of a refracted light (as shown in Figure 1.13). For such a case, on exchanging the angle of incidence φ_1 and the angle of refraction φ_2 and the refractive index n_1 and refractive index n_2 in Equations (1.90) through (1.93), the following equations are obtained:

$$r_s' = \frac{n_2 \cos \varphi_2 - n_1 \cos \varphi_1}{n_1 \cos \varphi_1 + n_2 \cos \varphi_2} = -r_s, \tag{1.97}$$

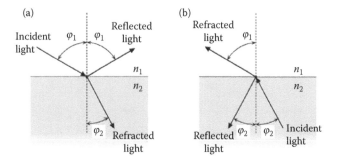

FIGURE 1.13　Reflection and refraction of a backward-traveling plane wave: (a) Refracted light from refractive index n_1 to n_2. (b) Backward-traveling wave from n_2 to n_1.

$$r'_p = \frac{n_1 \cos \varphi_2 - n_2 \cos \varphi_1}{n_2 \cos \varphi_1 + n_1 \cos \varphi_2} = -r_p. \tag{1.98}$$

That is to say, the amplitude reflectance for a light that is beamed in the reverse direction may differ in phase by only π, but the absolute values are equal. Consequently, the power reflectivity will be the same as described in the next exercise, Exercise 1-9.

On the other hand, the amplitude transmittance is obtained as follows:

$$t'_s = \frac{2n_2 \cos \varphi_2}{n_1 \cos \varphi_1 + n_2 \cos \varphi_2}, \tag{1.99}$$

$$t'_p = \frac{2n_2 \cos \varphi_2}{n_2 \cos \varphi_1 + n_1 \cos \varphi_2}. \tag{1.100}$$

This may be difficult to intuitively understand as the absolute value of the amplitude transmittance for the backward traveling light will be different, but as will be described in Solution 1-9, the power transmittance of the backward traveling wave will be the same. The equation

$$t_i t'_i + r_i^2 = t_i t'_i + r_i'^2 = 1 \quad (i = \text{s or p}) \tag{1.101}$$

is derived from Equations (1.99) and (1.100) and Equations (1.97) and (1.98). Equations (1.97), (1.98), and (1.101) are called the *Stokes' theorem*.

PROBLEM 1-9

Find the formula that would give the power reflectivity and the power transmittance of a plane wave at the dielectric boundary surface.

SOLUTION 1-9

Because Equations (1.90) through (1.93) give the electric field reflection coefficient (amplitude reflectance) and the electric field permeability coefficient (amplitude transmittance), it is necessary to use a complex Poynting vector to determine the power reflectivity and the power transmittance. From Equation (1.53), the power flow densities of the incident wave, the reflected wave, and the refracted wave are obtained as follows:

$$|\tilde{S}^{(1)}| = \frac{1}{2}\sqrt{\frac{\varepsilon_1}{\mu_0}}|A|^2 = \frac{n_1}{2Z_0}|A|^2, \tag{1.102}$$

$$|\tilde{S}^{(3)}| = \frac{n_1}{2Z_0}|B|^2 = \frac{n_1}{2Z_0}|A|^2|r_i|^2 \quad (i = s \text{ or } p), \tag{1.103}$$

$$|\tilde{S}^{(2)}| = \frac{n_2}{2Z_0}|C|^2 = \frac{n_2}{2Z_0}|A|^2|t_i|^2 \quad (i = s \text{ or } p). \tag{1.104}$$

However, the power flow density is the energy flow density per unit area; hence, one must be careful in that there is no change in the cross-sectional area of the energy flux for the reflection as shown in Figure 1.14, whereas there is a change in the cross-sectional area for refraction. Because the energy flux is equal in the projected cross-section in the boundary surface, the equality below must be satisfied.

$$|\tilde{S}^{(1)}|\cos\varphi_1 = |\tilde{S}^{(3)}|\cos\varphi_1 + |\tilde{S}^{(2)}|\cos\varphi_2. \tag{1.105}$$

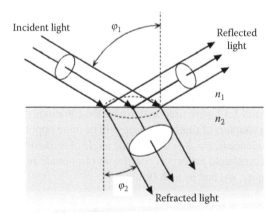

FIGURE 1.14 Change in the cross-sectional area of the energy flow during refraction and reflection.

From this relational expression, the *power reflectivity* R_i and the *power transmittance* T_i can be expressed with the following equation using r_i and t_i ($i = s$ or p):

$$R_i = \frac{|\tilde{S}^{(3)}|}{|\tilde{S}^{(1)}|} = |r_i|^2 \quad (i = s \text{ or } p), \qquad (1.106)$$

$$T_i = 1 - R_i = \frac{n_2 \cos \varphi_2}{n_1 \cos \varphi_1} |t_i|^2 = \frac{\sin \varphi_1 \cos \varphi_2}{\sin \varphi_2 \cos \varphi_1} |t_i|^2 \quad (i = s \text{ or } p). \qquad (1.107)$$

[End of Solution]

Now, from the power reflectivity equations (1.106), (1.97), and (1.98), the power reflectivity R_i' for the backward traveling light as shown in Figure 1.13b is seen to be equal to R_i. On the other hand, as seen from Equations (1.91), (1.93), (1.99), and (1.100), the absolute values of the amplitude transmittances of the backward traveling light are different from those of t_s and t_p. However, when Equations (1.99) and (1.107) are substituted to Equation (1.100)

$$T_i' = \frac{n_1 \cos \varphi_1}{n_2 \cos \varphi_2} |t_i'|^2 = \frac{n_2 \cos \varphi_2}{n_1 \cos \varphi_1} |t_i|^2 = T_1 \quad (i = s \text{ or } p) \qquad (1.108)$$

is obtained, and the power transmittance T_i' of the backward traveling light becomes equal to T_i. The reason why the amplitude transmittance of the forward traveling light is different from that of the backward traveling wave is that the electric field amplitude in media with different dielectric constants differ even if the power flow density is the same. (This can be understood if one compares Equations (1.102) and (1.104).)

Incidentally, the electric field reflection coefficient Equations (1.90) and (1.92) contain the angle of incidence φ_1 and the angle of refraction φ_2. The angle of refraction φ_2 can be obtained from Snell's law, given the refractive indices n_1 and n_2 and the angle of incidence φ_1. These equations can be written as a function involving n_1, n_2, and φ_1. Nevertheless, the angle of incidence φ_1 and the angle of refraction φ_2 measured from the normal line of the boundary surface in Figures 1.11 and 1.12 were used in deriving the equations thus far; however, in the following chapters of this book, there will be more opportunities for using the angle of incidence θ_1 ($= \frac{\pi}{2} - \varphi_1$) and the angle of refraction θ_2 ($= \frac{\pi}{2} - \varphi_2$). For this, the electric field reflection coefficient (amplitude reflectance) can be expressed using n_1, n_2, and θ_1 as follows:

$$r_s = \frac{n_1 \sin \theta_1 - \sqrt{n_2^2 - n_1^2 \cos^2 \theta_1}}{n_1 \sin \theta_1 + \sqrt{n_2^2 - n_1^2 \cos^2 \theta_1}}, \qquad (1.109)$$

$$r_p = \frac{n_2 \sin \theta_1 - \frac{n_1}{n_2} \sqrt{n_2^2 - n_1^2 \cos^2 \theta_1}}{n_2 \sin \theta_1 + \frac{n_1}{n_2} \sqrt{n_2^2 - n_1^2 \cos^2 \theta_1}}. \tag{1.110}$$

1.4.3 REFLECTION FROM A METAL SURFACE

Let us consider the reflection of a plane wave that was beamed from a dielectric medium with a refractive index n_1 to the surface of an ideal conductor with an electric conductivity of $\sigma = \infty$ as shown in Figure 1.15. Because there is no electric field inside and on the surface of the conductor, the tangential component of the electric field of the incident wave and the tangential component of the electric field of the reflected wave should mutually cancel each other.

First, let us consider a TE-polarized light (s wave). Granting that Equation (1.85) on the law of reflection holds true, when formulas similar to Equations (1.73) and (1.74) are used, the solitary tangential component $E_y = E_y^{(1)} + E_y^{(3)} = 0$ of the electric field on the surface of the conductor must be satisfied, and so

$$E_y^{(1)} = A \exp[-jk_0 n_1(-x \cos \varphi_1 + z \sin \varphi_1)], \tag{1.111}$$

$$E_y^{(3)} = -A \exp[-jk_0 n_1(x \cos \varphi_1 + z \sin \varphi_1)], \tag{1.112}$$

are obtained with A as the amplitude. Consequently, the amplitude reflectance is $r = -1.0$ (in the case of an ideal conductor with $\sigma = \infty$). On the other

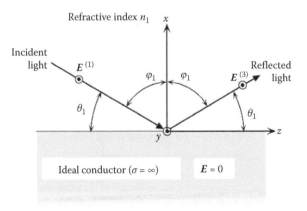

FIGURE 1.15 Reflection of the plane wave on a metal surface.

hand, the tangential component H_z of the magnetic field will be obtained from Equation (1.75) as follows:

$$H_z^{(1)} = -\frac{n_1}{Z_0}A \cos\varphi_1 \exp[-jk_0 n_1(-x\cos\varphi_1 + z\sin\varphi_1)], \quad (1.113)$$

$$H_z^{(3)} = -\frac{n_1}{Z_0}A \cos\varphi_1 \exp[-jk_0 n_1(x\cos\varphi_1 + z\sin\varphi_1)]. \quad (1.114)$$

For these equations, when the angle of incidence φ_1 is considered to be zero, that is, a vertical incidence, strange facts are discovered. As shown in Figure 1.16a, the phase of the electric field will invert upon reflection on the surface of the conductor (will be off by π), whereas the phase of the magnetic field will not shift. Consequently, for standing waves during a vertical incidence, the electric field will become the node on the surface of the conductor and, conversely, the magnetic field will become the antinode. Moreover, as the magnetic field exists at the surface of the conductor ($H = 0$ inside the conductor), surface current will flow. From Equations (1.113) and (1.114), the surface current density K_y is equal to

$$K_y = [H_z^{(1)} + H_z^{(3)}]_{x=0} = -2\frac{n_1}{Z_0}A. \quad (1.115)$$

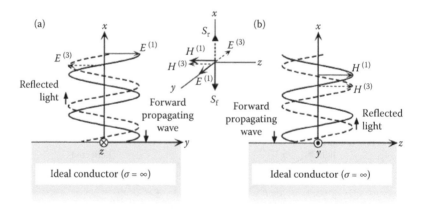

S_f: Poynting vector of forward propagating wave

S_r: Poynting vector of reflected wave

FIGURE 1.16 Phase changes during the reflection of the plane wave at the surface of a perfect conductor: (a) Phase relationship of the incident electric field and the reflected electric field. (b) Phase relationship of the incident magnetic field and the reflected magnetic field.

At the surface of the conductor, pressure will act according to this surface current density and the magnetic flux density. The time-averaged value p_x [Pa] of this pressure is given by

$$p_x = \frac{1}{2} \boldsymbol{K} \times \boldsymbol{B}^* = 2\varepsilon_0 n_1 A^2. \tag{1.116}$$

1.4.4 TOTAL INTERNAL REFLECTION

In cases where $n_1 > n_2$ as in Equations (1.109) and (1.110), when the angle of incidence θ becomes small and the term inside the square root becomes zero, $R_s = R_p = 1$ and a *total internal reflection* occurs. For such a condition, the angle of incidence θ should be smaller than the *critical angle* θ_c (*total supplementary angle of reflection*) which is defined by

$$\theta_c = \cos^{-1}\left(\frac{n_2}{n_1}\right) = \sin^{-1}\sqrt{1 - \frac{n_2^2}{n_1^2}} = \sin^{-1}\sqrt{2\Delta} \quad \text{[rad]}. \tag{1.117}$$

Here,

$$\Delta = \frac{n_1^2 - n_2^2}{2n_1^2}, \tag{1.118}$$

which is called the *relative index difference*. This will be used as the basic parameter to be used frequently in subsequent chapters on optical waveguide (normally, Δ is expressed in %). Equation (1.117) matches with Equation (1.59) for Snell's law, if the angle of refraction $\theta_2 = 0$. A real number will result if $n_2 \leq n_1$ in Equation (1.117), conversely, if $n_2 > n_1$, the reflectance will be less than unity $\theta_1 = 0$, and total internal reflection will not occur. In such a case, the power reflectivity is a positive real number; however, one must note that the electric field reflection coefficient (amplitude reflectance) of TE polarization will become negative (i.e., the phase will shift by π) if $n_2 > n_1$. The changes in R_s and R_p for the angle of incidence θ_1 are shown in Figure 1.17a and b, where the cases $n_2 \leq n_1$ and $n_2 > n_1$ are drawn separately.

On looking at the figure in detail, we can see that in the range of θ wherein total internal reflection does not occur, the reflectance of the TM-polarized light (p wave) is always smaller than that of the TE-polarized light (s wave) and that there is an angle where the reflectance of the TM-polarized light is zero. This angle is obtained by equating the numerator in Equation (1.110) to zero as follows:

$$\theta_B = \cot^{-1}\frac{n_2}{n_1} = \tan^{-1}\frac{n_1}{n_2} \quad \text{[rad]}. \tag{1.119}$$

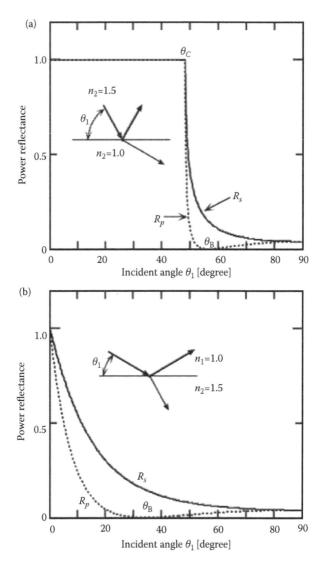

FIGURE 1.17 Dependence of power reflectivity on the angle of incidence: (a) $n_2 \leq n_1$ and (b) $n_2 > n_1$.

This is called the *Brewster angle*. If the total internal reflection does not occur, the reflectance of the TM-polarized light is always smaller than the reflectance of the TE-polarized light, thus reflected light from the boundary surface with different refractive indices such as the surface of water or glass will contain more TE-polarized components.* Moreover, one should note that the sign of the electric field reflection coefficient (amplitude reflectance) inverts before and after the Brewster angle. (That is, the phase will shift by π.)

Meanwhile, total internal reflection occurs if the angle of incidence θ_1 is smaller than the critical angle given by Equation (1.117), so the power reflectivity becomes 1.0 as seen in Figure 1.17. In such cases, the electric field reflection coefficients given by Equations (1.109) and (1.110) become complex numbers. When the electric field reflection coefficient is expressed in polar form, the absolute value becomes 1.0 and can thus be written as

$$r_i = \exp(j\Phi_i) \qquad (i = \text{TE or TM}) \qquad (1.120)$$

The phase term in the above equation for the TE-polarized light (s wave) is given by[†]

$$\Phi_{\text{TE}} = 2\tan^{-1}\left[\sqrt{\frac{2\Delta}{\sin^2\theta_1} - 1}\right] \quad \text{[rad]}, \qquad (1.121)$$

and the TM-polarized light (p wave) is given by

$$\Phi_{\text{TM}} = 2\tan^{-1}\left[\left(\frac{n_1}{n_2}\right)^2 \sqrt{\frac{2\Delta}{\sin^2\theta_1} - 1}\right] \quad \text{[rad]}. \qquad (1.122)$$

The phase change is π when the angle of incidence is zero (when incident parallel on the boundary surface), and the phase charge is zero when the angle of incidence θ_1 is just equal to the critical angle as shown in Figure 1.18.

The propagation of an obliquely incident light can be resolved into the normal component and tangential component of the boundary surface as shown in

---

* This phenomenon is useful when the reflection on the glass surface obstructs looking at an object through a glass, one can then see the object clearly by eliminating the TE-polarized light using a polarizing filter described in Section 1.5. This technique is common knowledge for photographers.
[†] Just because the term inside the $\sqrt{}$ of Equations (1.109) and (1.110) becomes negative, it does not simply mean that $\sqrt{-A} = jA$. When $\sqrt{-1}$ is square it becomes -1, so $\pm j$ is a mathematical solution. Physically, one needs to examine which sign is meaningful for $\sqrt{-A} = \pm jA$. For the total internal reflection, the $\cos\varphi_2$ in Equation (1.78) becomes $\cos\varphi_2 = \pm j\sqrt{\frac{n_1^2}{n_2^2}\sin^2\varphi_1 - 1}$ using Snell's law. However, the electric field in medium 2 during total internal reflection, which has an exponential decay solution in the negative x-direction (solution with $\exp(\alpha x)$) has a physical meaning. Accordingly, if Equation (1.121) is substituted to Equation (1.78), the solution for $\cos\varphi_2 = -j\sqrt{\frac{n_1^2}{n_2^2}\sin^2\varphi_1 - 1}$ is seen to satisfy the above condition. Consequently, it is seen from Equation (1.90) that the phase of the electric field reflection coefficient during a total internal reflection is positive.

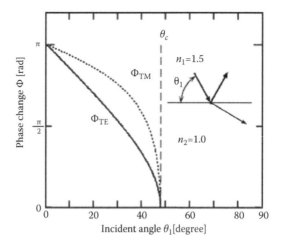

FIGURE 1.18 Dependence on the angle of incidence of the phase change of the electric field reflection coefficient during total reflection.

Figure 1.19, and the physical meaning of this phase delay is easy to understand when the normal component of the reflection is taken into consideration. In the case of a zero phase change when light is incident at exactly the critical angle, the boundary surface will become the antinode of the standing wave as shown in Figure 1.19b. On the other hand, if light is incident parallel on the boundary surface (angle of incidence $\theta_1 = 0$), the phase will shift by π, and the boundary surface will become the node of the standing wave. Consequently, for the angle of incidence in a range of $0 < \theta_1 \leq \theta_c$, the phase change will be between 0 and π, and the standing wave occurring at the boundary surface will be between a node and an antinode as shown in Figure 1.19c. When the state of Figure 1.19b is compared to a reflection from a metal surface, the electric field on the surface of the metal becomes zero, so the standing wave becomes a node. Consequently, the state of Figure 1.19b is equivalent to the reflection from the metal surface, which is located at a position secluded from the boundary surface of the dielectric medium by half wavelength of a standing wave (a quarter wavelength of a propagating wave). That is, the position of a reflection surface, which is equivalent to the reflection from the surface, is given by

$$\delta x_i = \frac{\pi - \Phi_i}{2k_0 n_1 \sin \theta_1} \quad (i = \text{TE or TM}), \qquad (1.123)$$

where δx is the distance from the boundary surface. This phase change will appear as the penetration distance of the electromagnetic field from the core layer (the layer where light is confined) to the cladding layer, and the field distribution of guided mode in optical waveguide is determined in a later

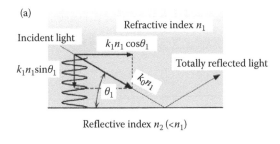

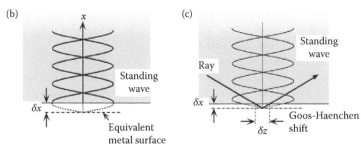

FIGURE 1.19   The standing wave formed by the incident wave and the reflected wave when the propagation vector is resolved into the normal component and the tangential component of the boundary surface: (a) normal component of the propagation vector; (b) standing wave when the phase change is zero; (c) standing wave when $-\pi < \Phi_i \leq 0$ ($i = $ TE or TM).

chapter 5. Furthermore, the shift of reflection point in the z-direction, and not in the x-direction, as shown in Figure 1.19c, is called the *Goos-Haenchen shift*.

PROBLEM 1-10

Let us consider a plane wave that is incident on a flat boundary surface of a dielectric material 1 with a refractive index $n_1 = 1.500$ and a dielectric material 2 with a refractive index $n_2 = 1.485$ at an angle of incidence θ (angle measured from the boundary surface). What will be the critical angle of total internal reflection θ_c for this?

SOLUTION 1-10

For Equation (1.118), if the difference between the refractive indices n_1 and n_2 is small enough relative to n_1, then an approximation of $n_1 + n_2 \simeq 2n_1$ can be applied, and the equation can be modified to

$$\Delta = \frac{n_1^2 - n_2^2}{2n_1^2} = \frac{(n_1 - n_2)(n_1 + n_2)}{2n_1^2} \simeq \frac{n_1 - n_2}{n_1}. \tag{1.124}$$

According to this equation, $\Delta = 0.01 = 1.0\%$. Consequently,

$$\theta_c = \sin^{-1}\sqrt{2\Delta} \simeq \sqrt{2\Delta} = 0.14 \text{ [rad]} = 8.0° \qquad (1.125)$$

is obtained using Equation (1.117).   **[End of Solution]**

PROBLEMS

1. What is the percentage of reflectance when a plane wave is incident perpendicular on a glass layer ($n = 1.5000$) of air?
2. What is the dB for this reflectance? (The amount of reflectance expressed in dB is called the return loss.)
3. A plane wave is incident perpendicularly on a medium with a refractive index of n_2 from a transparent medium with a refractive index of 1.5000. What is the needed range for the value of n_2 to have a return loss of less than -40 dB?
4. Consider a plane wave that is incident on a boundary dielectric material with $n_1 = 1.50, n_2 = 1.0$ (air) at the angle θ. Such a plane wave includes a 1:1 amplitude ratio of TE and TM polarization. At what angle of incidence θ will there be a phase change of 22.5 degrees between the TE and TM polarization due to a one-time total internal reflection?

   Using this principle, a quarter-wave plate and a half-wave plate can be configured using a prism with parallel oblique sides. Such a prism is called a Fresnel rhomb.

1.5   POLARIZATION AND BIREFRINGENCE

In Section 1.4, concerning the reflection and refraction of the plane wave, TE-polarized light (s wave) and TM-polarized light (p wave) were defined from the relationship of the electric field vector and the dielectric boundary surface. However, the phenomenon that has a dependence on the direction of polarization is not limited to reflection and refraction. In this section, the state of the polarization is derived from its relationship with the refractive index, and the functions of a quarter-wave plate (called a $\frac{\lambda}{4}$ plate) and a half-wave plate (called a $\frac{\lambda}{2}$ plate), which are commonly used as polarization state control devices for free-space optical beam, are explained. It is more convenient to use Stokes' parameter and Jones matrix for a more unified mathematical explanation of the polarization state, and this will be derived again in detail in Section 8.6.

In Maxwell's equation (1.15), the relationship between the electric field E and the electric flux density D is given by the auxillary equation

$$D = \varepsilon E, \qquad (1.126)$$

where ε is assumed to be a scalar quantity. However, there may be cases when the dielectric constant (refractive index) of a specific crystal axis direction for a type of crystal is different from the dielectric constant of another direction. For example, if light is propagated toward the z-axis direction, and dielectric constant for the x-axis direction and the y-axis direction are different, Equation (1.126) can be expressed as

$$
\begin{bmatrix} D_x \\ D_y \\ D_z \end{bmatrix} = \begin{bmatrix} \varepsilon_x, & 0, & 0 \\ 0, & \varepsilon_y, & 0 \\ 0, & 0, & \varepsilon_y \end{bmatrix} \cdot \begin{bmatrix} E_x \\ E_y \\ E_z \end{bmatrix} = \varepsilon_0 \begin{bmatrix} n_x^2, & 0, & 0 \\ 0, & n_y^2, & 0 \\ 0, & 0, & n_y^2 \end{bmatrix} \cdot \begin{bmatrix} E_x \\ E_y \\ E_z \end{bmatrix}. \quad (1.127)
$$

(If the plane wave is propagated toward the z-axis direction, ε_z is irrelevant, so $\varepsilon_z = \varepsilon_y$.) A material wherein the refractive index is different depending on the direction of the electric field is called a birefringent material, and such a refractive index is called *birefringence*. As in the aforementioned example, the direction with a refractive index that is different from that of another direction as in the x-axis is called the principal *birefringent axis* (intrinsic polarization axis).*

If an x-polarized plane wave (amplitude constant A_x) shown in Figure 1.7, is incident on such a birefringent material and is made to propagate a distance z, substituting $k = k_0 n_x$ into Equation (1.37) and expressing the time and propagation distance dependency not by the phasor notation $e^{j(\omega t - kz)}$ but by a real number, the electric field amplitude can be expressed by

$$
E_x = A_x \cos(\omega t - k_0 n_x z). \quad (1.128)
$$

Similarly, if a y-polarized light with an amplitude constant of A_y propagates by a distance z, the electric field amplitude will be given by the following equation:

$$
E_y = A_y \cos(\omega t - k_0 n_y z). \quad (1.129)
$$

Here, let us consider a case where a linearly polarized light, of which oscillation plane of electric field is inclined at an angle θ from the x-axis as shown in Figure 1.20, propagates by a distance z in a birefringent material as expressed in Equation (1.127). Because such a plane wave can be resolved into x-polarized light and y-polarized light, it can be expressed as

$$
E = A \cos\theta \cos(\omega t - k_0 n_x z)e_x + A \sin\theta \cos(\omega t - k_0 n_y z)e_y, \quad (1.130)
$$

---

* In general, when material constants differ based on the crystal orientation, they are said to be anisotropic. However, in the case of Equation (1.127), as the refractive index is different only in one direction, it would be called a single axis anisotropy. On the other hand, a material with a fixed material constant regardless of orientation is called an isotropic medium (alternatively, one can say that the material is isotropic).

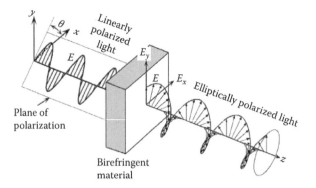

FIGURE 1.20 Propagation of a plane wave inclined from the main polarization axis of a birefringent material.

wherein the amplitude of each of the components of the polarized light in Equations (1.128) and (1.129) are $A_x = A \cos \theta$ and $A_y = A \sin \theta$, respectively. The trajectory of the temporal change of this electric field vector in the xy plane will become

$$\frac{E_x^2}{A_x^2} + \frac{E_y^2}{A_y^2} - 2 \cos \phi \frac{E_x E_y}{A_x A_y} = \sin^2 \phi, \qquad (1.131)$$

when the parameter ωt is eliminated from Equations (1.128) and (1.129). Here,

$$\phi = k_0(n_x - n_y)z. \qquad (1.132)$$

Equation (1.131) represents an ellipse, and at $\phi = 0$ (i.e., the incident end at $z = 0$) for the *planar polarization* or linear polarization inclined by an angle θ from the x-axis, in a range of $0 < |\phi| < \frac{\pi}{2}$, the major axis becomes an ellipse inclined from the x-axis to the θ direction. A polarization that depicts this kind of electric field vector trajectory is called an *elliptical polarization*. Then, at $|\phi| = \frac{\pi}{2}(z = \frac{\lambda}{4|n_x - n_y|})$, each of the major and minor axes become an ellipse at the x- and y-axis direction, respectively. In a range of $\frac{\pi}{2} < |\phi| < \pi$, the major axis becomes an ellipse inclined from the x-axis to the $-\theta$ direction. At $|\phi| = \pi$, it becomes a linearly polarized light inclined at an angle of $-\theta$ from the x-axis. Therefore, when the trajectory of this electric field vector is depicted inside the xy plane seen from the direction of the $+z$ axis to the $-z$ axis, eventually this ellipse will repeatedly change into a rectangle, which has a length of $2A_x$ along the x-axis and $2A_y$ along the y-axis, as shown in Figure 1.21.

In particular, when $\phi = \pm \frac{\pi}{2}$ at $A_x = A_y$ ($\theta = \frac{\pi}{4}$), the trajectory of the electric field vector will become a circle, and this will be called a circular

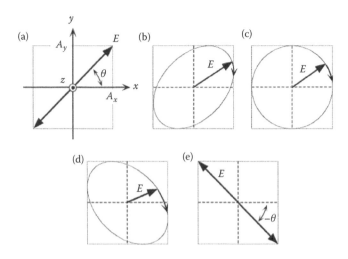

FIGURE 1.21 Changes associated with polarization state propagation in birefringent materials: (a) $\phi = 0$, (b) $0 < |\phi| < \frac{\pi}{2}$, (c) $\phi = \frac{\pi}{2}$, (d) $\frac{\pi}{2} < |\phi| < \pi$, and (e) $|\phi| = \pi$.

polarization. Here, if the phase difference ϕ is positive, the $(\phi = +\frac{\pi}{2})$ as shown in Figure 1.21 seen from the direction of the $+z$-axis to the $-z$-axis is called the right-handed *circular polarization* as the vector trajectory will rotate clockwise (right-handed rotation). Conversely, the $\phi = -\frac{\pi}{2}$ is called the left-handed circular polarization. The propagation distance (thickness of the birefringent material) $L_{\frac{\pi}{2}}$ for which a linearly polarized light inclined at 45 degrees from the principal birefringent axis that become circularly polarized is given by

$$L_{\frac{\pi}{2}} = \frac{\lambda}{4|n_x - n_y|}, \tag{1.133}$$

and the birefringent plate that was adjusted to this thickness is called a quarterwave plate (or a $\frac{\lambda}{4}$ plate) and is used when linear polarization is converted to circular polarization or conversely a circular polarization is converted to an arbitrary linear polarization.

On the other hand, if a linearly polarized light passes through a birefringent plate with a thickness twice as that indicated in Equation (1.133), the linearly polarized light of which polarization plane is inclined at an angle θ from the principal birefringent axis is converted to a linearly polarized light inclined at an angle θ in a direction opposite from the primary birefringent axis. That is, as the linearly polarized light will rotate by an angle of 2θ, if the angle is set to $\theta = 45°$, the linearly polarized light will be converted into an orthogonal linearly polarized light (e.g., from TE-polarized light to TM-polarized light). This kind of birefringent plate is called a half-wave plate (or a $\frac{\lambda}{2}$ plate). It is important to know the direction of a certain primary birefringent axis to be able to properly use $\frac{\lambda}{4}$ and $\frac{\lambda}{2}$ plates.

PROBLEMS

1. To convert a linearly polarized light of which polarization axis is inclined by an angle θ from x axis, how should the angle of the primary axis of the $\frac{\lambda}{2}$ plate be set up from the x-axis?
2. To convert any elliptically polarized light with the x component and the y component of the electric field expressed as

$$E_x = A_x \exp[j\omega t], \tag{1.134}$$

$$E_y = A_y \exp[j(\omega t - \phi)], \tag{1.135}$$

respectively, to a linearly polarized light with a certain direction used by a $\frac{\lambda}{4}$ plate, how should the angle be set up with respect to the x-axis of the primary axis of the $\frac{\lambda}{4}$ plate?

3. How should the $\frac{\lambda}{4}$ and $\frac{\lambda}{2}$ plates be combined to convert any elliptically polarized light similar to that in Question 2 to a linearly polarized light in an x-axis direction?

   [Supplement] In this way, any elliptically polarized light can be converted to any linearly polarized light by combining $\frac{\lambda}{4}$ and $\frac{\lambda}{2}$ plates. Moreover, any polarization state can be created using the $\frac{\lambda}{4}$ and $\frac{\lambda}{2}$ plates.*

1.6   PROPAGATION OF A PLANE WAVE IN A MEDIUM WITH GAIN AND ABSORPTION LOSS

In the previous sections, the discussion has proceeded with the assumption that the refractive index is a real number; however, for a plane wave propagated in a medium with a refractive index that is expressed as a complex number, a gain or absorption loss will appear. Let us consider an x-polarized plane wave propagating on the positive z-axis in a medium of which refractive index

---

* Incoherent light such as natural light can produce polarized components of a specific direction using a polarization filter; however, for this, there will be no correlation between the orthogonal polarization components so the polarization filter would suffer a 3 dB (50%) loss. On the other hand, a coherent light can be converted to any polarization state by combining $\frac{\lambda}{4}$ and $\frac{\lambda}{2}$ plates, and so if it is converted to a plane polarized light of a specific direction, it can pass through a polarization filter of that direction and there would be no loss.

is isotropic and homogeneous, independent of position and is expressed as a complex number 1*

$$n = n_r + jn_i. \tag{1.136}$$

Substituting this to Equations (1.136) and (1.37) gives

$$E_x = A \exp[j\{\omega t - k_0(n_r + jn_i)z\}]$$
$$= Ae^{j(\omega t - k_0 n_r z)} \exp(k_0 n_i z). \tag{1.137}$$

The first exponential function in Equation (1.137) is a term that represents ordinary propagation of a sinusoidal wave. However, if $n_i > 0$ in the second exponential function, the plane wave will be exponentially amplified, whereas if $n_i < 0$, the plane wave will be exponentially attenuated. Consequently, in each of these cases, the state of the propagation can be illustrated as shown in Figure 1.22.

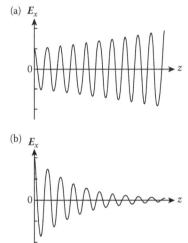

FIGURE 1.22 Amplification and attenuation of a plane wave in a medium with a complex refractive index. (a) Amplification of a plane wave (when $n_i > 0$). (b) Attenuation of a plane wave (when $n_i < 0$).

---

* In optics textbooks, the complex refractive index of an absorbing medium, such as metals, is sometimes written as

$$n = n + ik \quad (i \text{ is an imaginary unit})$$

The real part of the complex number n is the refractive index, and k is called the extinction coefficient. First of all, in this case, the time dependency is often assumed to be $e^{-i\omega t}$; consequently, when $k > 0$, the light will be attenuated.

Here, when the plane wave is amplified, the medium is said to have a *gain* (or is called a gain medium), whereas when a plane wave is attenuated, the medium is said to have an *absorption loss* (also called an absorbing medium).*

PROBLEM 1-11

When light of wavelength 1.0 μm is incident on a medium where the imaginary part of the refractive index $n_i = -0.005$, what will be the distance when the intensity is attenuated to 1/10?

SOLUTION 1-11

Equation (1.137) is an equation for the electric field amplitude, and as light intensity is proportional to $|E|^2$, the attenuation of light intensity I is given by the following equation:

$$I \propto \exp(2k_0 n_i z) = \exp\left(\frac{4\pi n_i z}{\lambda}\right). \tag{1.138}$$

Therefore, the attenuation coefficient of light per unit propagation distance α is expressed by

$$\alpha = \frac{4\pi n_i}{\lambda} = -\frac{4\pi \times 0.005}{1.0 \times 10^{-6}} = -6.28 \times 10^4 \text{ [Neper/m]}$$

$$= -2.73 \times 10^5 \text{ [dB/m]}. \tag{1.139}$$

Therefore, the propagation distance at which the loss is 10 dB is $\frac{10}{2.73 \times 10^5} = 36$ μm. Actually, the imaginary part of the refractive index of metals is much bigger than in this example, so light will not pass if the film is not considerably thin.[†]   **[End of Solution]**

PROBLEMS

1. Show that when a plane wave is beamed perpendicularly from vacuum with refractive index $n = 1.0$ to the flat surface of a medium (which is assumed to be thick enough) with a complex refractive index $n_r + jn_i$, the power reflectivity R is given by

$$R = \frac{(n_r - n)^2 + n_i^2}{(n_r + n)^2 + n_i^2}. \tag{1.140}$$

---

* Attenuation of light in a propagation area is called loss; however, there are two main causes for this phenomenon: an absorption loss resulting from the conversion of light energy to thermal energy due to absorption, and scattering loss or radiation loss resulting from the exit of light from the target area due to scattering and radiation.

† Traditional Kanazawa gold leaf is made by beating a lump of gold with a hammer bit-by-bit, until one can see through the leaf. This requires a thickness of less than several μm.

TABLE 1.2

Complex Refractive Indices of Some Metals at Wavelength $\lambda = 1.55$ μm

Name of Metal	n_r	n_i
Au	0.176	−10.1
Ag	0.368	−11.1
Al	1.42	−15.7
Cu	0.607	−8.26
Cr	4.19	−4.93

2. The values of the complex refractive indices in some metals at wavelength $\lambda = 1.55$ μm are shown in Table 1.2. Calculate the reflectivity when a plane wave of wavelength 1.55 μm is perpendicularly incident on these metals. (As seen from the results of the calculation for Equation (1.140) and this problem, the reason why the reflectivity of gold and silver is near 1.0 is explained by the fact that the real part of the complex refractive index of these metals is small and the imaginary part is large.)

1.7 WAVE FRONT AND LIGHT RAYS

We have considered plane waves in the previous sections, and so in this section, we will further expand a general expression of plane waves Equation (1.22) to a case where the amplitude and phase have location dependency $U(r)$ and $k_0 S(r)$. Assuming a time dependency of $e^{j\omega t}$, the electric and magnetic fields can be expressed as

$$E = \mathcal{E}(r)\exp[-j\omega t], \qquad (1.141)$$

$$H = \mathcal{H}(r)\exp[-j\omega t]. \qquad (1.142)$$

When these equations are substituted to Equations (1.14) and (1.15),

$$\nabla \times \mathcal{E} = -j\omega\mu_0\mathcal{H}, \qquad (1.143)$$

$$\nabla \times \mathcal{H} = j\omega\varepsilon\mathcal{E}, \qquad (1.144)$$

are obtained. Modeling after Equation (1.22) and omitting $e^{j\omega t}$, both electric and magnetic fields can be expressed as

$$\mathcal{E} = e(r)\exp[-jk_0 S(r)], \qquad (1.145)$$

$$\mathcal{H} = h(r)\exp[-jk_0 S(r)]. \qquad (1.146)$$

Substituting these equations to Equations (1.143) and (1.144) and using Equations (1.31) and (1.36)

$$\nabla \times e - jk_0(\nabla S \times e) = -j\omega\mu_0 h = -jk_0 Z_0 h, \qquad (1.147)$$

$$\nabla \times h - jk_0(\nabla S \times h) = j\omega\varepsilon e = j\frac{k_0 n^2}{Z_0} e, \qquad (1.148)$$

are obtained. Here, when we consider the limit where the wavelength becomes sufficiently short ($\lambda \to 0$), then $k_0 \to \infty$ according to Equation (1.41), Equations (1.147) and (1.148) will become

$$\nabla S \times e = Z_0 h, \qquad (1.149)$$

$$\nabla S \times h = -\frac{n^2}{Z_0} e, \qquad (1.150)$$

where it can be seen that the three vectors e, h, and ∇S are orthogonal to each other. Because the Poynting vector is also orthogonal to the electric field E and magnetic field H and propagates toward the propagation direction of light, its proportionality to ∇S is easily predicted. In fact, when Equations (1.150) and (1.149) are substituted to the complex Poynting vector Equation (1.52), the following equation is obtained, which will verify if the above prediction is correct:

$$\tilde{S} = \frac{1}{2}(e \times h^*) = \frac{1}{2Z_0} e \times (\nabla S \times e^*)$$

$$= \frac{1}{2Z_0}[(e \cdot e^*)\nabla S - (e \cdot \nabla S)e^*] = \frac{1}{2Z_0}|e|^2 \nabla S. \qquad (1.151)$$

Here, the set of points that satisfy $k_0 S(r) = $ const. is a set of points with the same phase, so an equiphase plane (phase front) is formed. Consequently, the $\nabla S(r)$ vector is orthogonal to this equiphase plane and the direction wherein $S(r)$ increases is the forward direction of the wave propagation, so the direction is the forward direction of the wave front. This state is represented by Figure 1.23. The lines of the gradient vector $\nabla S(r)$ (a curve that can connect the direction of the vector) are called *rays*, and $S(r)$ is called an *Eikonal*.

Next, h is expressed in terms of e using Equation (1.149), and then substituted in Equation (1.150). Using the vector formula,

$$\nabla S \times (\nabla S \times e) = (\nabla S \cdot e)\nabla S - (\nabla S \cdot \nabla S)e \qquad (1.152)$$

and the orthogonality e and ∇S, then

$$|\nabla S|^2 = n^2 \qquad (1.153)$$

is obtained. This equation is called an *Eikonal equation* and represents the propagation of a wave front in a medium with a refractive index that is not uniform and has a spatial distribution.

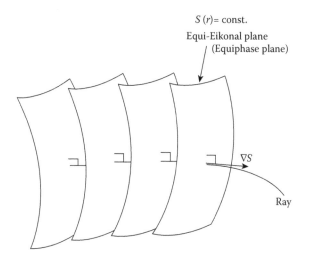

$S(r)$= const.
Equi-Eikonal plane
(Equiphase plane)

∇S

Ray

FIGURE 1.23   The wave front and the direction of the ray.

Now, when Equations (1.145) and (1.146) are compared with Equation (1.61), the optical path length S_{PQ} in Equation (1.61) is actually the difference of the Eikonal S_Q and S_P when a light wave propagates from point P to point Q and is expressed as

$$S_{PQ} = S_Q - S_P. \tag{1.154}$$

2 Fundamentals of Optical Waveguides

2.1 FREE-SPACE WAVES AND GUIDED WAVES

In contrast to a free-space wave that propagates in a free space (or in a uniform medium) as described in Chapter 1, light that propagates in a confined space limited to a cross-section perpendicular to the direction of propagation is called a guided wave, and the medium to which such light is confined is called an optical waveguide. For example, in a *slab waveguide* in which a cross-sectional structured transparent dielectric plate of high refractive index is sandwiched by dielectric materials with low refractive index, as shown in Figure 2.1, when light is beamed at an incident angle smaller than the total internal reflection angle from the high refractive index medium to a low refractive index medium, it will propagate the waveguide, repeating the total internal reflection at the upper and lower dielectric boundary surfaces. In this case, the layer of the high refractive index medium where the light is primarily guided is called the *core layer*, whereas the surrounding low refractive index layer is called the *cladding layer*. In this slab waveguide structure, the light is confined to the x-axis direction but not to the y-axis direction. Here, as shown in Figure 2.1b, if the size of the light beam is smaller than the thickness of waveguide, ray ① in the incident light will be totally reflected at point A at the upper side of the boundary surface and becomes ray ② propagating to a lower right direction and will further be totally reflected at point B at the lower side of the boundary surface and becomes ray ③ propagating to an upper right direction. In the case of a total internal reflection at point A, a standing wave S_A can be produced by the interference of the wave front belonging to incident ray ① and the wave front belonging to reflected ray ②. The node of the standing wave that is nearest to the boundary surface is produced at a position that is shifted from the boundary surface by a distance δx given by Equation (1.123). Similarly, the wave fronts belonging to ray ② and ray ③ will interfere at point B, forming a standing wave S_B.

Here, the smaller the width of the light beam compared to the thickness of the waveguide, the greater the chance of this standing wave not reaching the opposite side of the boundary surface, and so if the thickness of the guided wave becomes the same as that of the light beam, or if the light beam spreads over the entire waveguide due to the diffraction phenomenon that will be described in Chapter 3, standing wave S_A will be extended to the bottom and will be continuously linked to standing wave S_B. Then, as the wave fronts advancing diagonally

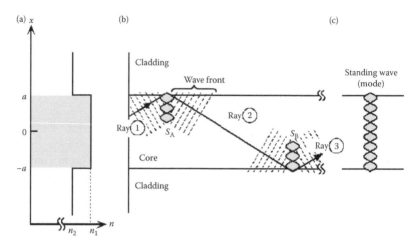

FIGURE 2.1 Cross-sectional structure of optical slab waveguide and a guided wave. (a) Cross-sectional refractive index distribution. (b) Beam propagation and standing interference wave inside the guided wave. (c) Mode.

toward the upper right and the lower right exist all over the waveguide, these will form standing waves that are the same as all those shown in Figure 2.1c, and these will spread throughout the waveguide. Because the standing wave that spreads throughout the core layer must have an integer number of nodes and antinodes in between the upper and the lower boundary surfaces, the wavelength of the light, the core thickness of the waveguide, the refractive indices of the core and the cladding, and the angle of the incident light must satisfy a particular conditional expression. Consequently, if the core width and the refractive indices of the core and cladding are given, and the wavelength of the light is also determined, the electromagnetic field distribution of the light that is guided inside that waveguide will have several nodes, and a fixed number of standing waves will propagate toward the forward direction. Thus, only some specific electromagnetic field distribution of the light waveguide is allowed to be guided inside the waveguide with different propagation constants. This particular electromagnetic field distribution is called "mode," and the concept of a mode is the most distinguishing characteristic between a guided wave and a free-space wave.

2.2 GUIDED MODE AND EIGENVALUE EQUATIONS

To form a guided mode, the angle of propagation, refractive index, wavelength, and waveguide size must satisfy certain conditions. Here, let us have a more detailed look into the propagation of light in a waveguide. As shown

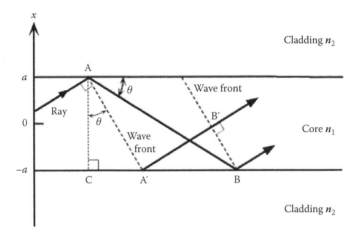

FIGURE 2.2 Propagation and interference conditions of the wave front in the slab waveguide.

in Figure 2.2, a plane wave is guided inside a slab waveguide leading to total internal reflection. As the waveguide structure in Figure 2.2 has the same refractive index for the upper and lower cladding layers, it is called a *symmetrical three-layer slab waveguide*. The basic parameters of this waveguide are the core refractive index n_1, cladding refractive index n_2, and core width $2a$ (or the core half-width a^*). The angle between the direction of propagation and the boundary of the core and cladding is defined as θ.

Here, let us consider a ray passing through two points A and A' on the same wave front. Point A is on the boundary surface of the core and the upper cladding, whereas point A' is on the boundary surface of the core and the lower cladding. Because these two points are on the same wave front, among the rays belonging to such wave front, the ray that passes through point A will be the one just before the reflection at the upper boundary surface, and the ray that passes through point A' will be the one immediately after the reflection at the lower boundary surface. After this situation, the ray that passes through point A will be reflected and will propagate toward the lower right direction, and will further be reflected at point B. At this time, it will undergo a phase change corresponding to Equations (1.121) and (1.122), along with total internal reflections at point A and point B. Then, the ray that passed by point A' will arrive at point B' and will interfere with the plane wave, which had an upper right direction immediately after undergoing reflection at point B on the lower boundary. If the phase is different from the phase at point B by integer multiples of 2π,

---

* This is called a core radius in optical fibers.

then a wave front will be formed. Similar phase relationships will continue infinitely even after subsequent reflections, and so plane waves that are repeatedly reflected will not cancel each other out by interference. Let us express such a relationship with an equation.

When a plane wave propagates a medium with a refractive index n_1 by a distance L, the phase factor will be $\exp[-jk_0n_1L] (= \exp[-jk_0S]$, where S is the optical path length) according to Equation (1.60). When the phase change corresponding to this phase factor is defined as $-k_0n_1L$,* the conditions described earlier will have the following equation:

$$-k_0n_1\overline{AB} + \Phi_A + \Phi_B = -k_0n_1\overline{A'B'} - 2\pi N. \qquad (2.1)$$

In this equation, $\overline{AB}$ and $\overline{A'B'}$ are the respective distances between points A and B and points A' and B', Φ_A and Φ_B are the respective quantities of phase changes for the total internal reflection at points A and B, and N is an integer (positive integer). Now, when simple geometry (like a quiz) is used, $\overline{AB}$ and $\overline{A'B'}$ can each be expressed, as follows, using the core half-width a and the angle of propagation θ:

$$\overline{AB} = \frac{2a}{\sin\theta}, \qquad (2.2)$$

$$\overline{A'B} = \frac{2a}{\tan\theta} - 2a\tan\theta, \qquad (2.3)$$

$$\overline{A'B'} = \left(\frac{2a}{\tan\theta} - 2a\tan\theta\right)\cdot\cos\theta$$

$$= \frac{2a\cos^2\theta}{\sin\theta} - 2a\sin\theta$$

$$= \frac{2a}{\sin\theta} - 4a\sin\theta. \qquad (2.4)$$

Because the refractive indices of the upper and the lower cladding are the same, the phase change due to the total internal reflection is $\Phi_A = \Phi_B = \Phi$. When Equations (2.2) and (2.4) are substituted to Equation (2.1), and after some calculations,

$$2k_0n_1a\sin\theta = \Phi_i + \pi N \quad (i = \text{TE or TM}) \qquad (2.5)$$

---

* When the starting point electric field represented by E_i, and the electric field after propagation is represented by E_o, when time is stopped and both are compared at the same time, $\frac{E_o}{E_i} = \exp[-jk_0n_1L]$, so the quantity of the phase change is defined as $-k_0n_1L$. The time dependency of wave motion has been defined as $\exp[j\omega t]$ in accordance to the customary practice of electrical engineering, but in optics, which is a branch of physics, this has been conventionally defined as $\exp[-i\omega t]$. Thus, in accordance to the definition used in optics, the quantity of the phase change is k_0n_1L. On the other hand, since the phase change due to the total internal reflection is $r = \frac{E_o}{E_i} = |r|\exp(j\Phi)$, the quantity of the phase change is Φ.

is obtained. In Equation (2.5), the phase change is given by Equation (1.121) for TE polarization and Equation (1.122) for TM polarization. Then, using TE polarization as an example, substituting Equation (1.121) into Equation (2.5),

$$k_0 n_1 a \sin \theta = \tan^{-1} \left[\sqrt{\frac{2\Delta}{\sin^2 \theta} - 1} \right] + \frac{\pi}{2} N \quad (N = 0, 1, 2, \dots) \qquad (2.6)$$

is obtained. When the waveguide structure (refractive indices of the core and cladding n_1 and n_2 and the core half-width a) and the wavelength $\lambda (k_0 = \frac{2\pi}{\lambda})$ of the light source are given, the solution will be $\sin \theta$. The angle of propagation is thus determined. In other words, the angle of propagation of the light in the waveguide can only be a certain angle determined by Equation (2.6). This specific value is called *eigenvalue* in mathematics. An equation just like Equation (2.6) that gives an eigenvalue is called an *eigenvalue equation*.

Here, when the electromagnetic distribution of the guided wave is expressed by

$$E = E^0(x, y) \exp[j(\omega t - \beta z)] \qquad (2.7)$$

and using the angle of propagation θ, the propagation coefficient β is expressed as

$$\beta = k_0 n_1 \cos \theta. \qquad (2.8)$$

2.3   EIGENMODE AND DISPERSION CURVES

Equation (2.6) has four parameters, namely, the core and cladding refractive indices n_1 and n_2, the core half-width a, and the wavelength λ of the light source. The equation will have to be solved if any of these parameters change. In such a case, when the parameters are lumped together and the number of parameters is reduced, the outlook will be much better. This operation is called normalization.

Here, when we consider the possible range of propagation constants, it is easy to see that $\beta < k_0 n_1$ from Equation (2.8). In addition, for guided modes, the angle of propagation must satisfy the condition $\theta < \theta_c$, wherein it must be smaller than the angle of total internal reflection. With this condition and Equation (1.117), it is seen that propagation constant will lie in the following range:

$$k_0 n_2 \leq \beta < k_0 n_1. \qquad (2.9)$$

In other words, electromagnetic field of the *guided mode* propagates almost confined inside the core, and because a portion of the light is leaked to the cladding layer corresponding to the phase delay at the core and the cladding boundaries, it is thought to propagate across both the core and the cladding. Consequently, the propagation constant is regarded as an intermediate value between the propagation constant of the plane wave propagating inside the core and the propagation constant of the plane wave propagating in the cladding. Because the possible range of the propagation constant is already known, the propagation constant can be normalized using the following equation:

$$b = \frac{(\beta/k_0)^2 - n_2^2}{n_1^2 - n_2^2} = \frac{n_1^2 \cos^2 \theta - n_2^2}{n_1^2 - n_2^2} = 1 - \frac{\sin^2 \theta}{2\Delta}. \tag{2.10}$$

This normalized parameter is called the *normalized propagation constant*, where corresponding to Equation (2.9), its range for the guided mode is

$$0 \le b < 1. \tag{2.11}$$

Here, the ratio of the propagation constant compared with the propagation constant of the plane wave in vacuum is

$$n_{eq} = \frac{\beta}{k_0}, \tag{2.12}$$

which is called the *equivalent index*.* This can be interpreted as the equivalent refractive index that the guided mode feels as propagating across both the core and the cladding.

In addition, the waveguide parameter is also defined as

$$V = k_0 n_1 a \sqrt{2\Delta}. \tag{2.13}$$

Here, Δ is the relative index difference defined in Equation (1.118) of Section 1.4.4. This waveguide parameter is called the *V parameter*.† When Equations (2.10) and (2.13) are used, the eigenvalue Equation (2.6) can only be written with the normalized parameters V and b, giving the following [1]:

$$V = \frac{1}{\sqrt{1-b}} \left[\tan^{-1} \sqrt{\frac{b}{1-b}} + \frac{\pi}{2} N \right] \quad (N = 0, 1, 2, \ldots). \tag{2.14}$$

---

* In microwaves, Equation (2.12) squared is called the *effective dielectric constant*, so Equation (2.12) is sometimes called the effective index. However, in the field of optics, as will be described in Section 4.2, $n_{eff} = n - \lambda \frac{dn}{d\lambda}$, which includes the wavelength dependence of the refractive index, appears in the equation that gives the resonant wavelength interval for resonators such as semiconductor lasers, leading to confusion as even this is called the effective refractive index. This is the reason why here the author decided to use the term, equivalent index.
† This was previously called the *normalized frequency*, but actually being a dimensionless quantity, this term is recently not often used.

Ideally, this eigenvalue equation may have already been solved if b is expressed as a function of V; however, as this is not such an equation analytically, it is written as above. Consequently, to draw a graph with V in the horizontal axis and b in the vertical axis, assuming N to be an integer, V is determined from Equation (2.14) for a given value of b. Then, the vertical and horizontal axes are exchanged. The graph drawn this way is shown in Figure 2.3. Here, the graph drawn when the V parameter is assigned to the horizontal axis and the normalized propagation constant b is positioned to the vertical axis is called the *dispersion curve*. Figure 2.3 is a dispersion curve for a symmetrical three-layer slab waveguide. If the waveguide structure is different, the dispersion curve will also be of a different shape. However, if the waveguide structure is the same, even with a change in the refractive index of the core or the wavelength, only the value of the V parameter determined from Equation (2.13) will change. Thus, one will have to change the horizontal axis value in Figure 2.3, read the value of b from the vertical axis, and then determine the actual propagation coefficient from Equation (2.10). In other words, by just drawing the dispersion curve once, the method for determining the propagation constant for each mode when given the waveguide structure and wavelength has been simplified, as shown in Figure 2.4.

A few points can be grasped from Figure 2.3. First, it is possible for the integer N to start from zero and become a positive value; however, the propagation constant is biggest at $N = 0$ mode, and every time N increases, the propagation constant will decrease. This N is called the mode order or the mode

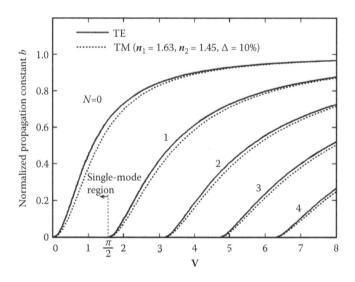

FIGURE 2.3 Dispersion curve for symmetrical three-layer slab waveguide.

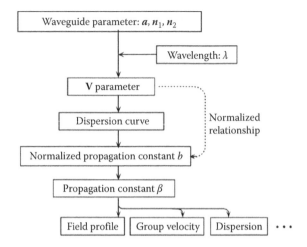

FIGURE 2.4 Flowchart for determining the propagation constant.

number, and the term used will be zero-order mode, first-order mode, and so on.* Moreover, as Equation (2.14) and Figure 2.3 are the eigenvalue equation and dispersion curve, respectively, for the TE-polarized light, the mode order is written as a subscript for the symbol that represents TE-polarized light; thus, it is written as TE_0 mode, TE_1 mode, and so on. As shown in Figure 2.3, when $V = \frac{N\pi}{2}$, $b = 0$, the dispersion curve of the TE_N mode will not exist for a region where the V parameter is smaller than $V = \frac{N\pi}{2}$. The point just as $b = 0$ is called the *cutoff*. Moreover, the number of guided modes M possible to be supported in a waveguide with a V value is given by

$$M = \left[\frac{2}{\pi}V\right] + 1. \tag{2.15}$$

Here, the function $[x]$ is called the Gaussian bracket and represents the largest integer that does not exceed the argument x.

Now, the cutoff V value for TE_1 mode is $\frac{\pi}{2}$, and if the V value is smaller than this value, only one mode will exist. The region for this kind of V is called the single-mode region, and the waveguide in this state is called the *single-mode waveguide*. In a single-mode region, only the zero-order mode (TE_0 and TM_0) will exist and this is called the *fundamental mode*. The cut-off for the fundamental mode does not exist in a symmetrical three-layer

---

* In regions where multiple modes exist, there is a rule that the mode number should be sorted in descending order of the propagation constant. In a slab waveguide, N is a mode number and N will start from 0. However, even in optical fibers, the mode number starts from 1, so care should be taken to avoid confusion.

slab waveguide*. The condition when the waveguide becomes a single-mode waveguide is called the single-mode condition and is given by Equation (2.16) for the step-index symmetrical three-layer slab waveguide

$$V \le \frac{\pi}{2}.$$ (2.16)

Conversely, when the V value is larger than the cutoff V value for the TE_1 mode, the region where multiple guided modes exist is called the multimode region, and such a waveguide is called a *multimode waveguide*.

A TE-polarized mode is assumed for Equations (2.14) and (1.121) and was used for the phase change Φ of the total internal reflection in Equation (2.5). For the TM-polarized mode, the following equation is derived using Equation (1.122):

$$V = \frac{1}{\sqrt{1-b}} \left[\tan^{-1} \left\{ \left(\frac{n_1}{n_2} \right)^2 \sqrt{\frac{b}{1-b}} \right\} + \frac{\pi}{2} N \right].$$ (2.17)

The TE mode eigenvalue equation (2.14) does not include the core and cladding refractive indices; however, the TM mode eigenvalue equation (2.17) involves n_1 and n_2. Consequently, the shape of the dispersion curve will change depending on the refractive index difference of the core and the cladding, and the propagation constant will be smaller than that of the TE mode. However, if the difference between n_1 and n_2 is small and Δ is at most about 1%, the difference between the dispersion curve of the TE and the TM modes is about 1 pencil line. Moreover, if the mode order is the same, the cutoff V value will also be the same.

2.4 ELECTROMAGNETIC DISTRIBUTION AND EIGENMODE EXPANSION

The general shape of the electromagnetic distribution in Equation (2.14) and Figure 2.3 is puzzling. So, returning to Equation (2.6), let us draw the general shape of the electromagnetic distribution. The relationship between the z-direction propagation constant β and the angle of propagation θ is shown by Equation (2.8). The transverse (x-direction) propagation constant is defined by

$$\kappa = k_0 n_1 \sin \theta = \sqrt{k_0^2 n_1^2 - \beta^2}.$$ (2.18)

Because the waveform of the transverse standing wave can be written in the form of a $\cos(\kappa x)$ or $\sin(\kappa x)$, it is possible to evaluate the number of standing waves by the magnitude relationship between κa and $\frac{\pi}{2}$.

---

* As discussed in Chapter 5, a cutoff exists in the fundamental mode for asymmetrical waveguides.

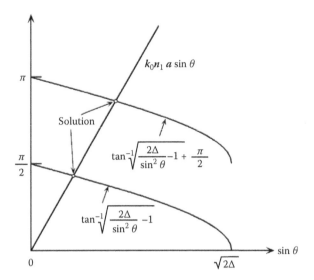

FIGURE 2.5   Left side and right side of Equation (2.6).

When the left side and the right side of Equation (2.6) are each drawn in a graph with $\sin\theta$ as a variable in the horizontal axis, it will look like Figure 2.5. The slope of the straight line that represents the left side is $k_0 n_1 a$ and the curve that represents the right side is a multiple-valued curve because each of them is obtained by shifting the fundamental mode by an integer multiple of $\frac{\pi}{2}$, which is $\frac{\pi}{2}$ when $\sin\theta = 0$ and is zero when $\sin\theta = \sqrt{2\Delta}$. When the slope representing the left side is large, many intersection points with the multiple-varied curve of the right side will exist, and it will become a multimode waveguide. However, when the value of the left side for $\sin\theta = \sqrt{2\Delta}$ is smaller than $\frac{\pi}{2}$ ($k_0 n_1 a\sqrt{2\Delta} = V \leq \frac{\pi}{2}$), the straight line of the left side will only have one intersection point with the multiple-varied curve. That is, it will become a single-mode waveguide. This condition is consistent with the single-mode condition of Equation (2.16) and is also in agreement with the total internal reflection condition of Equation (1.117).

Now, the intersection point in Figure 2.5 gives the solution for the guided mode. The vertical axis coordinate of this intersection is given by the value of $k_0 n_1 a \sin\theta = \kappa a$ and is within a range of $0 < \kappa a < \frac{\pi}{2}$ in the fundamental mode. The core width of the waveguide is $2a$, and consequently even if it propagates the distance $2a$ with the transverse propagation constant κ inside the core, the range of the phase change is $0 < 2\kappa a < \pi$. Therefore, only one transverse standing wave will be formed inside the core. If $2\kappa a$ is exactly equal to π, only one standing wave will exist, and a node will be formed at the boundaries of the core and the cladding. However, $2\kappa a$ is actually smaller than π, and the difference between $2\kappa a$ and π is equal to the phase change due to the total internal reflection. So, there appears to be a node at a position recessed from the boundary surface by the distance given by Equation (1.123). By adding this distance, which reaches equivalent reflection surface inside the cladding to a core half-width a, the equivalent core half width $a + \delta x_i = a_{\text{eff}}$ can be represented as follows from Equations (1.123), (2.5), and (2.18). Actually,

$$\frac{a_{\text{eff}}}{a} = \frac{(N+1)\pi}{2V\sqrt{1-b}} \tag{2.19}$$

the electric field distribution inside the cladding will decay exponentially as discussed in Section 5.1.2. Moreover, when the mode order N becomes large, the equivalent core half width in Equation (2.19) will increase. That is, the penetration depth of the electromagnetic field into the cladding will increase. The electric field profile is drawn, as shown in Figure 2.6, taking into account all these relations.

The dependence of electric field profile on the mode order N has already been discussed. The phase change that should be substituted to Equation (2.5) is Equation (1.121) for TE polarization and Equation (1.122) for TM polarization. Then, even if the mode order N is the same, the electromagnetic field of the mode will be a bit different depending on the polarization (this difference will be discussed in detail in Chapter 5). Light propagates in waveguides by virtue of repeated total internal reflections at the boundary surface, and

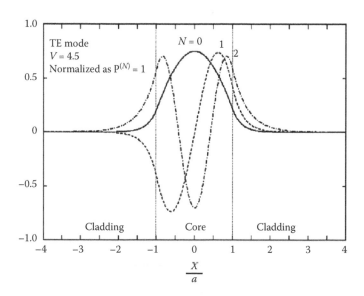

FIGURE 2.6 Electric field profile of the symmetrical three-layer slab waveguide.

as described in Exercises 1-7 of Section 1.4.2, the polarization state can be resolved into TE and TM polarizations. Thus, the light with any polarization state in the optical waveguide can be expressed as the sum of TE and TM polarizations. This difference in the polarization state is also called a mode, and these will be referred to as the *TE mode* (transverse electric mode) or the *TM mode* (transverse magnetic mode) (see Figure 2.7).

Finally, let us consider the relation between the propagation constant β of the mode, the transverse propagation constant κ, and the angle of propagation θ of the ray. As described in Section 1.7, the ray can be considered as a trajectory of a line (or an envelope curve) normal to the wave front. On the other hand, the propagation vector of the plane wave (propagating obliquely at an angle θ), which constitutes the mode, is also facing the direction perpendicular to the wave front, the magnitude of which is $k_0 n_1$. Consequently, if this propagation vector is drawn to correspond to Figure 2.2, then a vector with a magnitude of $k_0 n_1$ in the direction of the ray must first be depicted, as shown in Figure 2.8.

Because the plane wave forming the mode propagates in right upward and right downward directions, the propagation vector can also be drawn in right upward and right downward directions. Here, let us first consider a propagation vector with a right upward direction. From each of the Equations (2.8) and (2.18), the z and the x component of this propagation vector can be written as

$$\beta = k_0 n_1 \cos \theta, \tag{2.8}$$
$$\kappa = k_0 n_1 \sin \theta. \tag{2.18}$$

The β and κ expressed by Equations (2.8) and (2.18) will satisfy the $\beta^2 + \kappa^2 = k_0^2 n_1^2$ relationship. As is also apparent from the figure, β is the propagation constant of the guided mode (generally speaking, the propagation constant usually refers to the z-direction propagation constant of the propagation direction) and κ is the x-direction propagation constant. Because the right downward propagation vector of the plane wave with repeated total internal reflections

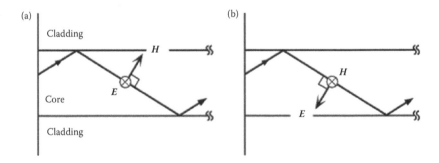

FIGURE 2.7 Polarization mode of the slab waveguide: (a) TE mode; (b) TM mode.

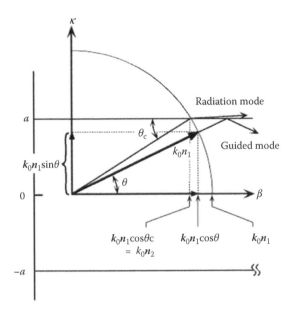

FIGURE 2.8 Correspondence between the propagation vector of the plane wave forming a mode and the angle of propagation of the ray.

has the x component $-\kappa$, interference (expressed as $\cos \kappa x$ and $\sin \kappa x$ as the sum and difference of $\exp(-j\kappa x)$ and $\exp(j\kappa x)$) with the x component κ of a propagation vector with a right upward direction occurs in the x-direction, resulting in the formation of a mode pattern. With this, we can imagine that the propagation will be toward the z-direction at a propagation constant β. When we think this way, we can easily see that the angle of propagation θ will become a critical angle and $\beta = k_0 n_2$ $(b = 0)$, when the mode reaches a cutoff. That is, the following equation will be obtained from Figure 2.8:

$$\theta_c = \cos^{-1} \left(\frac{\beta}{k_0 n_1} \right)_{\beta = k_0 n_2} = \cos^{-1} \left(\frac{n_2}{n_1} \right), \qquad (2.20)$$

and this is consistent with Equation (1.117). If the propagation constant of a guided mode becomes smaller than $k_0 n_2$, the angle of propagation θ of the mode will become larger than the angle of total internal reflection θ_c, and the rays will never undergo total internal reflection at the boundary surface of the core and the cladding. In such a case, a portion of light will be refracted and then radiated into the cladding. Such a mode, wherein the angle of propagation exceeds the angle of total internal reflection that results in radiation to the cladding, is called the *radiation mode*. In contrast to this, the mode wherein the rays are just confined to the core is called the *guided mode*. The radiation mode has a sinusoidal wave-like electric field profile toward the x-direction inside the core, but as the cladding also has the same sinusoidal wave-like electric field profile toward the x-direction, there is no need to form a standing wave inside the core. Consequently, there is no need for the propagation constant β of the radiation mode to have a discrete eigenvalue, so it will become a continuous value. When this state is shown on the β axis, it will look like the lower part of Figure 5.5. Because both the guided mode and the radiation mode are electromagnetic distributions that are determined based on the structure of the optical waveguide, these are generally called the *eigenmode*.

The major difference between guided waves and free-space waves is that the light propagating inside the waveguide allows only a particular form of electric field distribution (the magnetic field distribution can be calculated from the electric field distribution) represented by the eigenmode. However, an electromagnetic field incident on the input end of the waveguide may not have exactly the same electromagnetic field distribution as that of the waveguide eigenmode. So, if any electromagnetic field distribution is beamed, how will it be converted to a light propagating inside a waveguide? The answer to this involves the expansion of any electric field distribution in terms of the waveguide eigenmode, and then the electromagnetic field distribution of each of the eigenmodes propagates separately.

This is expressed mathematically by an orthogonal function expansion[*] and is called an *eigenmode expansion*. That is, if the scalar component[†] of the transverse electromagnetic field distribution of the guided mode is expressed as $f_j(x)$[‡] (j is a mode order), generally, for guided modes with different mode orders, the overlap integral given in the next equation becomes zero

$$\int_{-\infty}^{\infty} f_i(x) f_j(x)\, dx = \int_{-\infty}^{\infty} f_i^2(x)\, dx \cdot \begin{cases} \delta_{ij} & : \text{guided mode} \\ \delta(\beta_i - \beta_j) & : \text{radiation mode.} \end{cases}$$

$$(2.21)$$

Here, δ_{ij} is called a Kronecker's delta and is a function for integers expressed by

$$\delta_{ij} = \begin{cases} 1 & : \text{when } i = j \\ 0 & : \text{when } i \neq j, \end{cases} \qquad (2.22)$$

---

[*] The well-known Fourier series expansion is also one type of orthogonal function expansion. In addition, there are also a lot of orthogonal function systems in existence.

[†] While primarily a vector quantity, all of the other electromagnetic components are expressed in terms of E_y for the TE mode and H_y for the TM mode, so the field profile can be expressed as a scalar quantity.

[‡] In the slab waveguide, there is no confinement of light in the y-direction, so this is only a function of x. In the channel waveguide where confinement of light occurs in both the x- and y-directions, this should be written as $f_j(x, y)$ as will be discussed in Chapter 5. Moreover, the electromagnetic field should be expressed as a vector.

whereas $\delta(x)$ is a Dirac delta function and satisfies the following formula*
(refer to Appendix B) for an argument of a continuous number

$$\int_{-\infty}^{\infty} f(x)\delta(x)\,dx = f(0) \tag{2.23}$$

$$\text{and} \quad \delta(x) = 0 \quad \text{at } x \neq 0$$

where $f(x)$ is an arbitrary function, which is continuous at $x = 0$. Equation (2.21) should accurately be expressed using the vector electromagnetic field as will be discussed in Chapter 5; however, it can be simply expressed as a scalar function in slab waveguides. Then, if Equation (2.21) holds, the scalar function $g(x)$ of the incident electromagnetic field can be expanded as

$$g(x) = \sum_{j=0}^{N_{max}} C_j f_j(x) + \int_0^{k_0 n_2} C(\beta) f(\beta, x)\,d\beta. \tag{2.24}$$

Here, N_{max} is the maximum order of the guided mode. Meanwhile, for the radiation mode, the propagation constant is a continuous number, so the expansion coefficient and the scalar electromagnetic field function are expressed as a function of β, where they will be expressed not as a sum but as an integral. Moreover, the guided and the radiation modes are assumed to be the sum of both the TE mode and the TM mode. The expansion coefficient C_i is obtained by multiplying $f_i(x)$ to both sides of Equation (2.24), and using the relationship with Equation (2.21) as follows:

$$C_i = \frac{\int_{-\infty}^{\infty} f_i(x)g(x)\,dx}{\int_{-\infty}^{\infty} f_i^2(x)\,dx}. \tag{2.25}$$

Any electromagnetic field can be expanded into an eigenmode this way, and the component that has been expanded to an eigenmode will propagate inside the waveguide at a propagation constant β_j corresponding to each mode order j. The phase difference between each of the modes and the electromagnetic field distribution after the superposition of guided modes will change along with the propagation. Then, as shown in Figure 2.9, the incident light with an offset that is axially deviated to the lower side of the incident end will, along with the propagation, gradually shift to the upper side of the waveguide and then will meander back to the lower side. This will occur repeatedly. For an optical waveguide with a waveguide width of several centimeters, the mode number will be from several ten thousands to several hundred thousands,

---

* Generally speaking, a hypermetric function has a different definition, but remembering this definition will be helpful in calculations.

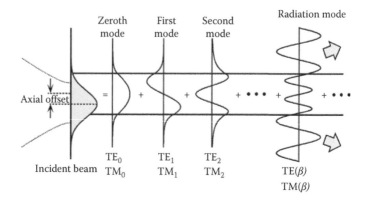

FIGURE 2.9   Eigenmode expansion of incident light beam.

so when a light beam with a diameter of a few millimeters is beamed obliquely, the meandering of the beam can be observed as a repeated total internal reflection of the light beam.* On the other hand, the component that was expanded into a radiation mode at the incident end will not be confined inside the waveguide but will be radiated to the cladding and will sufer radiation loss. This loss wherein the light is not absorbed but is rather radiated outside the core and is lost is called *radiation loss*. In particular, this is called a *coupling loss* as the loss occurs due to the poor coupling with the waveguide at the incident end. Because reflection occurs at the incident end, another loss called the *return loss* or the *Fresnel loss* occurs. Moreover, when light propagates through a waveguide, a *coupling loss* (as a phenomenon, it will be referred to as a radiation loss) occurs if offset or angular misalignment of the central axis of the waveguide happens midway to the coupling point. *Scattering loss* occurs when the boundary surface between the core and the cladding involves roughness or there is a fluctuation of the refractive index inside the waveguide. *Absorption loss* occurs when there are absorbers such as metallic ions (actually, the OH^- ion becomes an absorber in optical fibers), and if the waveguide bends (details will be discussed in Section 8.5), a *bending loss* (or bend loss) will occur. A summary of these phenomena is shown in Figure 2.10.

On the other hand, as only the fundamental mode exists in a single-mode waveguide, the component that was not expanded to the fundamental mode appears as the coupling loss. Consequently, to lower the coupling loss as much as possible (that is, to make the coupling efficiency for the fundamental mode

---

* If one buys an acrylic rod several centimeters in size from a DIY shop, one can observe this phenomenon with a simple experiment, wherein a laser pointer light is passed through the material.

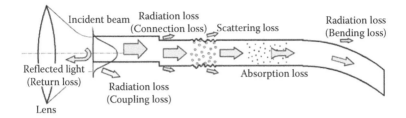

FIGURE 2.10 Various losses occurring in the waveguide.

as large as possible), the expansion coefficient C_0 for the zero-order mode given by Equation (2.25) should be made bigger. To realize this, $f_0(x)$ and $g(x)$ in the numerator should approach the same form as much as possible for the coefficient to become bigger ($g(x) = f_0(x)$, then $C_0 = 1$), and so the form of the incident light beam and the form of the fundamental mode of the waveguide should be the same.

2.5 FUNDAMENTAL PROPERTIES OF MULTIMODE WAVEGUIDES

In multimode waveguides where the number of guided modes are numerous ("numerous" means several hundreds to more than several thousands, wherein the angle of propagation can almost be considered as a continuous number), the angle of propagation of the highest order mode is considered to be almost equal to the angle of total internal reflection. In such a case, we can approximate that the angle of propagation of the guided mode will be continuously distributed in a range between $\theta = 0$ and $\theta = \theta_c$, and the highest order mode, which has an angle of propagation equal to the angle of total internal reflection, will be refracted at the incident end, which will have an angle of emergence of θ_{max}, as shown in Figure 2.11. If the medium outside the incident and exit ends is air ($n = 1.0$), Snell's law at the exit end can be expressed using Equation (1.86) (here, θ is the angle with respect to the normal line), so θ_{max} will be

$$2\theta_{max} = 2\sin^{-1}(n_1 \sin \theta_c) = 2\sin^{-1}(n_1 \sqrt{2\Delta}) = 2\sin^{-1}\sqrt{n_1^2 - n_2^2},$$
(2.26)

where Δ is the relative index difference defined by Equation (1.118) of Chapter 1. For optical waveguides where n_1 and n_2 are used as the refractive indices of the core and the cladding, it is defined exactly in the same way as

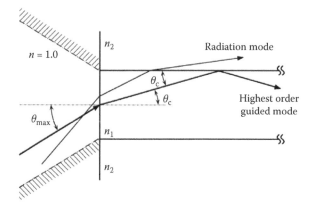

FIGURE 2.11 Maximum incident angle and maximum emergence angle of multimode waveguides.

$$\Delta = \frac{n_1^2 - n_2^2}{2n_1^2} \simeq \frac{n_1 - n_2}{n_1} \text{ (approximating, when } n_1 + n_2 \simeq 2n_1) \quad (2.27)$$

and is usually expressed in %. This angle is called the *maximum acceptance angle* because, conversely, at the incident end, only rays that are incident at an angle of incidence that is smaller than this angle will be guided at an angle of propagation smaller than the angle of total internal reflection inside the waveguide. Moreover,

$$\text{NA} = \sin \theta_{max} = n_1 \sqrt{2\Delta} = \sqrt{n_1^2 - n_2^2} \quad (2.28)$$

is called the *numerical aperture* and its acronym *NA* is also used to refer to the term.*

PROBLEM 2-1

Given a core half-width $a = 1.55$ μm, a wavelength of the ray $\lambda = 1.55$ μm, $n_1 = \frac{5}{3}$, and $n_2 = 2\sqrt{\frac{2}{3}}$, what will be the values of Δ and the V parameter? Here, express the V parameter as a constant containing π.

---

* In general, convergent and divergent light beams can be defined by NA, where $\text{NA} = \sin \theta$ and θ is the angle of convergence or the angle of divergence measured from the central axis. Moreover, for lenses, the θ used in $\sin \theta$ is the convergence angle θ of rays in a parallel pencil of rays converging from the outermost periphery of the lens toward the focus point on the central axis of the lens. See Chapter 3 for details.

SOLUTION 2-1

In this sample problem, the refractive indices of the core and the cladding were given unconventional numbers. The values $n_1 \simeq 1.667$ and $n_2 \simeq 1.633$ correspond to glass materials with fairly high refractive indices. The refractive index of silica glass used in optical fiber is about 1.45 at a wavelength region of 1.55 μm. The deliberate use of $\sqrt{}$ to designate a value was to obtain a clean number without using approximation for Equation (2.27), and so as defined by Equation (2.27), $\Delta = 0.02 = 2\%$. Moreover, substituting the value of wavelength to $k_0 = \frac{2\pi}{\lambda}$ in defining Equation (2.13) and then calculating, we will obtain $V = \frac{2\pi}{3}$.

[End of Solution]

PROBLEMS

1. For the waveguide in Exercise 2-1, how many TE- and TM-guided modes can propagate?
2. For this waveguide to become single mode, what is the minimum wavelength in micrometers of light that should be beamed? Answer with a significant three-digit number.
3. The normalized propagation constant for the TE zero-order mode (TE$_0$) of this waveguide is $b = 0.75$. What will be the value of the V parameter for this condition? Here, express the V parameter as a constant containing π.
4. Consider a metal waveguide with an air layer as the core and a metal as the cladding.
5. In metal waveguides, only one of either the TE or the TM mode can be propagated. Which kind of polarization can possibly be propagated?
6. The node of the electric field distribution can be formed roughly near the surface of the metal, so when the thickness of the core is set as D, the angle of propagation can be approximated as $\sin \theta \simeq \theta = \frac{\lambda}{2D}$. Assuming that the metal is gold, find the reflectance R at an angle of incidence θ to the metal surface at a wavelength of $\lambda = 1.55$ [μm].
7. Express the number of reflections M per unit guided length of the waveguide as a function of θ and D. Then, determine the approximate value of the propagation loss (in this case, the absorption loss) in decibels per unit length from $\alpha = 10 \log_{10}(R^M)$.

2.6   TRANSMISSION BAND OF MULTIMODE WAVEGUIDE

In the sections 2.1–2.5, we have discussed the characteristics of the propagation of light within an optical waveguide. However, when using a signal transmission line such as optical fiber, the characteristics of the signal propagation in a time domain and frequency domain are important. Thus, in this section, we will mainly derive the signal transmission characteristics of the multimode waveguide (or the multimode fiber). Because a detailed analysis of single-mode fibers requires the treatment of wave behavior, the derivation of such will be done in Part II.

2.6.1 PHASE VELOCITY AND GROUP VELOCITY

The phase velocity for a plane wave was derived in Exercise 1-5 of Section 1.3.3. As shown in Figure 2.12a, this phase velocity can be defined when the single-frequency (monochromatic) waves continue infinitely; however, information cannot be sent as in the same case of direct current. To transmit information using an electromagnetic wave, the amplitude and the phase (or the frequency) must be changed to correspond to the signal, as shown in Figure 2.12b. That is, modulation has to be applied. The propagation velocity of the modulated signal waveform is called a group velocity and can be calculated as follows:

As an example of the modulated signal, let us consider the simplest sinusoidal wave amplitude modulation. Now, if sinusoidal waves with the angular frequencies $\omega_0 + \Delta\omega$ and $\omega_0 - \Delta\omega$ are superimposed, the beatwave at the origin of propagation $z = 0$ can be expressed as

$$f(0,t) = \exp[j(\omega_0 + \Delta\omega)t] + \exp[j(\omega_0 - \Delta\omega)t]$$
$$= 2\cos(\Delta\omega t)e^{j\omega_0 t}. \qquad (2.29)$$

This waveform is an amplitude-modulated waveform, which is formed by applying an amplitude modulation of a sine wave with angular frequency $\Delta\omega$ as an amplitude envelope on the sine wave with angular frequency ω_0. When this waveform propagates by a distance z, the time waveform will be

$$f(z,t) = \exp[j\{(\omega_0 + \Delta\omega)t - \beta z\}] + \exp[j\{(\omega_0 - \Delta\omega)t - \beta z\}]; \qquad (2.30)$$

however, the value of the propagation constant at an angular frequency $\omega_0 + \Delta\omega$ does not have to be the same value at an angular frequency $\omega_0 - \Delta\omega$. Rather, the refractive index of a medium other than vacuum is generally frequency-dependent (i.e., wavelength-dependent) and the propagation constant in optical waveguides is wavelength-dependent due to the guided-wave phenomenon. Here, the angular frequency dependency of the propagation constant is expressed as

$$\beta(\omega) = \beta(\omega_0) + \left(\frac{d\beta}{d\omega}\right)_{\omega=\omega_0} (\omega - \omega_0) + \cdots, \qquad (2.31)$$

and if up to the second term of the above expansion is substituted to Equation (2.30), given that $\omega - \omega_0 = \Delta\omega$,

$$f(z,t) = 2\cos\left[\Delta\omega\left\{t - \left(\frac{d\beta}{d\omega}\right)_{\omega=\omega_0} z\right\}\right] \exp[j\{\omega_0 t - \beta(\omega_0)z\}] \qquad (2.32)$$

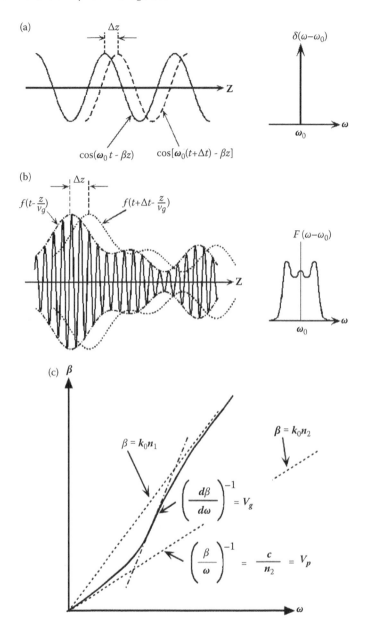

FIGURE 2.12 Phase velocity and group velocity. (a) Spectra and phase velocity of the single-frequency sinusoidal wave. (b) Spectra and group velocity of the amplitude-modulated sinusoidal wave. (c) Dispersion relationship of the fundamental mode of the waveguide.

will be obtained. Equation (2.32) shows that the signal waveform of the envelope oscillating at an angular frequency $\Delta\omega$ will propagate at a velocity of

$$v_g = \left(\frac{d\beta}{d\omega}\right)^{-1}. \tag{2.33}$$

This velocity is called the *group velocity*. As already mentioned, the propagation velocity of the signal waveform is given by the group velocity. Moreover, when the signal waveform is thought to be an isolated pulse waveform, as the isolated pulse is an energy packet, the propagation velocity of the energy of light is given by the group velocity.

From Equations (1.30), (1.45), and (1.58), the propagation constant of the plane wave in a medium or in vacuum with a refractive index n that is not frequency-dependent is $\beta = \frac{\omega n}{c}$. Consequently, the group velocity is $\frac{c}{n}$ and is equal to the phase velocity. However, in the optical waveguide (including optical fiber), since the refractive index of the medium depends on frequency[*] and the equivalent index n_{eq} is also frequency dependent resulting from the waveguide structure, β will not be a linear function of ω. Consequently, the group velocity will be different from the phase velocity.

Now, how do we consider the group velocity of the mode propagated inside the optical waveguide? It can be obtained by differentiating the propagation constant β of the optical waveguide with respect to ω in accordance to Equation (2.33); however, the exact calculation will be discussed in Section 6.2, and here let us consider the ray and plane waves to obtain an intuitive understanding. That is, the ray and plane waves are not expressed with modulated optical signals, so it is necessary to consider the propagation velocity of the energy to derive the group velocity. The phase velocity of the plane wave obliquely propagating at the angle of propagation θ in the core medium with refractive index n_1 is $\frac{c}{n_1}$, and as shown in Figure 2.13a, the propagation velocity of the equiphase plane becomes

$$\frac{c}{n_1}\cdot\frac{1}{\cos\theta} = \frac{\omega}{k_0 n_1 \cos\theta} = \frac{\omega}{\beta} = v_p, \tag{2.34}$$

making it equal to the phase velocity. On the other hand, the energy as shown in Figure 2.13b will be slower in the z-direction by a factor of $\cos\theta$ and is given by

$$v_g = \frac{c}{n_1}\cdot\cos\theta. \tag{2.35}$$

---

[*] Wavelength and frequency dependency of physical quantities such as these are called dispersion. However, dispersion in optical fiber refers to the second derivative of ω of the propagation constant β, and it is an important physical quantity that is used to determine the pulse width broadening of the optical pulse. Its detailed derivation is described in Chapter 6.

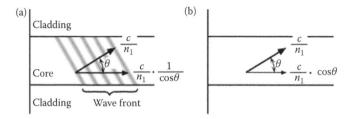

FIGURE 2.13 Propagation velocity of the equiphase plane and propagation velocity of the energy flow. (a) Propagation velocity (phase velocity) toward the z-direction of the equiphase plane. (b) Propagation velocity (group velocity) toward the z-direction of the energy flow.

Here, when the product of Equations (2.34) and (2.35) are taken, we can see that

$$v_p v_g = \frac{c^2}{n_1^2} = v^2 \tag{2.36}$$

holds. Here, v is the phase velocity of the plane wave in a medium with a refractive index of n_1, and this equation holds if the wavelength dependency of the refractive index is not taken into consideration.

2.6.2 Pulse Propagation and Frequency Response in Multimode Waveguides

Because the angle of propagation θ is larger for higher order modes, it can be seen from Equation (2.35) that the group velocity will be delayed for higher order modes. Consequently, the lower order mode will propagate faster, whereas the higher order mode will propagate slower in the multimode optical waveguide and multimode optical fiber with homogeneous refractive indices inside the core, so even if a short optical pulse is beamed at the incident end, it will become a pulse with a broadened temporal waveform at the exit end. In other words, it will not be suitable for high-speed transmission as the pulse interval cannot be shortened.

Let us first discuss the state of this pulse propagation on the time axis. As the angle of propagation of the lowest order mode (fundamental mode) is extremely small and propagates approximately parallel to the optical axis of the waveguide, it will arrive fastest at the exit end. On the other hand, the angle of propagation of the highest order mode is almost equal to the angle of total internal reflection and propagates at a distance longer by a factor of $\frac{1}{\cos \theta_c}$ than the fundamental mode, as shown in Figure 2.14. As a result, the difference in the propagation time Δt between the fundamental mode that was propagated to a distance L and the highest order mode is

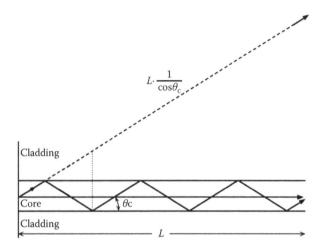

FIGURE 2.14 Difference in the propagation time between the fundamental mode and the highest order mode.

$$\Delta t = \frac{n_1 L}{c}\left[\frac{1}{\cos\theta_c} - 1\right]$$
$$= \frac{n_1 L}{c}\frac{n_1 - n_2}{n_2} \simeq \frac{n_1 L}{c}\Delta. \tag{2.37}$$

This time difference is surprisingly large for $n_1 = 1.50$ and $\Delta = 1.0\%$; the pulse broadening per unit length will be $\frac{\Delta t}{L} = 5 \times 10^{-8}$ [s/km]. Because this value is about 50 ns/km, we may feel it is small, but as will be understood after converting to the frequency characteristic, it is a surprisingly big value.

Next, let us calculate the frequency characteristic. The frequency characteristic is calculated using the output waveform when the signal given by the Dirac δ function* is beamed, that is, by the Fourier transform of the impulse response. First, it is necessary to determine the impulse response.[†] If all the modes at the incident end are assumed to be excited at the same proportion, each of the modes will arrive at the exit end at slightly different times. If the

---

* The Dirac δ function or the impulse response is familiar to us because it appears in the discussion of the circuit theory in the curriculum of the Department of Electrical and Electronic Engineering for the lower years; however, for readers who want to check this out again, please refer to "Appendix C."

[†] Because light is an electromagnetic wave, an impulse containing a high frequency that is much higher than the carrier frequency and reaches infinity is improbable, so this was considered for convenience in the analysis. The transmission function obtained from this analysis only includes a frequency component that is much lower than the carrier frequency, so this is not a problem.

number of modes is assumed to be very large and to arrive at the exit end in a continuous manner, the output pulse will be a square waveform with width Δt, as shown in Figure 2.15.

For the signal amplitude detected by a photodetector, the incident waveform (here, it is the intensity waveform and not the electric field amplitude) is an impulse, so the energy E_{in} of the incident light will become*

$$E_{in} = \int_{-\infty}^{\infty} \delta(t)\, dt = 1 \tag{2.38}$$

On the other hand, if the signal amplitude of the exiting square pulse is set to A_{out}, the exiting light energy E_{out} will become

$$E_{out} = \int_{-\infty}^{\infty} A_{out}[U(t) - U(t - \Delta t)]\, dt$$
$$= A_{out}\Delta t. \tag{2.39}$$

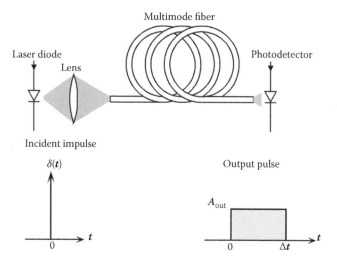

FIGURE 2.15 Impulse response in multimode fibers.

---

* Normally, a modulation of signal is applied on the voltage or current, so the power is square of the amplitude. Thus, the signal's energy is the time integration of the square of the amplitude. Here, the energy of the light inside the optical waveguide is considered. However, assuming the intensity modulation used in the On–Off Keving pulse transmission, the signal at the modulator of the incident end and the photodetector of the exit end is given by the voltage or the current, and the voltage amplitude or the current amplitude of the electric signal is converted into the intensity signal of the light in the optical waveguide. Therefore, the calculation has to be this way as the light energy (photon number) is conserved inside the optical waveguide.

When the waveguide loss (this corresponds to the loss of the direct current component) is expressed by α [dB], the A_{out} is given by

$$A_{\text{out}} = \frac{10^{\frac{-\alpha}{10}}}{\Delta t}. \tag{2.40}$$

Then, the Fourier transform (Refer to Appendix A for the Fourier transform formula) of the rectangular pulse with width Δt and amplitude A_{out} is given by

$$\begin{aligned}
F(\omega) &= \int_{-\infty}^{\infty} A_{\text{out}}[U(t) - U(t - \Delta t)]e^{-j\omega t}\, dt \\
&= \frac{10^{\frac{-\alpha}{10}}}{\Delta t}\left[\left(\frac{1}{j\omega} + \pi\delta(\omega)\right) \cdot \left(1 - e^{-j\omega\Delta t}\right)\right] \\
&= \frac{10^{\frac{-\alpha}{10}}}{\Delta t}\left[\left(\frac{1}{j\omega} + \pi\delta(\omega)\right) \cdot 2j \sin\frac{\omega\Delta t}{2}e^{-j\frac{\omega\Delta t}{2}}\right]. \tag{2.41}
\end{aligned}$$

Here, as the product of the δ function and the sine function is zero because of the nature[*] of the δ function, when the absolute value of Equation (2.41) is calculated, the sinc function (a function of the form $\frac{\sin x}{x}$) will be obtained as follows:

$$|F(\omega)| = 10^{\frac{-\alpha}{10}}\frac{\sin\frac{\omega\Delta t}{2}}{\frac{\omega\Delta t}{2}}. \tag{2.42}$$

The -3 dB bandwidth B of the transmission line is given by the value obtained by dividing the angular frequency ω, which becomes $|F(\omega)| = \frac{1}{2}|F(0)|$ with 2π. Then, using the approximation of $\sin x \simeq x - \frac{1}{6}x^3$ and further substitution of Equation (2.37) to Δt,

$$\text{BL} = \frac{\sqrt{3}}{\pi}\frac{c}{n_1\Delta} \simeq 0.55\frac{c}{n_1\Delta} \tag{2.43}$$

is obtained. In this way, the distance characteristics of the transmission bandwidth of the multimode fiber is given by the product of the -3 dB bandwidth and the distance L, which is called the BL product.[†]

---

[*] Because the value of the δ function is determined by integration, the reader can consider that for the formula containing the δ function, the operation that will be integrated later will be implicitly omitted. Consequently, as $\int_{-\infty}^{\infty} \delta(x-c) f(x)\, dx = f(c)$ holds true with the arbitrary function as $f(x)$, so $\delta(t)\sin(t) = 0$.

[†] Because the traversal of power between the guided modes (called mode coupling or mode transition) arises due to the structural imperfections actually existing in the fiber, the band distance product is represented as $\text{BL}^\gamma (0.5 \le \gamma \le 1.0)$. The value of γ is dependent on the size of the structural imperfections, but the value is usually about 0.8–0.9.

PROBLEM 2-2

Determine the BL product of step-index multimode fiber for a core with a refractive index of $n_1 = 1.500$ and a relative index difference of $\Delta = 1.0\%$.

SOLUTION 2-2

Eleven MHz·km is obtained when the given numbers are substituted to Equation (2.43). As this can only be broadcasted to a few television channels within a radius of about 1 km, when using a multimode fiber, a distributed-index (graded-index) multimode fiber is necessary.

[End of Solution]

3 Propagation of Light Beams in Free Space

3.1 REPRESENTATION OF SPHERICAL WAVES AND THE DIFFRACTION PHENOMENON

As described in Section 1.3, in the propagation of plane waves, the plane wave will spread infinitely in a cross-section that is perpendicular to the direction of propagation (that is to say, the amplitude is the constant regardless of the location). However, in beam waves, the amplitude is distributed to an area in the cross-section that is perpendicular to the direction of propagation. For such a localized amplitude distribution, the amplitude distribution (and the distribution of the phase) changes with the propagation. Such a phenomenon is called a diffraction phenomenon. This section will derive, in a strict sense, the equation that would describe such a diffraction phenomenon (although this is done by using the scalar wave approximation). However, as the concept of diffraction phenomenon can also be qualitatively understood using Huygen's Principle, readers who want to know only the concept can take into mind Equation (3.15), which expresses the diffraction phenomenon, and think of the physical meaning of the equation using Exercise 3-1.

In Section 1.3, the solution to the wave equation (1.20) in the Cartesian coordinate system led to the plane wave; however, the *spherical wave* can be derived via the polar coordinates, whereas the cylindrical wave can be derived from the cylindrical coordinate system. So, before specifying the coordinate system, the position-dependent amplitude function $U(r)$ and the unit polarization vector e_p are used instead of the position-independent amplitude A in Equation (1.22) for plane waves, to express the electric field as

$$E = U(r)e_p e^{j\omega t}. \tag{3.1}$$

Substituting Equation (3.1) to Equation (1.20), we will obtain

$$\nabla^2 U(r) + k^2 U(r) = 0. \tag{3.2}$$

Such a wave equation corresponding to the amplitude function of the scalar quantity is called a scalar wave equation.

When the polar coordinate system is used to write Equation (3.2), it becomes

$$\frac{1}{r^2}\frac{\partial}{\partial r}\left(r^2\frac{\partial U}{\partial r}\right) + \frac{1}{r^2 \sin\theta}\frac{\partial}{\partial \theta}\left(\sin\theta\frac{\partial U}{\partial \theta}\right) + \frac{1}{r^2 \sin^2\theta}\frac{\partial^2 U}{\partial \phi^2} + k^2 U = 0. \tag{3.3}$$

Here, the position vector r for $U(r)$ is represented as polar coordinates. The general solution of this equation is expressed using the spherical Bessel function and the spherical Neumann function with respect to the coordinate r and the associated Legendre function and the sine (and cosine) function with respect to coordinate θ and ϕ. However, as the simplest solution here, an isotropic solution depending only on r and not on both the coordinates θ and ϕ is assumed (similar to that of a plane wave). In such a case, Equation (3.3) becomes

$$\frac{1}{r^2}\frac{\partial}{\partial r}\left(r^2\frac{\partial U}{\partial r}\right) + k^2 U = 0. \tag{3.4}$$

This general solution can be expressed as

$$U(r) = A\frac{\exp(-jkr)}{r} + B\frac{\exp(jkr)}{r} \tag{3.5}$$

with A and B as constants [omitting the term $\exp(j\omega t)$]. The first term of Equation (3.5) is the *divergent spherical wave* emanating from the coordinate origin to the total solid angle, whereas the second term is the *convergent spherical wave* conversely converging to the origin.

Incidentally, the solution can certainly be verified if Equation (3.5) is substituted to Equation (3.3); however, only the origin actually becomes a singularity. This is because the wave source for plane wave is set at $z = -\infty$, with only the propagating wave being considered at the spatial region being contemplated. However, for the spherical wave, the wave source is positioned at $r = 0$, which is involved in the area of interest. To be accurate, for the divergent spherical wave caused by the unit wave source existing at position r', Equation (3.4) can be written using the Dirac δ function (refer to Appendix B) as

$$\frac{1}{r^2}\frac{\partial}{\partial r}\left(r^2\frac{\partial U}{\partial r}\right) + k^2 U = -\delta(r - r'). \tag{3.6}$$

Then, the particular solution (the general solution is given by the sum of particular solution of Equation (3.6) and Equation (3.5)) is given by

$$G(r) = \frac{1}{4\pi}\frac{\exp(-jk|r - r'|)}{|r - r'|}, \tag{3.7}$$

which can be proven by substituting to Equation (3.6).* Here, $|r - r'|$ is the distance between points r and r', so when written using the Cartesian coordinate

---

* Even if $r' = 0$ is set as the origin of the wave source position for simplicity, the generality is not lost. From Gauss theorem on vector analysis, when volume integration is done in the spatial domain V, including the wave source position $r' = 0$ on both sides of Equation (3.6), the integral of the δ function becomes 1, so $\iiint_V [\nabla^2 G + k^2 G]dr^3 = \iint_S \frac{\partial G}{\partial r}\boldsymbol{n}da + \iiint_V k^2 G dr^3 = -1$. Here, $\boldsymbol{n}$ is the unit vector from the origin facing the exterior of the region. When the region V, as microspheres of infintesimal radius δr, is substituted to $G(r) = A\frac{\exp(-jkr)}{r}$, the second term will become 0 at a limit of $\delta r \to 0$. Moreover, the first term will become $\iint_S \frac{\partial G}{\partial r}da = -A(jk + \frac{1}{\delta r})\frac{e^{-jk\delta r}}{\delta r} \cdot 4\pi\delta r^2$ and will be $-4\pi A$ at a limit of $\delta r \to 0$. Consequently, $A = \frac{1}{4\pi}$.

system, it will become $|r - r'| = \sqrt{(x - x')^2 + (y - y')^2 + (z - z')^2}$. Equation (3.7) is a function of the position vector r, but it is also actually a function of r', so this can also be written as $G(r, r')$. (When this is used in integral representation, $G(r, r')$ is called the Green's function.)

Here, let us explain the *diffraction* phenomenon using spherical waves. Diffraction is a phenomenon originating from the wave nature of light that occurs in a light beam of which amplitude distribution is localized in a certain area in a cross-section perpendicular to the direction of propagation, different from plane waves. The amplitude distribution in the cross-section changes along with the propagation due to the diffraction. First, when the amplitude distribution is assumed to be localized approximately in the vicinity of the origin in the xy plane, as shown in Figure 3.1, let us consider the amplitude distribution after the propagation at a distance of d in the z-direction of a free-space wave (here, there is no need for the wave front to be a plane, and the direction of propagation represented by the normal direction of such wave front does not have to be completely along the z-axis) propagating approximately in the z-direction as an amplitude distribution in the $x'y'$ plane. This phenomenon can be phenomenologically (or intuitively) explained using Huygen's principle. However, for a more precise mathematical representation, we can explain the phenomenon by deriving the diffraction integration equation, although this is somewhat a more circuitous route.

Let us take into consideration the two independent solutions $u(r)$ and $v(r)$ that will satisfy Equation (3.2). Then, in general, *Green's theorem* below for the closed region V surrounded by a closed surface S holds true (refer to Appendix C for proof).

$$\iiint_V [u\nabla^2 v - v\nabla^2 u]dr^3 = \iint_S \left(u\frac{\partial v}{\partial n} - v\frac{\partial u}{\partial n}\right)n \cdot da. \qquad (3.8)$$

Here, n is the unit normal vector standing on the closed surface S, and the outgoing direction is the positive direction. Moreover, $\frac{\partial}{\partial n}$ represents the differential (grad) toward the direction of n on the closed surface S. Because u

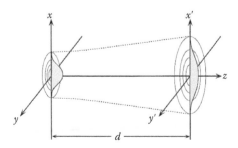

FIGURE 3.1 Incident wave front coordinates and exit wave front coordinates taking into consideration the diffraction phenomenon.

and v are arbitrary, with u as the amplitude distribution in the xy plane of Figure 3.1, v is the amplitude distribution (that is, the amplitude distribution given by Equation (3.7)) of the spherical wave emanating from a point r' in the $x'y'$ plane. In addition, the integral region of the surface integral and the volume integral of Equation (3.8) will be separated into two, assuming that the amplitude distribution of the incident wave front in Figure 3.1 is limited to the radius R_0 from the origin on the xy plane, with the outside amplitude set to 0, as shown in Figure 3.2. That is, V is set as the free space surrounded by S_1 and S_2, with the surface area of the sphere with radius R and point r' as the center as S_2 and the circular integral region including the circular region with a radius of R_0 in the xy plane that includes the incident wave front as S_1. (The area S_1 is a circular region with a center at the intersection of the xy plane and the perpendicular line taken down the xy plane from r' and assumed to completely encompass the area where the amplitude distribution of the incident wave front exists.)

In such an integral region, when Equation (3.7) with $v(r) = G(r)$ is substituted to Equation (3.8), the volume integral on the left side will become $-u(r')$ due to the nature of the δ function*, so it will become

$$-u(r') = \frac{1}{4\pi} \iint_{S_1} \left[u \frac{\partial}{\partial n} \left(\frac{e^{-jkr}}{r} \right) - \frac{e^{-jkr}}{r} \frac{\partial u}{\partial n} \right] n \cdot da$$

$$+ \frac{1}{4\pi} \iint_{S_2} \left[-u \left(jk + \frac{1}{R} \right) \frac{e^{-jkR}}{R} - \frac{e^{-jkR}}{R} \frac{\partial u}{\partial r} \right] n \cdot da. \quad (3.9)$$

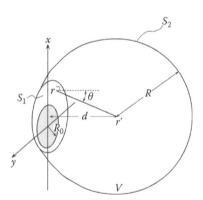

FIGURE 3.2 Integral region applied with Green's theorem to derive the diffraction integral.

---

* From Equation (3.2), u will become $\nabla^2 u(r) + k^2 u(r) = 0$, and from Equation (3.7), v will become $\nabla^2 v(r) + k^2 v(r) = -\delta(r - r')$, so from the nature of the δ function, $\int_{-\infty}^{\infty} f(x)\delta(x - x')dx = f(x')$, the volume integral on the left side will become $-u(r')$.

Here, r in the first integral on the right side of Equation (3.9) is the distance between the point on the area S_1 and r' ($r = |r - r'| = \sqrt{(x - x')^2 + (y - y')^2 + d^2}$), whereas r in the $\frac{\partial u}{\partial r}$ term in the second integral on the right side is the distance between point on the area S_2 and r', and so it should be written as R. However, it is written as r to express clearly that it is an argument of differentiation.

When the incident wave propagates in approximately the positive z-direction, the $\frac{\partial}{\partial n} n \cdot da$ in the first integration on the right side of Equation (3.9) is expressed as

$$
\frac{\partial}{\partial n}\left(\frac{e^{-jkr}}{r}\right) n \cdot da = \frac{\partial}{\partial r}\left(\frac{e^{-jkr}}{r}\right) \frac{\partial r}{\partial n} n \cdot da
$$

$$
= -\left(jk + \frac{1}{r}\right) \frac{e^{-jkr}}{r} \frac{(-e_z) \cdot (r - r')}{|r - r'|} da
$$

$$
= -\left(jk + \frac{1}{r}\right) \frac{e^{-jkr}}{r} \cos\theta\, da, \tag{3.10}
$$

$$
\frac{\partial u}{\partial n} n \cdot da = -\frac{\partial u}{\partial z} da = jku(r)da. \tag{3.11}
$$

Here, as shown in Figure 3.2, the θ in Equation (3.10) is the angle between the straight line drawn from the point r on the incident plane S_1 to the observation point r' and the z-axis (normal line of area S_1 with an inward direction). Consequently, the first integral on the right side of Equation (3.9) will be as follows:

$$
-\frac{1}{4\pi} \iint_{S_1} u(r)\frac{e^{-jkr}}{r} \left[jk + \left(jk + \frac{1}{r}\right)\cos\theta\right] da. \tag{3.12}
$$

On the other hand, the second integral on the right side of Equation (3.9) can be written as

$$
-\frac{1}{4\pi} \iint_{S_2} \left[\left(jku + \frac{\partial u}{\partial r}\right)\frac{e^{-jkR}}{R} + \frac{e^{-jkR}}{R^2}u\right] n \cdot da, \tag{3.13}
$$

when organized in the order of terms. Because in the infinity of $R \to \infty$ both u and $\frac{\partial u}{\partial r}$ become asymptotic to 0, the integral of the second term inside [ ] in Equation (3.13) will be 0. In the same way, the integral of the first term will also become 0.*

---

* To be precise, it is necessary for

$$
\lim_{r \to \infty} r\left(jku + \frac{\partial u}{\partial r}\right) = 0
$$

to hold true. This is called the radiation condition. The spherical wave certainly satisfies this condition.

Consequently, Equation (3.9) will become as follows:

$$u(\mathbf{r}') = \frac{1}{4\pi} \iint_{S_1} u(\mathbf{r}) \frac{e^{-jkr}}{r} \left[jk + \left(jk + \frac{1}{r} \right) \cos\theta \right] da. \qquad (3.14)$$

This equation is called the *Fresnel-Kirchhoff integral*. When the distance d between the xy plane (the plane that takes into consideration the incident wave) and the $x'y'$ plane (the plane that takes into consideration the diffracted wave) in Figure 3.1 is sufficiently large compared to the wavelength, and the diffraction wave is localized almost in the vicinity of the z-axis, we can approximate $1 \ll kr$ and $\cos\theta \simeq 1$, so using $k = \frac{2\pi}{\lambda}$, Equation (3.14) can be simplified as shown in Equation (3.15). However, as the amplitude distribution of the incident wave and the diffracted wave used the same symbol u, this has become confusing. To avoid this, the amplitude distribution of the incident wave is expressed as $\psi(x, y)$, and the amplitude distribution of the diffracted wave is expressed as $\Psi(x', y')$, and so the equation becomes

$$\Psi(x', y') = \frac{j}{\lambda} \iint_{S_1} \psi(x, y) \frac{e^{-jkr}}{r} dx\, dy. \qquad (3.15)$$

Here, $r = \sqrt{(x - x')^2 + (y - y')^2 + d^2}$.

PROBLEM 3-1

Explain the physical meaning of Equation (3.15).

SOLUTION 3-1

The readers may have already learned Huygen's principle in senior high school. If we use this, the integrand of Equation (3.15) expresses a newly generated spherical wave from point P in the incident wave coordinates $(x, y, z = 0)$ in Figure 3.1. The amplitude of this spherical wave is proportional to $\psi(x, y)$ and as $\psi(x, y)$ is a complex number, its phase distribution will also be expressed as the wave front shape (phase distribution) of the incident wave in the xy plane. Here, $r = \sqrt{(x - x')^2 + (y - y')^2 + d^2}$ is the distance between the center of the wave source (point P) of that spherical wave and point Q (coordinates are (x', y')) on the $x'y'$ plane of the diffracted wave being considered. Consequently, Equation (3.15) will determine the amplitude at point Q by summing up the spherical wave that is generated from point P and arrives at point Q over the xy planes for point P. However, Huygen's principle cannot explain the constant term in front of the integration of Equation (3.15), and the spherical wave will also propagate in the $-z$-direction. Therefore, the diffraction phenomenon cannot be fully described by Huygen's principle alone.

[End of Solution]

3.2 FRESNEL DIFFRACTION AND FRAUNHOFER DIFFRACTION

When the propagation distance d is sufficiently large compared to the areas of amplitude distribution of the incident wave and the diffracted wave, that is, when the following inequalities are satisfied in the areas where $\psi(r)$ and $\Psi(r')$ have appreciable value, as shown in Figure 3.1,

$$(x - x')^2 \ll d^2, \quad (y - y')^2 \ll d^2 \tag{3.16}$$

the denominator r in the integrand of Equation (3.15) can be approximated by d. On the other hand, as e^{-jkr} will change significantly with the change in the wavelength of r, when closely approximated as

$$r \simeq d \left[1 + \frac{(x - x')^2 + (y - y')^2}{2d^2} \right], \tag{3.17}$$

Equation (3.15) can be written as follows:

$$\Psi(x', y') = \frac{j}{\lambda d} e^{-jkd}$$
$$\times \iint_{S_1} \psi(x, y) \exp\left[-jk \frac{(x - x')^2 + (y - y')^2}{2d} \right] dx\, dy. \tag{3.18}$$

This kind of approximation is called the *Fresnel approximation*.

Moreover, at a sufficiently far distance, as the broadening of $\Psi(x', y')$ due to the diffraction will be greater than that of the $\psi(x, y)$ and d is also large, ignoring the last term inside the [ ] in the following expanded form of Equation (3.17):

$$r \simeq d \left[1 + \frac{x'^2 + y'^2}{2d^2} - \frac{xx' + yy'}{d^2} + \frac{x^2 + y^2}{2d^2} \right]. \tag{3.19}$$

A further approximated form of Equation (3.18) can be obtained as follows:

$$\Psi(x', y') = \frac{j}{\lambda d} \exp\left[-jk \left(d + \frac{x'^2 + y'^2}{2d} \right) \right]$$
$$\times \iint_{S_1} \psi(x, y) \exp\left[jk \frac{xx' + yy'}{d} \right] dx\, dy. \tag{3.20}$$

This kind of approximation is called the *Fraunhofer approximation*. Now, when the exponent term of the integrand of Equation (3.20) undergoes a conversion of variables such as

$$k\frac{x'}{d} = -\Omega_x(= -2\pi\nu_x), \quad k\frac{y'}{d} = -\Omega_y(= -2\pi\nu_y). \quad (3.21)$$

Equation (3.20) can be written as*

$$\Psi(x', y') = \frac{j}{\lambda d}\exp\left[-jk\left(d + \frac{x'^2 + y'^2}{2d}\right)\right]\mathcal{F}[\psi(x, y)]. \quad (3.22)$$

Here, $\mathcal{F}[f(x, y)]$ represents the Fourier transform of the function $f(x, y)$ and is defined by Equation (A.2) of Appendix A. Ω_x and Ω_y are called *spatial angular frequencies* (ν_x and ν_y are called *spatial frequencies*).[†]

Because the Fresnel approximation and Fraunhofer approximation are dependent on the propagation distance d, their applicable region should be discriminated. The border for discriminating the region is based on the magnitude of the last term inside the [ ] in Equation (3.19). This term is involved in the exponent r of e^{-jkr}, if

$$k\left[\frac{x^2 + y^2}{2d}\right] \ll \frac{\pi}{4}. \quad (3.23)$$

Then this term can be practically neglected. That is, when the diameter of the region where the amplitude of the incident wave is primarily distributed (diameter of the beam) is designated by D, and if

$$\frac{D^2}{\lambda} \ll d \quad (3.24)$$

holds true, then that region is the Fraunhofer approximation region. When d approaches $\frac{D^2}{\lambda}$ or if it is smaller than $\frac{D^2}{\lambda}$, then it will be the Fresnel approximation region. Moreover, when d is approximately several times to 10 times that of the wavelength, the amplitude distribution after propagation is close

---

* The negative sign attached during the changing of variables in Equation (3.21) is for Equation (3.22) to become a Fourier transform. If a negative sign in not attached, it will become an inverse Fourier transform. Fourier transform and inverse Fourier transform are mutually reciprocal transformations, so either expression is fine. However, for the Fourier transform to be defined in the form of Equation (A.2) of Appendix A, a negative sign was used for the changing of variables in Equation (3.21). Yet, in textbooks on optics, a positive sign is used for the changing of variables in Equation (3.21) and a lot of textbooks describe Fourier transform using Equation (3.22). This is not incorrect; however, $e^{j2\pi\nu t}$ is used instead of $e^{-j2\pi\nu t}$ in the defining equation of Fourier transform (Equation (A.3) of Appendix A).

[†] In the field of optics, the conversion formula (A.3), which uses frequency instead of angular frequency, is often used for Fourier transform. In such cases, the spatial frequencies $\nu_x = \frac{\Omega_x}{2\pi}$ and $\nu_y = \frac{\Omega_y}{2\pi}$ are used instead of the spatial angular frequencies.

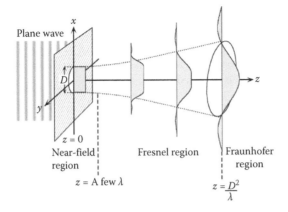

FIGURE 3.3 Applicable approximate regions with regard to diffraction.

to that of the incident wave because the diffraction of the incident wave is not sufficiently generated. Such a region is called a near-field region.*

In this near-field region, it is necessary to perform calculations in accordance with Equation (3.14).

As an example of the discrimination of this approximate region, when a plane wave is beamed to a partitioning screen with a circular aperture of diameter D, each region is illustrated in Figure 3.3.

3.3 FRAUNHOFER DIFFRACTION OF A GAUSSIAN BEAM

As the incident wave $\psi(x, y)$ of the Fraunhofer diffraction integral in Equation (3.20), let us consider the following amplitude distribution at $z = 0$

$$\psi(x, y) = A \exp\left(-\frac{x^2 + y^2}{w_0{}^2}\right) = A \exp\left[-\left(\frac{r}{w_0}\right)^2\right], \qquad (3.25)$$

(here $r^2 = x^2 + y^2$). Because this amplitude distribution is represented by a Gaussian function, it is called a *Gaussian beam*. Here, w_0 is called the *spot size* and is primarily the parameter that represents the radius of the Gaussian beam. However, as a lot of the electromagnetic distribution of the propagation mode of laser beams and single-mode optical waveguides can be

---

* In recent years, a new field in optics called near-field optics has received much attention. In English, this is of course referred to as near-field, but in the field of optics, the propagating wave is not dealt with, rather it is the portion of the evanescent wave effusing from the boundary surface of the electromagnetic field due to the total internal reflection that has been described in Section 1.4.4.

approximated using the Gaussian function, spot size is also used as a parameter to represent the spread of the electromagnetic field distribution in these areas.[*]

When Equation (3.25) is substituted to Equation (3.22), the Fourier transform of the Gaussian function is also a Gaussian function, so the following equation (refer to Appendix A) is obtained.

$$\Psi(x', y') = \underbrace{\frac{\pi w^2 A}{\lambda d} \exp\left[-jk\left(d + \frac{x'^2 + y'^2}{2d}\right) + j\frac{\pi}{2}\right]}_{\text{phase term}}$$

$$\times \underbrace{\exp\left[-\frac{w^2}{4}\left(\Omega_x^2 + \Omega_y^2\right)\right]}_{\text{amplitude term}}. \tag{3.26}$$

When Equation (3.21) is used, the amplitude term of the Gaussian function in the second row of Equation (3.26) is rewritten as

$$\text{Amplitude term} = \exp\left[-\frac{x'^2 + y'^2}{\left(2\frac{d}{kw}\right)^2}\right] = \exp\left[-\frac{r'^2}{\left(2\frac{d}{kw}\right)^2}\right]. \tag{3.27}$$

Equation (3.27) means that the spot size increases in proportion to the propagation distance d, so the beam diameter will broaden along with the propagation distance. Then, when the *beam broadening angle* $2\theta_D$ is defined as that in Figure 3.4, $2\theta_D$ can be determined as[†]

$$2\theta_D = 2\tan^{-1}\left(\frac{2\frac{d}{kw_0}}{d}\right) = 2\tan^{-1}\left(\frac{\lambda}{\pi w_0}\right) \simeq 1.273\frac{\lambda}{2w_0}. \tag{3.28}$$

---

[*] In some of the references, the portion of the Gaussian function is expressed as

$$\exp\left[-\frac{1}{2}\left(\frac{x}{w'}\right)^2\right].$$

The above definition is convenient because the form will not change even if Fourier transform is done during the calculation of the diffraction integral, and others. According to standards, such as JIS and ISO, spot size is defined as "the distance between the point where the electric field amplitude becomes $\frac{1}{e}$ and the center of the beam," or "the distance between the point where the light intensity becomes $\frac{1}{e^2}$ of that at the beam center," so it is necessary to be cautious regarding the definition of spot size. Therefore, the relationship between the definition of spot size w, which is used in this book, and that of the spot size w' in the above equation is given simply as $w = \sqrt{2}w'$.

[†] In general, the beam that has a diameter of about D with a wave front that is almost flat will have a broadening angle that is almost $\theta_D \simeq \frac{\lambda}{D}$ regardless of the amplitude distribution of the beam. When the amplitude distribution of the beam is exactly calculated, the definition of the beam diameter will differ depending on the amplitude distribution, so the coefficient will slightly deviate from 1.0 (by about 20%).

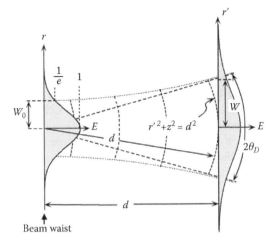

FIGURE 3.4 Broadening due to the Fraunhofer diffraction of the Gaussian beam.

Using the half-angle θ_D of the broadening angle, the numerical aperture (NA) of the Gaussian beam based on the same definition as Equation (2.28) is expressed as*

$$\mathrm{NA} = \sin \theta_D = \frac{\lambda}{\sqrt{\lambda^2 + (\pi w_0)^2}}. \tag{3.29}$$

On the other hand, in determining the form of the equiphase plane from the phase term of Equation (3.26), the equiphase plane is a set of points at which the phase is equal to that at the positions $x' = 0$ and $y' = 0$ for $z = d$, and so the equality

$$k\left(z + \frac{x'^2 + y'^2}{2z}\right) - \frac{\pi}{2} = kd - \frac{\pi}{2} \tag{3.30}$$

holds true. Here, when the approximation of $r' \ll d$ is used, the following equation gives the equiphase plane:

$$r'^2 + z^2 = d^2. \tag{3.31}$$

---

* In general, the NA of the beam converging or diverging at an angle of θ can be expressed based on

$$\mathrm{NA} = \sin \theta.$$

On the other hand, when a parallel beam of light is beamed parallel to the optical axis, the NA of the lens is defined based on $\mathrm{NA} = \sin \theta$, using the angle θ relative to the optical axis of the ray that is focused (diverged in the case of a concave lens) from the outermost periphery of the lens.

This is an equation of a spherical surface. Consequently, at a sufficiently far distance of $\frac{w^2}{\lambda} \ll z$, as shown in Figure 3.4, the intensity distribution of the Gaussian beam wave is expressed by the Gaussian function, wherein size increases (with a constant angle of divergence) in proportion to the distance of propagation and the wave front will broaden like that of a spherical wave with a center at about the origin $z = 0$.

On the other hand, at $z = 0$, as seen from Equation (3.25), which is expressed as a real function, the wave front becomes a plane (x, y plane) that is perpendicular to the direction of propagation. Then, the spot size becomes the smallest at the portion where the wave front has become a plane (this is symmetrical to the plane at $z = 0$ at the region where $z < 0$, so it will become a convergent Gaussian beam). Consequently, this position where the spot size is the smallest and the wave front is a plane is called the *beam waist*.

Here, we have considered diffraction at the Fraunhofer region, but actually the propagation of the Gaussian beam preserves the Gaussian form even in the Fresnel region (the detailed derivation will be described in Chapter 7). Then, the following equation uses the *Fresnel number*, which is defined as

$$N_F = \frac{w_0^2}{\lambda d} \tag{3.32}$$

the classification that separates the Fresnel region from the Fraunhofer region (or rather the rough estimate) is as follows:

$$N_F \ll 1 \quad \text{Fraunhofer region,}$$
$$N_F \gtrsim 1 \quad \text{Fresnel region.}$$

PROBLEM 3-2

Show the spot size w_f at the focal point when a Gaussian beam is focused by a convex lens using parameters such as diameter of the lens D, the focal length f, and the wavelength λ.

SOLUTION 3-2

Let us assume that the Gaussian beam has a beam waist just in front of a convex lens. The spot size at that beam waist is designated as W. If the diameter of the lens is not much larger than the diameter of the beam, then the peripheral part of the Gaussian distribution will be rejected depending on the finite aperture of the lens, and the shape of the beam at the focal point will deviate from the Gaussian distribution. Therefore, $2W < D$ is necessary, and normally D is set up at about twice the beam diameter of $2W$. Let us consider the focusing of light with these conditions being satisfied.

When a Gaussian beam is focused by a convex lens, a beam waist will form at the focal point as shown in Figure 3.5a (lens aberrations are not

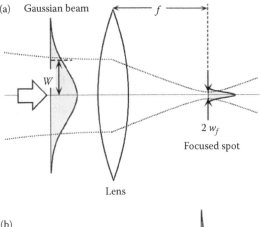

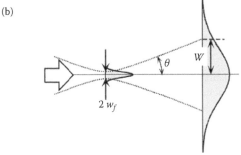

FIGURE 3.5 Focusing of Gaussian beam by a convex lens and the spot size. (a) Focusing of light by a lens. (b) Emission and diffraction of a microscopic spot beam.

considered). Then, broadening will occur because of diffraction when the focal point is passed by. This broadening due to diffraction is symmetric to the focusing of light with respect to the focal plane. That is, if the broadening is considered to be because of the diffraction of a beam that is emitted from a light source, which is in the focal point as shown in Figure 3.5b, the problem of the focusing of the light will be solved.

The spot size at the focal point will broaden at a broadening angle of the Gaussian beam w given by Equation (3.28), so from the condition that the spot size will become W after the propagation by a distance f

$$\tan \theta = \frac{W}{f} = \frac{\lambda}{\pi w_f} \tag{3.33}$$

will hold true. Consequently,

$$w_f = \frac{\lambda f}{\pi W} \tag{3.34}$$

will be obtained. That is, if the light is focused by a lens (large NA) with a short focal length having an aperture that is large enough to cover W, then it can be focused at a small spot.* Then, as W and f are almost of the same order, the spot size of the focusing spot will be comparable to the wavelength.

Incidentally, as Equation (3.28) is primarily a Fraunhofer diffraction equation, it is necessary to verify the scope at which Equation (3.34) is applicable. If the Fresnel number defined by Equation (3.32) is used for Equation (3.34), then

$$N_F = \frac{w_f^2}{\lambda f} = \frac{w_f}{\pi W} \ll 1 \qquad (3.35)$$

is obtained, where it can be seen with certainty that it has become a Fraunhofer region. That is, we know that for diffraction from a focused beam at a steep angle or from a beam diameter that is approximately the same size as the wavelength, the Fraunhofer diffraction formula can be used even for a propagation distance that is several times that of the wavelength.

PROBLEM 3-3

Determine the relationship of the electric field amplitude to the beam power and the intensity at the center of the Gaussian beam.

SOLUTION 3-3

The scalar electromagnetic field distribution function $\psi(x, y)$ given by Equation (3.25) is considered to be the amplitude portion of the electric field distribution corresponding to $U(r)$ of Equation (3.1). That is,

$$E = \psi(r)e_p \exp[j(\omega t - k \cdot r)]. \qquad (3.36)$$

On the other hand, the intensity distribution is given by the real part of the complex Poynting vector of Equation (1.52), so it is necessary to express the H in Equation (1.52) as E. The equation that expresses H in terms of E is given by Equation (1.25) for the plane wave; however, as the plane wave amplitude is a constant value A regardless of the location, only the $e^{-jk \cdot r}$ term can be differentiated in $\nabla \times E$. However, when the curl operation in Equation (3.36) is performed, and $\nabla \times$ is operated from the left of both sides of the equation, and the vector analysis formula (D.7) of Appendix D is used, then

$$\nabla \times E = -j(k \times E) + \psi(\nabla \times e_p) \exp[j(\omega t - k \cdot r)]$$
$$+ (\nabla \psi) \times e_p \exp[j(\omega t - k \cdot r)] \qquad (3.37)$$

is obtained. The first term at the right side of Equation (3.37) is an expression that is obtained for a plane wave. Because e_p in the second term is a

---

* The size of the focus spot is given as almost $w_f \simeq \frac{\lambda}{NA}$, regardless of the shape of the beam.

unit vector, $\nabla \times e_p = 0$. As the third term is not 0, when Equation (3.37), including the third term, is substituted to Maxwell's equation (1.24),

$$H = \frac{1}{\omega\mu_0}\left[(k \times E) + (\nabla\psi) \times e_p \exp[j(\omega t - k \cdot r)]\right] \qquad (3.38)$$

is obtained. When Equation (3.38) is substituted to the H of Equation (1.52), and the formula (D.1) of Appendix D and Equation (D.6) are used in the triple vector product, then

$$\tilde{S} = \frac{1}{2}(E \times H^*)$$

$$= \frac{1}{2\omega\mu_0}\left[E \times (k \times E)^* + E \times \{(\nabla\psi) \times e_p \exp[j(\omega t - k \cdot r)]\}^*\right]$$

$$= \frac{1}{2\omega\mu_0}[(E \cdot E^*)k - (k \cdot E^*)E + \psi e_p \times (\nabla\psi^*) \times e_p]$$

$$= \frac{1}{2\omega\mu_0}[|E|^2 k + (e_p \cdot e_p)\psi\nabla\psi^* - (\nabla\psi^* \cdot e_p)\psi e_p] \qquad (3.39)$$

is obtained. Here, in the reduction process of Equation (3.39), as E and k are orthogonal, $(k \cdot E^*) = 0$ is used. Also, the second and third terms of Equation (3.39) represent the orthogonal components in the e_p of the vector $\psi\nabla\psi^*(= \psi \text{ grad } \psi^*)$. In general, as the $\nabla\psi^*$ faces the cross-section that is perpendicular to the direction of propagation (direction of the k vector) in the beam wave, only the first term in the right side of Equation (3.39) will be involved in determining the energy flow density toward the direction of propagation. Because the direction of the complex Poynting vector expresses the direction of the energy flow and magnitude expresses the energy flow density per unit area, the energy flow density toward the direction of propagation can be calculated by

$$\tilde{S} = \tilde{S} \cdot \frac{k}{k} = \frac{k}{2\omega\mu_0}|E|^2. \qquad (3.40)$$

Here, k becomes $k_0 n$ in a medium with a refractive index of n. Equation (3.40) expresses the light intensity distribution at the cross-section perpendicular to the direction of propagation.

Now, let us bring back the Gaussian beam of which the amplitude distribution at $z = 0$ is expressed by Equation (3.25). Assuming that the z-axis direction is the direction of propagation and the polarization is an x polarization, and then substituting Equation (3.25) to the ψ in Equation (3.39), the equation will become

$$\tilde{S} = \frac{1}{2\omega\mu_0}\left[|\psi(x, y)|^2 k e_z - 2\frac{y}{w^2}A^2 \exp\left(-2\frac{x^2+y^2}{w^2}\right)e_y\right]$$

$$= \frac{k}{2\omega\mu_0}A^2 \exp\left(-2\frac{x^2+y^2}{w^2}\right)\left[e_z - \frac{2y}{kw^2}e_y\right]. \qquad (3.41)$$

Here, the direction of propagation of the beam is the z-direction, and the second term inside the parenthesis at the right side is the y-direction component, so the second term has no relationship to the energy flow density toward the direction of propagation. (Moreover, when $\lambda \ll w$, the second term involving e_y is negligible compared to the first term involving e_z.) Therefore, the light intensity distribution is expressed up to the first term of Equation (3.41). Moreover, the energy (power flow) flowing per unit time through the x, y plane toward the z-direction is determined by integrating the z component of Equation (3.41) over the whole x, y cross-section (using Equation (E.1) of Appendix E) and then using the following equation to obtain:

$$
\begin{aligned}
P &= \frac{k}{2\omega\mu_0} A^2 \iint_{-\infty}^{\infty} \exp\left(-2\frac{x^2+y^2}{w^2}\right) dx\, dy \\
&= \frac{k}{2\omega\mu_0} A^2 \frac{\pi w^2}{2}.
\end{aligned} \tag{3.42}
$$

When the amplitude A at the center of the beam is expressed using beam power P from Equation (3.42), it becomes

$$
A = \frac{1}{w}\sqrt{\frac{4P}{\pi n}}\sqrt{\frac{\mu_0}{\varepsilon_0}} = \frac{1}{w}\sqrt{\frac{4Z_0 P}{\pi n}} \quad \text{[V/m]}. \tag{3.43}
$$

That is, when the value (e.g., $1 \text{ mW} = 1 \times 10^{-3}$ W) of the beam power is substituted with P of Equation (3.43), the electric field intensity at the center of such beam will be determined.   **[End of Solution]**

3.4 WAVE FRONT TRANSFORMATION EFFECT OF THE LENS

Before discussing wave front transformation by lenses, we must first describe simply the formula for imaging using rays. As shown in Figure 3.6a, convex lens possesses the effect of collecting parallel rays of light (i.e., a beam of coherent light) to a focal point. At this time, the distance f between the central plane of the lens and the focal point of the lens is called the *focal length*.*

---

* Because a lens actually possesses a finite thickness, the ray is focused by refracting it twice, once at the spherical surface at the anterior surface of the lens and then at the spherical surface at the posterior surface of the lens. Consequently, if the lens is thick, the focusing position (the position of intersection with the optical axis) of the ray passing through the center of the lens and that of the ray passing through the periphery of the lens will deviate, and this is called spherical aberration. In the actual imaging using a lens, there are other aberrations, such as Seidel's five aberrations (aberrations occur when rays cannot be focused at one point). However, a detailed explanation of aberrations has not been provided in the book, so please refer other reference books on optics.

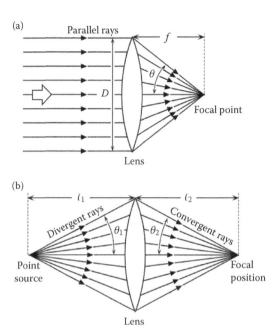

FIGURE 3.6 Focusing of coherent light by a convex lens. (a) Focusing of a parallel beam of light. (b) Focusing of a divergent beam of light.

Here, we assume that the thickness of the lens is negligibly thin (called a thin lens). In a thin lens, the focal distance of the anterior side of the lens is also f when parallel rays of light are beamed backwards from the posterior side of the lens.

As shown in Figure 3.6(b), let us suppose a case wherein not parallel rays of light but divergent rays that come out from one point on the optical axis far from the focal point in the anterior side of the lens is focused by the lens. The distance between the light source and the lens is designated as ℓ_1, whereas the distance between the lens and the focal position is designated as ℓ_2. Then, the well-known equation

$$\frac{1}{\ell_1} + \frac{1}{\ell_2} = \frac{1}{f}$$

(3.44)

is derived. Figure 3.6b was considered with the light source being at one point on the optical axis. However, as shown in Figure 3.7, when an object is located at a position with a distance ℓ_1 from the center of the lens on the optical axis and is illuminated by an incoherent light, rays of light will diverge from various points of the object, and each of the rays will be independently focused by the

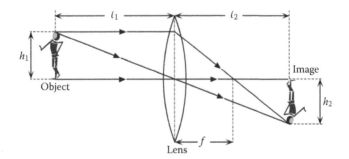

FIGURE 3.7   Imaging of incoherent light by a convex lens.

lens, so the image would be formed at a position with a distance ℓ_2 from the center of the lens. Then, the ratio of the height of the object h_1 and the height of the image h_2

$$M = \frac{h_2}{h_1} \tag{3.45}$$

is called the *lateral magnification*.

Here, as shown in Figure 3.6b, for a ray in the divergent light from the light source that is focused after passing through the outermost periphery of the lens, if the angle between the ray and the optical axis at the side of the light source is designated as θ_1 and the angle between the ray and the optical axis at the focal position side is designated as θ_2, then

$$\ell_1 \tan \theta_1 = \ell_2 \tan \theta_2 \tag{3.46}$$

is derived. This is called the *Liouvilles' theorem*.

An important parameter other than the focal distance that expresses the performance of the lens is *NA* (which has the same definition as the same parameter used for optical waveguides), which is defined by the following equation that makes use of the focusing angle (or the convergence angle) of Figure 3.6a:

$$\text{NA} = n \sin \theta \simeq n \tan \theta = n \frac{D}{2f}. \tag{3.47}$$

Here, n is the refractive index (1.0 if air) of the medium surrounding the lens, D is the diameter of the lens, and the second approximation is an approximation if θ is small.*

---

* NA expresses the brightness of the lens or its focusing capability. The bigger the NA, the brighter (or bigger the focusing capability of) the lens. In optics, the brightness of the lens is usually represented by NA; however, for camera lenses in photography, the F value defined as $F \equiv \frac{f}{D}$ is the parameter used to express brightness. The smaller (usually greater than 1.0 but near 1) the F value, the brighter the lens.

Up to now, we have reviewed the fundamental characteristics of lenses according to ray optics. However, saying that a ray is refracted is equivalent to saying that the wave front is bending, so when a coherent light passes through a lens, the shape of the wave front changes. Thus, when focusing of parallel rays of light in Figure 3.6a is considered by means of the wave front, as the ray will propagate perpendicular to the wave front of the optical wave, as shown in Figure 3.8, the parallel rays of light are considered to represent the propagation of the plane wave. On the other hand, the focused beam of rays corresponds to a spherical wave converging at one point, so the lens can be considered as an element that converts the wave front from a plane wave to a spherical wave (or in the case of Figure 3.6b, from a divergent spherical wave to a convergent spherical wave). Thus, as shown in Figure 3.8, when a ray that is beamed at a distance r from the optical axis (or the central axis of the lens) is taken into consideration, point A on the optical axis of Figure 3.8 and point B, where the light intersects with the center plane of the lens, are of the same phase as the ray belongs to a plane wave. On the other hand, as the focused rays will all gather at the focal point O, the intersection point C where the circle passing through point A with O as the center intersects with the straight line BO is also of the same phase with point A. Consequently, the phase corresponding to the optical path length S_{BC} is a phase change that is due to the lens. When the r-dependent phase change $\Delta\Phi(r)$ is determined, it becomes

$$\Delta\Phi(r) = k_0 S_{BC} = k_0\sqrt{f^2 + r^2} - f = k_0 f\left(\sqrt{1 + \frac{r^2}{f^2}} - 1\right). \qquad (3.48)$$

Here, only the ray that is near the optical axis is considered (this is called the *paraxial approximation*), and so using the approximation $r \ll f$,

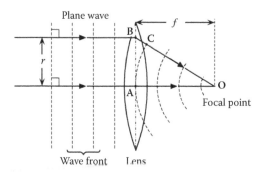

Plane wave

r

B
C
f

A

O

Focal point

Wave front   Lens

FIGURE 3.8 Wave front transformation from a plane wave to a spherical wave by a convex lens.

$$\Delta\Phi(r) \simeq k_0 f \left[1 + \frac{1}{2} \left(\frac{r}{f} \right)^2 - 1 \right] = k_0 \frac{r^2}{2f} \qquad (3.49)$$

is obtained. This is the phase change on the incident wave that is due to the convex lens. On the other hand, the phase change of a concave lens will have a sign that is opposite to that of Equation (3.49).

PROBLEM 3-4

Why will the phase change of Equation (3.49) have a positive sign for the convex lens?

SOLUTION 3-4

Equation (3.49) represents the phase difference between point A and point B in Figure 3.8. Because point A and point C are in phase, Equation (3.49) also represents the phase difference between point C and point B. Point C is the position after the propagation from point B, so according to Equation (1.60), the phase of point C at a fixed time is different by $\exp[-jk_0\overline{BC}]$ when compared with that of point B. Thus, conversely, the phase of point B is different by $\exp[jk_0\overline{BC}]$ when compared with that of point C, so the phase variation amount of Equation (3.49) is given a positive sign.

[End of Solution]

Now, an ideal lens will change the wave front from a plane wave (or a divergent spherical wave) to a convergent spherical wave; however, if the lens has aberrations, it will not result in an ideal convergent spherical wave. In such a case, as shown in Figure 3.9, the difference (not the phase amount but the distance) between the wave front after passing through the lens and the ideal spherical surface (with the center as the focal point) is normalized by the wavelength and is expressed as a function of the distance (called the height of the ray) from the optical axis of the lens. This kind of representation is called a wave aberration. In general, as the aberration in lenses is small at a distance near the optical axis and becomes larger at the periphery, for the wave front aberration, the portion that is about less than $\frac{\lambda}{4}$ (the effective diameter is designated as D_e) is the low aberration region, which can be effectively focused with a beam. Light that is beamed outside this region will have a different focal position along the optical axis due to aberration and it cannot be effectively focused at the focal point (in other words, it cannot form a small focusing spot). Consequently, the NA of the lens is defined by Equation (3.47), but if the portion of the lens, which can effectively focus light, is actually less than the lens aperture D, then the effective NA of the lens is given by the equation where D in Equation (3.47) is replaced with D_e.

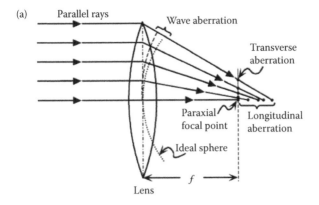

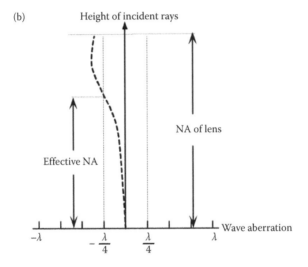

FIGURE 3.9 Definition of the wave front aberration of a convex lens and its effective NA.

PROBLEM

1. Consider a transparent plate that is placed perpendicular to the plane wave (would be the same also for a beam wave) that propagates toward the z-direction. Prove that the transparent plate has a lens effect, if this transparent plate has a refractive index distribution that is expressed by

$$n^2(x, y) = n(0)^2[1 - g^2(x^2 + y^2)] - n(0)^2(1 - g^2 r^2). \qquad (3.50)$$

(This type of medium is called a *lens-like medium* or *distributed index medium* and will be discussed in subsequent chapters. When the refractive index distribution inside the core of multimode fiber is set to this distribution shape, broadbanding becomes possible. g is called the focusing constant.)

3.5 FOURIER TRANSFORM WITH LENSES

In Section 3.2, the fact that a diffraction wave (Fraunhofer diffraction) at a distance that is sufficiently farther than a wavelength will become a Fourier transform of the original amplitude distribution was discussed. However, when a lens is used, the Fourier transform can be obtained even at a not so far distance. First, as shown in Figure 3.10, let us consider an amplitude distribution $\Psi(x', y')$ on the focal point at the rear of the lens when there is an amplitude distribution $\psi(x, y)$ (called an input image in optics) just in front of a lens with a focal point distance f. First, $\psi(x, y)$ will undergo a phase change by the lens as given by Equation (3.49). Then, it will become a spherical wave and reach the focal point. Because this can be considered as a Fresnel region, when $d = f$ in Equation (3.18),

$$\Psi(x', y') = \frac{j}{\lambda f} e^{-jk_0 f} \iint_{-\infty}^{\infty} \psi(x, y) \cdot \underbrace{\exp\left[jk_0 \frac{x^2 + y^2}{2f}\right]}_{\text{phase change due to lens}}$$

$$\times \exp\left[-jk_0 \frac{(x - x')^2 + (y - y')^2}{2f}\right] dx\, dy$$

$$= \frac{j}{\lambda f} \exp\left[-jk_0\left(f + \frac{x'^2 + y'^2}{2f}\right)\right]$$

$$\times \iint_{-\infty}^{\infty} \psi(x, y) \exp\left[jk_0 \frac{x'x + y'y}{f}\right] dx\, dy \qquad (3.51)$$

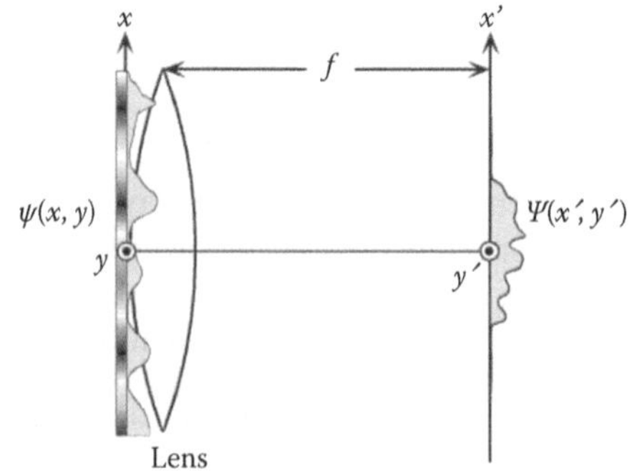

FIGURE 3.10   Fourier transform of a complex amplitude distribution in close proximity of the front surface of convex lens.

can be obtained. When this equation undergoes a conversion of variables in the same way as Equation (3.21) with

$$k_0 \frac{x'}{f} = -\Omega_x (= -2\pi \nu_x), \quad k_0 \frac{y'}{f} = -\Omega_y (= -2\pi \nu_y) \qquad (3.52)$$

then, Equation (3.51) can be written as

$$\Psi(x', y') = \frac{j}{\lambda f} \exp\left[-jk_0 \left(f + \frac{x'^2 + y'^2}{2f}\right)\right] \mathcal{F}[\psi(x, y)]. \qquad (3.53)$$

That is, we can see that the Fourier transform of the amplitude distribution just in front of the lens can be obtained at the focal point at the rear of the lens.

Next, as shown in Figure 3.11, let us consider an amplitude distribution $\Psi(x', y')$ on the focal plane at the rear side when there is an amplitude distribution $\psi(x_i, y_i)$ on a focal plane on the front side of the lens. In such a case, when the amplitude distribution after propagation at a distance of f from the focal plane on the front side to a plane just in front of the lens is set as $\Psi(x, y)$, $\Psi(x, y)$ can be obtained from the simple Fresnel diffraction Equation (3.18) as follows:

$$\Psi(x, y) = \frac{j}{\lambda f} e^{-jk_0 f} \iint_{-\infty}^{\infty} \psi(x_i, y_i)$$

$$\times \exp\left[-jk_0 \frac{(x - x_i)^2 + (y - y_i)^2}{2f}\right] dx_i dy_i. \qquad (3.54)$$

Here, we can see that the formula in Equation (3.54) is a convolution integral that is defined by Equation (A.23) in Appendix A. That is, when $g(x, y)$ is defined as

$$g(x, y) \equiv \exp\left[-jk_0 \frac{x^2 + y^2}{2f}\right] \qquad (3.55)$$

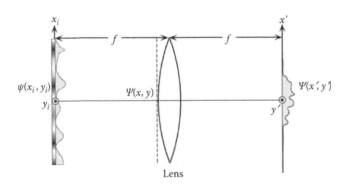

FIGURE 3.11   Fourier transform of a complex amplitude distribution at the focal point in front of a convex lens.

then, $\Psi(x, y)$ can be expressed as

$$\Psi(x, y) = \frac{j}{\lambda f} e^{-jk_0 f} [\psi(x, y) * g(x, y)]. \qquad (3.56)$$

Because this $\Psi(x, y)$ can then be substituted to $\psi(x, y)$ of Equation (3.53), using the convolution integral theorem (Equation (A.25) of Appendix A), the amplitude distribution on a focal plane of the lens can eventually become

$$
\begin{aligned}
\Psi(x', y') &= \frac{j}{\lambda f} \exp\left[-jk_0\left(f + \frac{x'^2 + y'^2}{2f}\right)\right] \\
&\quad \times \mathcal{F}\left[\frac{j}{\lambda f} e^{-jk_0 f} \{\psi(x, y) * g(x, y)\}\right] \\
&= \frac{j}{\lambda f} \exp\left[-jk_0\left(2f + \frac{x'^2 + y'^2}{2f}\right)\right] \\
&\quad \times \mathcal{F}[\psi(x, y)] \cdot \mathcal{F}\left[\frac{j}{\lambda f} g(x, y)\right]. \qquad (3.57)
\end{aligned}
$$

Here, when the formula in Equation (A.23) of Appendix A is used in the Fourier transform of $g = (x, y)$ in Equation (3.57), it will become

$$\Psi(x', y') = \frac{j}{\lambda f} \exp(-jk_0 2f) \cdot \mathcal{F}[\psi(x_i, y_i)]. \qquad (3.58)$$

This is of course a Fourier transform of the amplitude distribution at the focal point on the front side of the lens; however, we can see that it can also appear on the focal plane at the rear side of the lens.

Previously, the Fourier transform effect of the lens was used for the purpose of erasing the dots in half-tone dot-modulated photos* and in highlighting the contours of blurred photos, but as computer image processing technology has progressed in recent years, it has not been of much use. However, it is still an important concept in understanding the nature of lenses in terms of wave optics.

---

* When photos in newspapers are magnified and then closely observed, we can see that they consist of small dots and that the shading is exhibited by the size of the dots. This kind of photograph is called a half-tone dot-modulated photograph, which is suitable for economical printing, such as those in newspapers. It is characterized by its ease in reproduction as even when the photo is copied, the shading will remain the same.

4 Interference and Resonators

4.1 PRINCIPLE OF TWO-BEAM INTERFERENCE

Interference occurs when two free-space beams with the same direction of polarization overlap at the same frequency. Here, let us consider an intensity distribution formed on a screen by the interference of two plane waves (supposing that the y polarization is the direction of polarization) with angular frequency ω_0, which are propagating in the z-direction. The refractive index of the medium is designated as n, and the electric fields are designated as

$$E_y^{(1)} = A_1 \exp[j(\omega_0 t - k_0 n z + \Phi_1)], \tag{4.1}$$

$$E_y^{(2)} = A_2 \exp[j(\omega_0 t - k_0 n z + \Phi_2)]. \tag{4.2}$$

From Equation (3.40), the light intensity I_L on the screen is

$$
\begin{aligned}
I_L &= \frac{k_0 n}{2\omega\mu_0} \left| E_y^{(1)} + E_y^{(2)} \right|^2 \\
&= \frac{k_0 n}{2\omega\mu_0} \left[A_1{}^2 + A_2{}^2 + 2A_1 A_2 \cos(\Phi_1 - \Phi_2) \right] \\
&\propto (A_1 - A_2)^2 \\
&\quad + \left[(A_1 + A_2)^2 - (A_1 - A_2)^2 \right] \cos^2\left(\frac{\Phi_1 - \Phi_2}{2} \right).
\end{aligned}
\tag{4.3}
$$

Here, when the light intensity in Equation (4.3) is plotted on the vertical axis and $\Phi_1 - \Phi_2$ is plotted on the horizontal axis, the graph is drawn as shown in Figure 4.1. Let us suppose two beams (assuming plane waves) incident on a screen that are tilted at an angle of θ. If the direction of propagation of beam 1 is designated as the z-axis, and the z'-axis direction of propagation of beam 2 is tilted at an angle of θ around the y-axis, as shown in Figure 4.2, each of the electric fields can be expressed as

$$E_y^{(1)} = A_1 \exp[j(\omega_0 t - k_0 n z + \Phi_1)], \tag{4.4}$$

$$E_y^{(2)} = A_2 \exp[j\{\omega_0 t - k_0 n(z \cos\theta + x \sin\theta) + \Phi_2\}]. \tag{4.5}$$

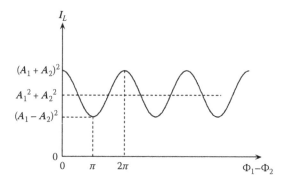

FIGURE 4.1 The relationship of the phase difference and intensity in the interference of two plane waves.

When the light intensity distribution $I_L(x)$ on the screen is determined in the same manner as Equation (4.3), the following equation is obtained:

$$I_L(x) = \frac{k_0 n}{2\omega\mu_0}\big[A_1{}^2 + A_2{}^2 + 2A_1A_2$$
$$\times \cos\{\Phi_1 - \Phi_2 + k_0 n(z - z\cos\theta - x\sin\theta)\}\big]$$
$$\propto (A_1 - A_2)^2$$
$$+ \big[(A_1 + A_2)^2 - (A_1 - A_2)^2\big]\cos^2\left(\frac{\delta\Phi - k_0 n x \sin\theta}{2}\right). \quad (4.6)$$

Here, as the screen is fixed perpendicular to the optical axis (z-axis), $\delta\Phi = \Phi_1 - \Phi_2 + k_0 n(z - z\cos\theta)$ is a constant, and the light intensity given by $I_L(x)$ can be changed depending on the x-coordinate. In other words, interference fringes appear in the y-axis direction of the screen. The interval D between the interference fringes is given by

$$D = \frac{\lambda}{n\sin\theta}. \quad (4.7)$$

When this relation is represented graphically as in Figure 4.2, it is seen that the relation is a periodic function of x, and the contrast of the interference fringes is clear when the amplitudes A_1 and A_2 of the two beams are the same. However, if there is a difference in the amplitudes, the contrast of the interference fringes is diluted. Here, the following measure defined by

$$I_V = \frac{I_L^{(max)} - I_L^{(min)}}{I_L^{(max)} + I_L^{(min)}} = \frac{2A_1A_2}{A_1^2 + A_2^2} \quad (4.8)$$

is called the *visibility* of the interference fringe. When $A_1 = A_2$, the visibility is 1.0.

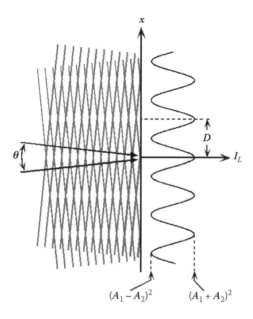

FIGURE 4.2   Interference fringe due to the interference of tilted incident plane waves.

PROBLEM 4-1

In Figure 4.2, for the interval between the interference fringes to be 1,000 times that of the wavelength, how many [rads] should be the angle of inclination θ of the wave front?

SOLUTION 4-1

From Equation (4.7), the angle of inclination is expressed as

$$\sin \theta = \frac{\lambda}{nD}, \tag{4.9}$$

so substituting the given values $\frac{D}{\lambda} = 1,000$ and $n = 1.0$ in Equation (4.9) will result in $\sin \theta \simeq \theta = 10^{-3}$ rad = 0.057 degrees. If the wavelength is 0.6 μm, then the interval between the interference fringes would be 0.6 mm, as the problem states that the interval of the interference fringes will be 1,000 times the wavelength. At such value, this interference fringe is barely discernible to the human eye. That is, if an interferometer is assembled, and the interference fringe is observed using such device, the inclination of the two light beams should be so small that they appear as parallel as possible. Otherwise, the interference fringe cannot be observed by the human eye.   **[End of Solution]**

PROBLEMS

1. In Equations (4.1) and (4.2) that represent the electric fields of two light waves, let us assume that the angular frequency of one of the light waves slightly shifted from ω_0 and became $\omega_0 + \Delta\omega$. When these two light waves are combined and then converted to electrical signals by a photodetector, what kind of electric signal will be obtained?

    [Hint]: An electromagnetic wave of optical frequency ω_0 can be converted to direct current in a photodetector, but note that the $\Delta\omega$ component of the constant frequency wave will become an alternating current of angular frequency $\Delta\omega$ in the same way as the electrical signals. This kind of photodection method is called the *heterodyne detection method*. When the intensity of one of the two waves is weak, it can be combined with the strong intensity of the other light wave (with only a slightly different frequency) so that it can be detected. Because of this combination, the amplitude of the alternating current component of the electrical signal will be the product of the amplitudes of the two light waves (the noise can be removed using an electrical signal filter), and the weak light signals can be detected.

2. Let us assume that of the electric fields of the two light waves expressed in Equations (4.1) and (4.2), the direction of polarization of one of them will be inclined at an angle of θ. Determine the intensity of interference of the light beams that are combined.

4.2 RESONATORS

As shown in Figure 4.3, let us consider a free-space beam that is beamed perpendicular to two translucent plane mirrors that are parallel to each other. The z-axis will be designated as the optical axis, the refractive index of the medium between the translucent mirror is n, and the electric field reflectance of the translucent mirror is designated as r (a complex number as it includes the case when the phase changes). A portion of the free-space beam is reflected on the translucent plane mirror #1 and the rest would reach translucent plane mirror #2, where the reflected component will begin to bounce back and forth between translucent plane mirrors #1 and #2. However, if the free-space beam is a coherent light wave, a standing wave is generated, if the interval L of the translucent mirrors is a half-integer multiple of the wavelength, and resonance occurs. This resonator is called a *Fabry–Perot resonator*.

Let us consider that the spot size of the space beam is large enough compared to the wavelength, and the diffraction of the beam is negligible. When diffraction can be ignored, as the transverse amplitude distribution of the beam remains unchanged in the direction of propagation, the problem becomes a one-dimensional problem that is dependent on only the z-axis. Assuming that the

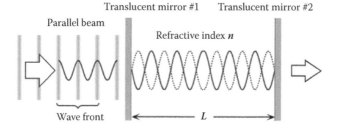

FIGURE 4.3 Basic configuration of the Fabry–Perot resonator.

electric field of the free-space beam is a y polarization and the position of the transluscent mirror #1 is the origin $z = 0$, the electric field is expressed by

$$E_y = A \exp[j(\omega_0 t - k_0 n z)], \qquad (4.10)$$

(strictly speaking, A is a function of x and y; however, Equation (4.10) deals with a one-dimensional problem and is assumed to be a constant). The light wave departs from the translucent mirror #1 (origin) and will repeatedly be reflected between the translucent mirrors #1 and #2. The electric field that is emitted from translucent mirror #2 is designated as $E_y^{(2)}$. The refractive index of the medium inside the resonator is designated as n, with the medium outside simply set as air (or vacuum, with a refractive index $n = 1.0$). The reflectance and transmittance of the light wave that is beamed from the air to the translucent mirror are expressed as r_i and t_i (where $i = 1$ or 2), and conversely, the reflectance and transmittance of the light wave that is beamed from the resonator to the translucent mirror are expressed as r_i' and t_i' (where $i = 1$ or 2), respectively. Consequently, as shown in Figure 4.4, as the sum of all the transmitted components at the translucent mirror #2 will be $E_y^{(2)}$, assuming the electric field ($z = 0$) of the incident beam as $E_y^{(1)}$, it can be expressed as

$$E_y^{(2)} = E_y^{(1)} \left[t_1 t_2' e^{-jk_0 n \cdot L} + t_1 t_2' r_1' r_2' e^{-jk_0 n \cdot 3L} + t_1 t_2' r_1'^2 r_2'^2 e^{-jk_0 n \cdot 5L} + \cdots \right]$$

$$= E_y^{(1)} \frac{t_1 t_2' e^{-jk_0 n L}}{1 - r_1' r_2' e^{-jk_0 n \cdot 2L}}. \qquad (4.11)$$

Therefore, using Equation (4.11), the amplitude transmittance t is given by $\dfrac{E_y^{(2)}}{E_y^{(1)}}$ and the power transmittance $T = |t|^2$ is expressed by

$$T = \frac{|t_1 t_2'|^2}{\{1 - |r_1'||r_2'| \cos \Phi\}^2 + |r_1'|^2 |r_2'|^2 \sin^2 \Phi}. \qquad (4.12)$$

Here, $r_1' = |r_1'| e^{j\phi_1'}$, $r_2' = |r_2'| e^{j\phi_2'}$, and $\Phi = k_0 n \cdot 2L - \phi_1' - \phi_2'$.

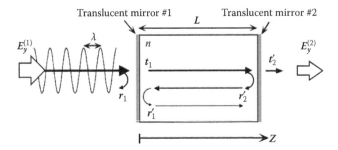

FIGURE 4.4 Relationship between light wave reflection and output field in Fabry–Perot resonator.

PROBLEM 4-2

Let us consider using the Fresnel reflection between air and a medium with a refractive index n for a translucent mirror. In such a case, determine the equation of the power reflectance of a Fabry–Perot resonator.

SOLUTION 4-2

In this case, the reflectance of the two translucent mirrors will be the same. First, from Equation (1.94), the amplitude reflectance when the light wave is beamed from air to a medium with a refractive index n will be

$$r_1 = r_2 = -r_1' = -r_2' = \frac{1.0 - n}{1.0 + n}, \tag{4.13}$$

so the power reflectance will be equal for all reflections (set as R). Moreover, from Stoke's Theorem equation (1.101), the amplitude transmittance will become

$$|t_1 t_2'| = |t_1 t_1'| = 1 - |r_1|^2 = 1 - R. \tag{4.14}$$

If these are substituted to Equation (4.12) and then $\phi_1' = \phi_2' = 0$ is used, the power transmittance can be expressed as*

$$T = \frac{1}{1 + \frac{4R}{(1-R)^2} \sin^2(k_0 n L)}. \tag{4.15}$$

---

* In addition, the characteristics of the resonance peak vicinity are approximated by $\sin(k_0 n L) \simeq k_0 n L$ and expressed as

$$T = \frac{1}{1 + \frac{4R}{(1-R)^2}(k_0 n L)^2}.$$

The function with this form is called the *Lorentzian function* and often appears along with the Gaussian function in optics. Moreover, the function $f(x) = \frac{1}{1+Ax^2}$ before the approximation is called the Airy function.

By substituting $k_0 = \frac{2\pi}{\lambda}$ to Equation (4.15), the transmittance characteristics for the wavelength can be calculated. If the absorption is included, n will become a complex number ($n_i < 0$, please refer to page 45).

[End of Solution]

When Equation (4.15) is drawn for $k_0 n L$ by setting the power reflectance R as the parameter, Figure 4.5 is obtained. The spectral characteristics will have periodic characteristics with a period of $k_0 n L = \pi$, and the wavelength region inside one period is called the free spectral range, which is abbreviated as *FSR*.* From the relationship of $k_0 = \frac{2\pi\nu}{c}$ (c is the speed of light), the horizontal axis in Figure 4.5 is the quantity that is proportional to the frequency. Consequently, the period of transmittance characteristics will be equally spaced in frequency. The peak frequency ν_0 or the peak wavelength λ_0, which give the maximum transmittance are expressed as

$$\nu_0 = \frac{Nc}{2nL} \quad (N \text{ is an integer}), \tag{4.16}$$

$$\lambda_0 = \frac{2nL}{N} \quad (N \text{ is an integer}). \tag{4.17}$$

At the peak frequency, the light that is repeatedly reflected in between the two translucent mirrors generates a reinforced interference at the exit end, and a standing wave is formed inside the resonator. The number of standing waves N is called the resonance order.

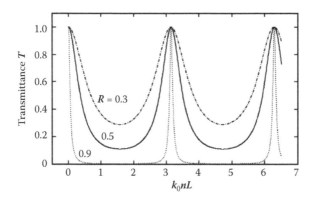

FIGURE 4.5 Transmittance wavelength characteristics of the Fabry–Perot resonator.

---

* There are a lot of cases where it is expressed using another term in a different field even for a physical quantity that is based on the same physical phenomenon. FSR is also referred to as the longitudinal mode spacing in Fabry–Perot semiconductor laser, which will be discussed later.

Incidentally, when Figures 4.5 and 4.1 are compared, the interference of the two beams becomes a cosine function, but in a Fabry–Perot interferometer, the transmittance will decrease rapidly with the wavelength deviation from the peak wavelength. In the Fabry–Perot interferometer, the emitted light is configured from the interference of a large number of reflected light components bouncing back and forth inside the resonator, so even a slight difference in the phases of each of the reflected light components will work strongly toward weakening the interference. For the sharp spectral characteristics shown in Figure 4.5, the interval (or the wavelength width) between the two frequencies wherein the transmittance becomes half that of the maximum transmittance is called the *full width at half maximum* (which is abbreviated as FWHM) bandwidth.[*] Moreover, the ratio of the FWHM bandwidth $\delta\nu_{\text{FWHM}}$ and FSR is

$$F = \frac{\text{FSR}_\nu}{\delta\nu_{\text{FWHM}}} = \frac{\frac{c}{2nL}}{\frac{c}{\pi nL}\sin^{-1}\frac{1-R}{2\sqrt{R}}} \simeq \frac{\pi\sqrt{R}}{1-R}, \tag{4.18}$$

which is called *finesse*. The last approximation is for the case when R is close to 1.0. The closer the reflectance is to 1.0, the larger the finesse, and the spectral characteristics will become sharper.

The FSR described earlier is the wavelength interval (or the frequency interval) between the resonance peaks. By transforming Equation (4.17) to an equation for N, differentiating it with respect to λ_0 and substituting $dN = 1$, the wavelength interval $\Delta\lambda$ of the two peak wavelengths whose resonance order N differs by 1 is expressed by the following equation:

$$\text{FSR}_\lambda = \Delta\lambda = -\frac{\lambda_0^2}{2n_{\text{eff}}L}. \tag{4.19}$$

Here, n_{eff} is given by

$$n_{\text{eff}} = n\left(1 - \frac{\lambda_0}{n}\frac{dn}{d\lambda}\bigg|_{\lambda=\lambda_0}\right), \tag{4.20}$$

which is a quantity that includes the wavelength dependency of the refractive index and is called the *effective index*.[†]

Among semiconductor lasers, there is the Fabry–Perot semiconductor laser (abbreviated as FP)[‡], wherein the cleavage facet of a semiconductor crystal is utilized to form a reflector on both ends of the active layer (the layer that has

---

[*] Because half the transmission is equivalent to a loss of 3 dB, this is the same as the 3 dB bandwidth.

[†] As discussed in Section 2.3, a lot of papers and textbooks confuse this effective index with the equivalent index, so it is necessary to be careful with the defining equation.

[‡] Often used as an optical disk pickup light source.

gain in the core layer of the waveguide) configuring a Fabry–Perot resonator. For this oscillation wavelength of the FP semiconductor laser, a large number of wavelengths that satisfies Equation (4.17) may be present in the gain wavelength region of the active layer, so the lasing spectrum will have multiple peaks that look like a tooth comb, as shown in Figure 4.6.* For such a case, the resonance order N or its corresponding spectral peak is called the *longitudinal mode*,[†] and the laser that is oscillating at multiple longitudinal modes is called a longitudinal multimode laser (lasers that have only one longitudinal mode is called a single longitudinal mode laser). The interval (or wavelength interval)[‡] of this longitudinal mode is given by Equation (4.19), and the effective index of Equation (4.20) has a value that is more than 10% bigger than the refractive index n in semiconductor lasers. Therefore, the wavelength dependency of the refractive index is not negligible.

If a semiconductor laser with the lasing spectrum shown in Figure 4.6 is used for optical fiber transmission, the transmission bandwidth of the single

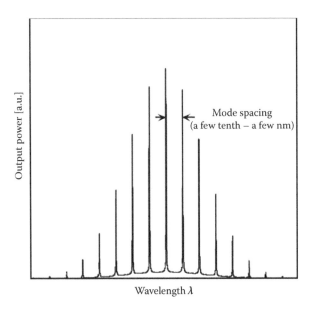

FIGURE 4.6 Typical oscillation spectrum of a Fabry–Perot semiconductor laser.

---

* The vertical axis unit (a.u.) stands for arbitrary unit. It means that the unit is calibrated arbitrarily or is nonstandard
† In contrast to this, the guided mode in the active layer (core) is a standing wave oscillating in the transverse direction and is called the *transverse mode*.
‡ This has the same physical quantity as that of FSR and is called such in semiconductor lasers.

mode fiber will deteriorate (refer to Chapter 6), so Fabry–Perot lasers cannot be used for optical communication. For optical communications, it is necessary to have only one lasing wavelength spectrum. Even if there is only one lasing spectrum at the steady state with an unmodulated injected current, once the injected current is modulated, the laser may oscillate in the multilongitudinal mode. Here, even with modulation, the laser that retains one lasing spectrum is called a dynamic single mode laser. One of such dynamic single mode laser is the DFB (distributed feedback) laser, a laser that forms a diffraction grating on the surface of the active layer, thereby configuring a resonator.

PROBLEM 4-3

Determine the lasing condition of an FP semiconductor laser.

SOLUTION 4-3

As mentioned in Section 1.6, the gain and absorption loss can be related to the imaginary part n_i of the refractive index. However, the imaginary part n_i of the refractive index itself does not represent the gain and absorption loss, and as known from Equation (1.139), the gain and absorption loss in terms of optical power are equal to $2k_0 n_i$. Moreover, in semiconductor lasers, gain occurs when current is injected; however, even if there is no injection of current, there is still a constant absorption loss. Here, when the gain (power amplification factor) per unit length of the semiconductor laser under current injection is set as g and the absorption loss coefficient is designated as α, the net gain is $g - \alpha$. When this is used to rewrite Equation (1.137) (the time dependent term $e^{j\omega t}$ is omitted), $E_y(z)$ is expressed by

$$E_y(z) = Ae^{-jk_0 n_r z} \exp\left[\frac{g-\alpha}{2}z\right]. \qquad (4.21)$$

When this electric field departs $z = 0$, makes a round trip inside the resonator and comes back, and then faces the same direction of propagation, it will undergo reflection on the translucent mirror (actually, at the boundary surface of the semiconductor and air) twice, so the propagation distance will be $2L$, resulting in

$$E_y(2L) = Ar^2 e^{-jk_0 n_r 2L} \exp\left[\frac{g-\alpha}{2}2L\right]. \qquad (4.22)$$

The interference condition of the Fabry–Perot interferometer is that the phase of the light that made a round trip and came back is different from the phase of the light at the starting point by multiple integers of 2π. Moreover, the amplitude condition is such that the amplitude after the round trip should be at least the same or more than that at the starting point for amplification to occur. Therefore, the lasing condition is represented by the following two conditions:

$$2k_0 n_r L = 2\pi N \quad (N \text{ is an integer}) \quad (: \text{phase condition}), \quad (4.23)$$
$$r^2 \exp[(g - \alpha)L] \geq 1.0 \quad (: \text{amplitude condition}). \quad (4.24)$$

From the phase condition, the lasing wavelength is determined by Equation (4.17). Moreover, the amplitude condition gives the necessary gain condition that would lead to oscillation, and the condition at which lasing will start is called the *threshold*. Because $r^2 = R$, from Equation (4.24), the threshold gain g_{th} is given by

$$g_{th} = \alpha + \frac{1}{2L} \log_e \frac{1}{R^2} = \alpha + \frac{1}{2L} \log_e \frac{1}{R_1 R_2}. \quad (4.25)$$

The second equation is the equation when the reflectance of the two reflectors is different. The gain of the semiconductor laser can be obtained by injecting a current in the forward-biased direction on the diode, so the oscillation of the semiconductor laser will occur above a certain current, with characteristics shown in Figure 4.7. These characteristics are called the current–light output characteristics of semiconductor lasers (also known as the $I-L$ characteristics), and the current value at which laser oscillation begins is called the threshold current I_{th}.

The threshold current of commercially available semiconductor lasers ranges from several mA to tens of mA, and lasers under 1 mA threshold have a very low threshold. However, at research levels, there are lasers

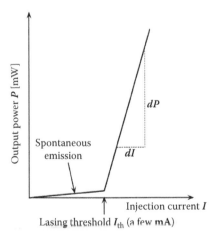

FIGURE 4.7 Current versus light output characteristics of semiconductor lasers.

with threshold currents much less than 1 mA. Moreover, the output of normal lasers (for optical disks and optical communications) is from several mW to tens of mW, but there are high-power semiconductor lasers exceeding 1 W for special applications.

If current that is larger than the threshold current I_{th} is injected to a semiconductor laser, the light output will increase in proportion to the amount of the increase of the current. The efficiency will be physically easy to understand when expressed as a ratio of the number of carriers injected to the number of photons generated. Here, when the amount of increase of the current exceeding the threshold is designated as dI, and the amount of increase in the light output is set as dP, the resulting equation

$$\eta_d = \frac{dP/\hbar\omega}{dI/e} \simeq \frac{dP}{dI \cdot E_g} \qquad (4.26)$$

is called external differential quantum efficiency. Here, E_g is the bandgap energy of the active layer.   **[End of Solution]**

PROBLEMS

1. Draw a figure of the transmittance wavelength characteristics when the reflectances of the two resonators in the Fabry–Perot resonator of Exercise 4-2 are not equal (e.g., they are 0.9 and 0.5). Moreover, think of a reason why the transmittance cannot be 1.0 at the peak wavelength.

2. For the set up in Exercise 4-2, assuming that the reflectances of the two reflectors are equal, calculate the finesse for $R = 0.95$ and $R = 0.5$.

3. For the set up in Exercise 4-2, determine the transmittance and reflectance when the medium has absorption.

4. Prove that the ratio of the electric field in the resonator and the incident electric field outside the resonator is $\frac{1+r}{1-r}$, when the electric field reflectance is r.

4.3 VARIOUS INTERFEROMETERS

As described in Section 4.1, the interferometer has to be configured to bring about the interference of two light beams.

When two light beams are separated on the translucent mirror as shown in Figure 4.8, the circuit that again merges these light beams on the translucent mirror after different optical paths, involving total reflection mirrors, is called the *Mach–Zehnder interferometer*.* As a translucent mirror, one side of a glass plate can be subjected to an antireflection coating, whereas the other side can be

---

* This can also be configured with an optical waveguide. In such a case, a directional coupler is used instead of a translucent mirror, as will be discussed in Chapter 8.

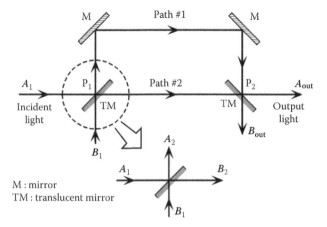

FIGURE 4.8 Mach–Zehnder interferometer.

used for Fresnel reflection; however, the reflectance will be different for TE and TM polarizations.[*] The refractive index of the medium constituting the optical path is set to n, and the amplitude reflectance and amplitude transmittance of the translucent mirror are set to r and t, respectively. As shown in the lower part of Figure 4.8, if the electric field amplitude of the incident light from the entrance ports (planes of incidence) 1 and 2 of the translucent mirror are defined as A_1 and B_1, and the electric field amplitude of the exiting light is defined as A_2 and B_2, respectively, the relation between them can be expressed as

$$\begin{bmatrix} A_2 \\ B_2 \end{bmatrix} = \begin{bmatrix} r & t' \\ t & r' \end{bmatrix} \cdot \begin{bmatrix} A_1 \\ B_1 \end{bmatrix}. \tag{4.27}$$

Here, similar to that in Equation (1.97) through Equation (1.100), r' and t' are the amplitude reflectance and amplitude transmittance seen from the opposite side of the translucent mirror.[†]

---

[*] In addition, by using a wedge plate to gradually change the thickness so that both sides of the glass plate will not be parallel, only one of the Fresnel reflections from both sides of the glass plate can be configured to have an interferometer optical path. Moreover, the difference of the reflectance between TE polarization and TM polarization can be decreased by forming a thin metal film; however, absorption loss may occur because of the metal film. To decrease the polarization dependence of the Fresnel reflection, it is best to have the direction of reflection as close to 90° as possible.

[†] When the determinant of Equation (4.27) is calculated, it becomes $rr' - tt'$. When this result is used together with the Stokes' theorem equation (1.101), the determinant becomes -1. This is because the exit port A_2 is taken from the same side of the entrance port A_1 with respect to the translucent mirror, so element 11 of the matrix will represent a reflectance. If the labels of exit ports A_2 and B_2 are exchanged, the determinant will become $+1$, resulting in a normal lossless matrix.

When the lengths of the two optical paths 1 and 2, from point P_1 just at the back of the branching translucent mirror until point P_2 in front of the merging translucent mirror, are set as L_1 and L_2, respectively, the optical path lengths defined in Equation (1.62) will be $S_1 = nL_1$ and $S_2 = nL_2$. Therefore, optical path 1 and optical path 2 will each undergo a phase change of $e^{-jk_0 S_1}$ and $e^{-jk_0 S_2}$, respectively. Consequently, the transmission matrix between these two points will be a diagonal matrix with these phase changes as the diagonal elements. After the propagation, they will again be merged at the translucent mirror, and finally, the outgoing light amplitudes A_{out} and B_{out} from the exit ports of the Mach–Zehnder interferometer circuit are given by

$$
\begin{bmatrix} A_{\text{out}} \\ B_{\text{out}} \end{bmatrix} = \begin{bmatrix} r & t' \\ t & r' \end{bmatrix} \cdot \begin{bmatrix} e^{-jk_0 S_1} & 0 \\ 0 & e^{-jk_0 S_2} \end{bmatrix} \cdot \begin{bmatrix} r & t' \\ t & r' \end{bmatrix} \cdot \begin{bmatrix} A_1 \\ B_1 \end{bmatrix}
$$
$$
= \begin{bmatrix} r^2 e^{-jk_0 S_1} + tt' e^{-jk_0 S_2}, & rt' e^{-jk_0 S_1} + r't' e^{-jk_0 S_2} \\ rt e^{-jk_0 S_1} + r't e^{-jk_0 S_2}, & tt' e^{-jk_0 S_1} + r'^2 e^{-jk_0 S_2} \end{bmatrix} \cdot \begin{bmatrix} A_1 \\ B_1 \end{bmatrix}. \quad (4.28)
$$

To further simplify, when the amplitude reflectance of the translucent mirror is assumed to be $r = \frac{1}{\sqrt{2}}$ (the power reflectance is $R = \frac{1}{2}$), it will result in $r' = -\frac{1}{\sqrt{2}}$ based on Equations (1.97) and (1.98). Moreover, the amplitude transmittance at the boundary surface of media with different refractive indices will not be equal to the amplitude reflectance; however, if one side of the glass plate is used as a Fresnel reflection translucent mirror and the other side is applied with an antireflection coating, the transmitted light will come out to the medium with the same refractive index as that of the incident side, so the amplitude transmittance will also become $t = t' = \frac{1}{\sqrt{2}}$. Consequently, Equation (4.28) will be further simplified and can be rewritten as

$$
\begin{bmatrix} A_{\text{out}} \\ B_{\text{out}} \end{bmatrix} = e^{-jk_0 \frac{S_1+S_2}{2}} \begin{bmatrix} \cos\left(k_0 \dfrac{S_1 - S_2}{2}\right) & j\sin\left(k_0 \dfrac{S_1 - S_2}{2}\right) \\ j\sin\left(k_0 \dfrac{S_1 - S_2}{2}\right) & \cos\left(k_0 \dfrac{S_1 - S_2}{2}\right) \end{bmatrix} \cdot \begin{bmatrix} A_1 \\ B_1 \end{bmatrix}.
$$
$$
(4.29)
$$

Here, when we consider a light that is beamed only at one side of the entrance port, i.e. $B_1 = 0$, and when the output light intensity P is determined from the exit port using $P = |A_{\text{out}}|^2$,

$$
P = |A_1|^2 \cos^2\left[k_0 \frac{S_1 - S_2}{2}\right] = |A_1|^2 \cos^2\left[\frac{\pi}{\lambda} S\right] \quad (4.30)
$$

will be obtained. Here, $S = n(L_1 - L_2)$ represents the difference in the optical path lengths of optical path #1 and optical path #2.

Equation (4.30) is illustrated as shown in Figure 4.9, and the light output of the Mach–Zehnder interferometer with an optical path length difference will

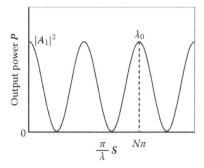

FIGURE 4.9   Wavelength filter characteristics of the Mach–Zehnder interferometer.

have a periodic dependence to the reciprocal of the wavelength. The transmittance peak wavelength λ_0 of the wavelength filter can be expressed using

$$\lambda_0 = \frac{1}{N\pi} S \qquad (N = 0, 1, 2, \ldots). \tag{4.31}$$

Because the Mach–Zehnder interferometer has such periodic wavelength filter characteristics, if several interferometers with an optical path length difference S of an integral multiple of two are cascaded, a band-pass filter can be configured to retrieve the desired wavelength from a particular wavelength band. Moreover, a Mach–Zehnder interferometer with the optical path length difference set to zero (the translucent mirror in the latter part of Figure 4.8 can be exchanged with the total reflection mirror) would not have wavelength filter characteristics. In such a symmetric Mach–Zehnder interferometer, if a sample medium is inserted into one arm of the optical path and the phase of the light that is transmitted through such a medium changes (because of the changes in the refractive index brought about by adding voltage or stress or changes in the refractive index resulting from temperature change), the output of the interferometer will also change, so the quantity of the phase change can be measured accurately.*

PROBLEMS

1. At the translucent mirror on the exit side of the Mach–Zehnder interferometer, the centers of the two beams that have to be merged are

---

* If two beams are slightly tilted so that they will not overlap at the exit end of the Mach–Zehnder interferometer, an interference fringe such as that in Equation (4.6) will appear, and the direction of the movement of such interference fringe will change depending on the sign (positive or negative) of the phase change. Consequently, one can measure the sign of the phase change in addition to the magnitude. Moreover, if one counts the number of interference fringes that moves, a phase change that exceeds 2π can also be measured.

slightly shifted and then merged and set up in such a way that the centers of the beams will coincide on the screen. In such a case, one should be able to observe the interference fringes. Let us assume that the distance between the exit side translucent mirror and the screen is 30 cm, and the wavelength is 0.6 μm. What micron should the spacing between centers of the two beams on the exit side of translucent mirror be, so that the interval of the interference fringes on the screen is more than 0.6 mm (i.e. the fringe interval that is observable to the human eye)?

[**Hint**]: This is easily determined from Exercise 4-1. Moreover, from this example, the reader will know how necessary it is to adjust the optical path of the Mach–Zehnder interferometer with high precision.

2. The interferometer shown in Figure 4.10 is called the *Michelson interferometer*, and it is used to measure the change in the distance (crustal deformation for earthquake prediction) of two remote spots. Find an equation that can express the intensity of light emitted from the interferometer.

3. The interferometer shown in Figure 4.11 is called the *Sagnac interferometer*, and it is used as a laser gyro (or an optical fiber gyro if the entire optical path is configured using fiber optics) as the amount of movement of the interference fringes that is proportional to the angular velocity can be measured, when the entire interferometer is rotated. Find an equation that can express the intensity of light emitted from the interferometer.

4. Figure 4.12 depicts a resonator that is called the *ring resonator*. It looks very similar to the Sagnac interferometer in Figure 4.11, with a difference of only 90° in the orientation of the translucent mirror. Unlike that in a Sagnac interferometer, light will not return to the incident direction. Find an equation that will express the intensity of the output light, when there is an absorbing medium with an amplitude transmittance t inside the ring resonator.

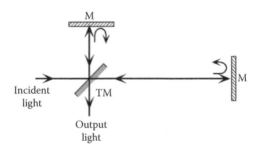

FIGURE 4.10 Michelson interferometer.

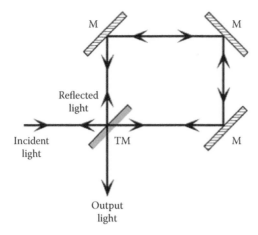

FIGURE 4.11   Sagnac interferometer.

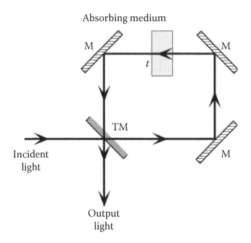

FIGURE 4.12   Ring resonator.

4.4   DIFFRACTION BY GRATINGS

So far, we have considered a plane (or a spherical surface for lenses) as the boundary surface where reflection and refraction occurs. However, if there is an object with a refractive index that is different from that of a surrounding medium with a uniform refractive index, and such object is smaller than a wavelength, light will be scattered. If these scatterers are arranged periodically at equal intervals, then each of the scattered light will form a wave front of a fixed direction and will be strongly radiated from such direction.

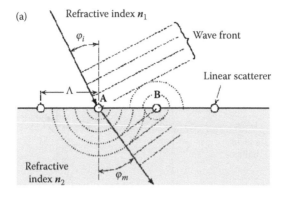

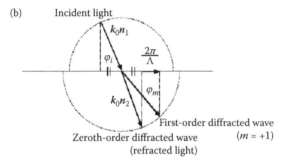

FIGURE 4.13 Plane wave diffraction by gratings.

Now, let us periodically arrange the linear scatterers at a period Λ on a plane. Such an element is called a *grating*. A plane wave is incident on the grating at an angle φ_i to the normal line of the surface of the grating and perpendicular to these linear scatterers. This situation can be thought of in a two-dimensional plane as shown in Figure 4.13.

When the angle of incidence is φ_i, the angle of diffraction is φ_m, the refractive index of the medium at the incident side is set as n_1, and the refractive index of the medium at the exit side is n_2, the light scattered at point A in the figure will become a cylindrical wave and will be radiated. However, the cylindrical wave that was scattered at point B will form a wave front, if the conditions of the following equation is satisfied:

$$k_0 n_2 \Lambda \sin \varphi_m = k_0 n_1 \Lambda \sin \varphi_i + 2\pi m \quad (m = 0, \pm 1, \pm 2, \ldots) \qquad (4.32)$$

or

$$n_2 \sin \varphi_m = n_1 \sin \varphi_i + \frac{\lambda}{\Lambda} m \quad (m = 0, \pm 1, \pm 2, \ldots). \qquad (4.33)$$

The integer m is called the *diffraction order*. If Equation (4.32) is further transformed, it will become*

$$k_0 n_2 \sin \varphi_m = k_0 n_1 \sin \varphi_i + \frac{2\pi}{\Lambda} m \quad (m = 0, \pm 1, \pm 2, \ldots). \tag{4.34}$$

That is, when the propagation vectors of the incident wave and the diffracted wave are defined as k_i and k_m, respectively, and the absolute value of the lattice vector of the grating is expressed as

$$|K| = \frac{2\pi}{\Lambda}, \tag{4.35}$$

taking a direction wherein φ_m increases, then Equation (4.33) can be written as

$$|k_m| \sin \varphi_m = |k_i| \sin \varphi_i + |K| m \quad (m = 0, \pm 1, \pm 2, \ldots). \tag{4.36}$$

In the case of normal incidence ($\varphi_i = 0$), the angle of diffraction φ_m-th of the m-order diffracted light becomes

$$\varphi_m = \sin^{-1} \left(\frac{m\lambda}{n_2 \Lambda} \right) \quad \left(\frac{m\lambda}{n_2 \Lambda} \leq 1.0 \right). \tag{4.37}$$

As seen in Equations (4.33) and (4.37), the angle of diffraction is dependent on the wavelength λ, so if light of wide spectrum is beamed to the grating, each of the wavelength components will be separated such that there will be a small angle of diffraction for the short wavelengths and a big angle of diffraction for the long wavelengths. Consequently, gratings find applications in wavelength filters, spectrometers (spectrum analyzers), and so on.

Then, let us consider that instead of linear scatterers, planar scatterers are laminated in layers, as shown in Figure 4.14, and light is gradually reflected from each of the boundary surfaces. The spacing between the scattering planes is defined as Λ, and the reflected light from each of the scattering planes are assumed to be weak.[†] In such a case, the angle of reflection θ_m will be equal to the angle of incidence θ_i, and the condition wherein the reflected light will be reinforced due to interference will be

$$\sin \theta_m = \sin \theta_i = \frac{m\lambda}{2\Lambda} \quad (m = 0, \pm 1, \pm 2, \ldots). \tag{4.38}$$

Or, if Equation (4.38) is rewritten in terms of the propagation vector and lattice vector, then the equation can be written as

$$k_m = k_i + mK \quad (m = 0, \pm 1, \pm 2, \ldots). \tag{4.39}$$

---

[*] If $m = 0$, then $n_2 \sin \varphi_m = n_1 \sin \varphi_i$, matching the equation for Snell's Law of refraction.
[†] If the reflections from each of the planes are not weak, then multiple reflections will occur between the laminated planes, resulting in reflectance of multilayer reflective films.

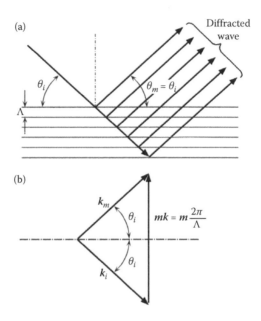

FIGURE 4.14   Bragg diffraction by planar scatterers.

This condition is called the *Bragg condition*, and the angle determined by Equation (4.38) is called the *Bragg angle*, with the reflection resulting from this phenomenon being referred to as the *Bragg reflection*. The condition of Equation (4.38) is different from that of Equation (4.33) in that only certain values are allowed for the angle of incidence. Bragg reflection is used in X-ray reflections of the crystal lattice and in other applications.

PROBLEMS

1. Assuming that the refractive indices of the medium at the incident side and the medium at the exit side are 1.0, determine the first-order angle of diffraction and the second-order angle of diffraction, when a plane wave of wavelength $\lambda = 0.5\ \mu m$ is beamed perpendicularly to a grating with period $\Lambda = 1.6\ \mu m$.

4.5   MULTILAYER THIN FILM INTERFERENCE

Multilayer films (multilayer optical thin films) with multiple stacked thin films of different refractive indices have a wide range of applications as wavelength filters, polarization splitters, antireflection coatings, and high-reflection mirrors. In these multilayer films, the Fresnel reflections generated at the boundary of multiple layers with different refractive indices are superimposed and cause interference. As a result, wavelength dependence and polarization dependence of the reflectance are generated, and the increase or decrease of the reflectance occurs.

In the multilayer film interference, the interference overlaps in between various boundaries, and the reflex pathway has to be considered one by one, making the analysis difficult. Therefore, the boundary conditions that hold true for the incident wave, reflected wave, and transmitted wave at each of the boundaries, and the phase change due to the propagation within each of these layers are all expressed using the matrix formalism. First, as shown in Figure 4.15, the boundary between the $i-1$th layer and ith layer is defined as the boundary surface $i-1$, where the forward propagating waves facing the positive z-axis direction are the incident wave and the transmitted wave, whereas the backward propagating wave facing the negative z-direction is the reflected wave. The electric and magnetic fields of the incident wave beamed from the ith layer to the boundary surface are expressed as E_i^+ and H_i^+, whereas the electric and magnetic fields of the reflected wave are expressed as E_i^- and H_i^-.

Let us consider a TE-polarized plane wave that is beamed to a multilayer film. If the tangential components E_i and H_i of the electric and magnetic fields in the ith layer of the boundary surface i plane will have an angle φ_i between the normal line of the boundary surface i plane and the propagation direction of the plane wave (direction of the ray), direction of the electric field E_i^+ and the magnetic field H_i^+ for the incident wave and direction of the electric field E_i^- and magnetic field H_i^- for the reflected wave is

$$E_i = |E_i^+| + |E_i^-|, \tag{4.40}$$

$$H_i = |H_i^+| \cos \varphi_i - |H_i^-| \cos \varphi_i. \tag{4.41}$$

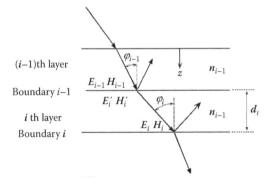

FIGURE 4.15 The relationship between the electric and magnetic fields of the $i-1$th layer and the ith layer in multilayer films.

Moreover, when a plane wave is propagated at the i layer between the boundary surface $i-1$ plane and i plane, the phase change quantity ϕ_i will be*

$$\phi_i = k_0 n_i d_i \cos \varphi_i = \frac{2\pi}{\lambda} n_i d_i \cos \varphi_i \qquad (4.42)$$

Consequently, the tangential components E_i' and H_i' of the electric and magnetic fields in the ith layer of the boundary surface $i-1$ plane are expressed by the following equations:

$$E_i' = |E_i^+|e^{j\phi_i} + |E_i^-|e^{-j\phi_i}, \qquad (4.43)$$

$$H_i' = |H_i^+|e^{j\phi_i} \cos \varphi_i - |H_i^-|e^{-j\phi_i} \cos \varphi_i. \qquad (4.44)$$

From the boundary conditions, the tangential component E_{i-1} of the electric field across the boundary surface $i-1$ plane is equal to E_i', whereas the tangential component H_{i-1} of the magnetic field is equal to H_i', so the following equations are derived:

$$E_{i-1} = E_i' = |E_i^+|e^{j\phi_i} + |E_i^-|e^{-j\phi_i}, \qquad (4.45)$$

$$H_{i-1} = H_i' = |H_i^+|e^{j\phi_i} \cos \varphi_i - |H_i^-|e^{-j\phi_i} \cos \varphi_i. \qquad (4.46)$$

On the other hand, when Equations (1.25) and (1.30) of Section 1.3.2 are used, the relationship between the electric and the magnetic fields of the plane wave is as follows:[†]

$$|H_i^{\pm}| = \sqrt{\frac{\varepsilon_0}{\mu_0}} n_i |E_i^{\pm}|. \qquad (4.47)$$

When Equation (4.47) is used in Equations (4.41) and (4.46), H_i and H_{i-1} can be represented as follows:

$$H_i = \sqrt{\frac{\varepsilon_0}{\mu_0}} n_i (|E_i^+| - |E_i^-|) \cos \varphi_i, \qquad (4.48)$$

$$H_{i-1} = \sqrt{\frac{\varepsilon_0}{\mu_0}} n_i (|E_i^+|e^{j\phi_i} - |E_i^-|e^{-j\phi_i}) \cos \varphi_i. \qquad (4.49)$$

---

* For the electric field E_i^+ that is progressing from the boundary surface $i-1$ plane to the i plane, this phase change will be $e^{+j\phi_i}$. For $e^{j(\omega t - k_0 n z)}$, the electric field at the boundary surface $i-1$ plane is one that occurs temporally, at the previous time. So, to use the electric field that temporally occurs afterwards (at the boundary surface i plane) to express such a term, it must be multiplied by $\exp[+jk_0 n z]$.

† If we think about the direction of propagation, to be exact, H_i^- propagates in the direction of the negative z-axis, so the sign of the vector equation is negative. However, as we are considering the amplitude here, they are expressed using Equation (4.47).

On the other hand, using E_i and H_i from Equations (4.40) and (4.48), $|E_i^+|$ and $|E_i^-|$ can be expressed as

$$|E_i^+| = \frac{1}{2}\left[E_i + \frac{H_i}{\sqrt{\frac{\varepsilon_0}{\mu_0}}n_i \cos \varphi_i}\right], \tag{4.50}$$

$$|E_i^-| = \frac{1}{2}\left[E_i - \frac{H_i}{\sqrt{\frac{\varepsilon_0}{\mu_0}}n_i \cos \varphi_i}\right]. \tag{4.51}$$

When these equations are substituted to Equations (4.45) and (4.46) and expressed in matrix form, the following equation is obtained:

$$\begin{bmatrix} E_{i-1} \\ H_{i-1} \end{bmatrix} = \begin{bmatrix} \cos \phi_i, & j\frac{1}{Y_i}\sin \phi_i \\ jY_i \sin \phi_i, & \cos \phi_i \end{bmatrix} \cdot \begin{bmatrix} E_i \\ H_i \end{bmatrix}. \tag{4.52}$$

Here,

$$Y_i = \sqrt{\frac{\varepsilon_0}{\mu_0}}n_i \cos \varphi_i \quad \text{(TE wave)} \tag{4.53}$$

is called the *optical admittance*.

The TE wave was derived earlier. However, the TM wave can be derived in the same way by replacing E and H in Equations (4.40) and (4.41), and then taking the reciprocal of $\sqrt{\frac{\varepsilon_0}{\mu_0}}n_i$ of Equation (4.47). This would result in the optical admittance of the TM mode to be defined as

$$Y_i = \sqrt{\frac{\varepsilon_0}{\mu_0}}n_i\frac{1}{\cos \varphi_i} \quad \text{(TM wave)}, \tag{4.54}$$

which can be expressed in matrix form in the same way as Equation (4.52). The matrix portion of Equation (4.52), which is

$$[M_i] = \begin{bmatrix} \cos \phi_i, & j\frac{1}{Y_i}\sin \phi_i \\ jY_i \sin \phi_i, & \cos \phi_i \end{bmatrix}, \tag{4.55}$$

is called the *interference matrix*. It can easily be seen from Equation (4.52) that the determinant of the interference matrix will be unity.*

Next, let us determine the reflectance from multilayer films. With the number of layers designated as N layers, the medium on the incident side (of course, this is the medium that spreads semi-infinitely in the negative z-direction of Figure 4.15) is set as the zeroth layer, and the medium on the exit side (the medium that spreads semi-infinitely in the positive z-direction) is set as the $(N + 1)$th layer. In such a case, the boundary surface exists from the 0 plane to the N plane. To determine the reflectance and transmittance, one should know the relationship between the electric and magnetic fields E_0 and H_0 of the zeroth layer of the medium on the incident side and the electric and magnetic fields E'_{N+1} and H'_{N+1} of the $(N + 1)$th layer of the medium on the exit side. Using Equation (4.52), this relationship can be calculated as follows:

$$
\begin{bmatrix} E_0 \\ H_0 \end{bmatrix} = [M_1] \begin{bmatrix} E_1 \\ H_1 \end{bmatrix} = [M_1][M_2] \begin{bmatrix} E_2 \\ H_2 \end{bmatrix} = \cdots
$$

$$
= [M_1][M_2]\cdots[M_N] \begin{bmatrix} E_N \\ H_N \end{bmatrix} = [M] \begin{bmatrix} E'_{N+1} \\ H'_{N+1} \end{bmatrix}. \qquad (4.56)
$$

Here, $[M_i]$ is the interference matrix of the ith layer represented by Equation (4.55), whereas

$$
[M_i] = [M_1][M_2]\cdots[M_N] = \begin{bmatrix} m_{11}, & m_{12} \\ m_{21}, & m_{22} \end{bmatrix} \qquad (4.57)
$$

is the interference matrix of the whole multilayer film, with m_{11} to m_{22} being the matrix elements. The angle of propagation φ_i in each of the layers is included in these matrix elements; however, this can be easily determined by $n_0 \sin \varphi_0 = n_1 \sin \varphi_1 = \cdots = n_i \sin \varphi_i$ from the angle of incidence and Snell's Law. When the electric and magnetic fields of the incident wave, reflected wave, and the transmitted wave are related in terms of the optical admittance, the following equations will be obtained (because both TE polarization and TM polarization are the same, only the example for TE polarization is presented below):

$$
E_0 = E_0^+ + E_0^-, \qquad (4.58)
$$

$$
H_0 = (H_0^+ - H_0^-) \cos \varphi_0 = (E_0^+ - E_0^-)Y_0, \qquad (4.59)
$$

$$
H'_{N+1} = E'_{N+1} Y_{N+1}. \qquad (4.60)
$$

---

* In other words, in the interference matrix of multilayer films without absorption loss, the determinant becomes 1.

Substituting these equations to Equation (4.56), eliminating E'_{N+1}, and determining the electric field reflectance, would result in

$$r = \frac{E_0^-}{E_0^+} = \frac{m_{11} - Y_0^{-1}Y_{N+1}m_{22} + Y_{N+1}m_{12} - Y_0^{-1}m_{21}}{m_{11} + Y_0^{-1}Y_{N+1}m_{22} + Y_{N+1}m_{12} + Y_0^{-1}m_{21}}. \tag{4.61}$$

The power reflectance is given by $R = |r|^2$ in the same way as Equation (1.106) of Section 1.4.2. When this interference matrix is used, the design and analysis of multilayer films will be possible.

PROBLEMS

1. Given a refractive index of the medium on the incident side as n_s, and a refractive index of the medium on the exit side as n_a, let us suppose a multilayer film, having films with two kinds of refractive indices n_1 and n_2, that is configured by stacking M sets alternatively (n_1 on the layer next to n_s, next is n_2, with n_a as the last). Here, the number of layers will be $2M$. Assuming the angle of incidence to be $\varphi_0 = 0$ (normal incidence) and designating the thickness d_i of each layer as

$$d_i = \frac{\lambda}{2n_i}, \tag{4.62}$$

prove that the reflectance of the multilayer film can be given by

$$R = \left[\frac{n_s \left(\frac{n_1}{n_2}\right)^{2M} - n_a}{n_s \left(\frac{n_1}{n_2}\right)^{2M} + n_a} \right]^2. \tag{4.63}$$

This multilayer film can be used in configuring high-reflection films.
[**Hint**]: Any matrix M squared to N will become

$$[M]^N = \begin{bmatrix} A, & B \\ C, & D \end{bmatrix}^N, \tag{4.64}$$

with the parameter defined as

$$\xi = \sqrt{\det[M]} = \sqrt{AD - BC} \tag{4.65}$$

$$\eta = \frac{A + D}{2}, \tag{4.66}$$

and using the matrix elements $A \sim D$, it can be calculated as follows:

(a) If $\xi \geq \eta$ and θ is defined as $\theta = \cos^{-1} \frac{\eta}{\xi}$,

$$[M]^N = \frac{\xi^N}{\sin \theta} \left[\frac{1}{\xi} \sin N\theta [M] - \sin(N-2)\theta [I] \right]. \quad (4.67)$$

(b) If $\xi \leq \eta$ and θ is defined as $\theta = \cosh^{-1} \frac{\eta}{\xi}$,

$$[M]^N = \frac{\xi^N}{\sinh \theta} \left[\frac{1}{\xi} \sinh N\theta [M] - \sinh(N-2)\theta [I] \right]. \quad (4.68)$$

Here, $[I]$ is the identity matrix.

2. Given a refractive index with the medium on the incident side as n_s and a refractive index with the medium on the exit side as n_a (assuming that $n_s > n_a$), wherein one layer with a refractive index n_{AR} and a thickness d_{AR} is sandwiched between the media. Assuming

$$n_{AR} = \sqrt{n_s n_a}, \quad d_{AR} = \frac{\lambda}{4n_{AR}}, \quad \lambda = \frac{2\pi}{k_0}, \quad (4.69)$$

prove that the reflectance for a normal incidence is zero. Here, lambda is the wavelength in vacuum. This kind of film is called an *antireflection coating* (or AR coating).

Part II

Description of Light
Propagation through
Electromagnetism

5 Guided Wave Optics

Optical waveguiding means confining light from a light source to a medium with a nearly uniform structure in the direction of propagation and then guiding it to a target region. This medium is called an *optical waveguide*, and the axis of the direction of propagation is called the optical axis. Although a spatially localized electromagnetic wave (in other words, a beam wave) has the property wherein it spreads by diffraction, the light diverging from the optical axis, as it propagates inside the optical waveguide, is pulled back to the direction of the optical axis because of reflection and refraction and is confined in the core thus waveguided. A transparent medium (dielectric materials and semiconductors with band gap energy that is larger than the photon energy) is generally used in the core and the cladding, although sometimes metals are used in the cladding.

Optical waveguides can be roughly classified by cross-sectional shapes, into cylindrical cross-sectional optical waveguides (primarily used as transmission lines in communications), such as optical fibers and rod lenses, and optical waveguides fabricated on a planar substrate (used in optical integrated circuits and waveguide-type optical devices). There are various cross-sectional structures as shown in Figure 5.1. When classified by material, optical fibers can be a silica optical fiber with silica glass (SiO_2 glass) as the main component and a plastic optical fiber (POF), which makes use of plastics. On the other hand, optical waveguides fabricated on a planar substrate can be formed from thin glass film (silica glass and multicomponent glass) on a silicon substrate or a silica glass substrate using the thin film deposition method and the thick film formation method, or formed by the diffusion of ions in electro-optic crystals ($LiNbO_3$ and $LiTaO_3$, which are used to use the electro-optical effect and the acousto-optical effect). For those using semiconductors, thin films made of material systems (GaAs/AlGaAs system, GaInAsP/InP system, etc.) that are nearly the same as that in semiconductor lasers are formed on lattice-matched (the crystal lattice constant is equal) substrates by the crystal growth method. On the other hand, a structure where light is confined in the direction parallel to the substrate surface in addition to the normal direction is called the *channel waveguide* (sometimes called the three-dimensional waveguide). Figure 5.1b through 5.1e corresponds to this category, and to process the core layer into a stripe pattern, photolithographic and etching techniques that are the same as the semiconductor process are used.

Reflection at the boundary surface between the core and the cladding and refraction inside the core are used to confine light to the core. As shown in

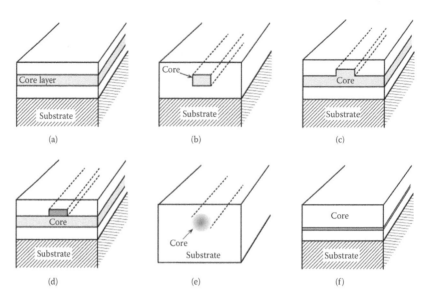

FIGURE 5.1 Various cross-sectional structures of optical waveguides. (a) Slab waveguide. (b) Rectangular (embedded) waveguide. (c) Ridge waveguide. (d) Strip-loaded waveguide. (e) Diffusion-type waveguide. (f) ARROW-type waveguide.

Figure 5.2a, when total internal reflection and refraction are used, all of the optical power can be confined to the core (and the cladding region in its vicinity), so in principle, optical power loss due to propagation will not occur. On the other hand, as shown in Figure 5.2b, when Fresnel reflection is used, reflectance will be less than 1.0, leading to transmittance to the cladding. Consequently, with each reflection, optical power will gradually escape from the core to the cladding, and optical power loss due to propagation will occur. This loss is called the radiation loss. Moreover, optical loss will also occur when the core and the cladding media absorb light (the refractive index is a complex number and is expressed as $n = n_r + jn_i$, absorption occurs when n_i is negative). This loss is called an absorption loss. The optical waveguide can be classified according to the waveguiding principle of confining light to the core and the materials of core and cladding as summarized in Table 5.1.

The wave theory, which treats light as an electromagnetic wave, is suitable for a rigorous analysis of optical waveguides, wherein one must start from Maxwell's electromagnetic equations. In Chapter 2, we derived the guided mode using the optical path length and the principle of interference; however, in this chapter, we will start with Maxwell's equations and use the wave theory to describe the basic characteristics of waveguides and the method for analyzing them.

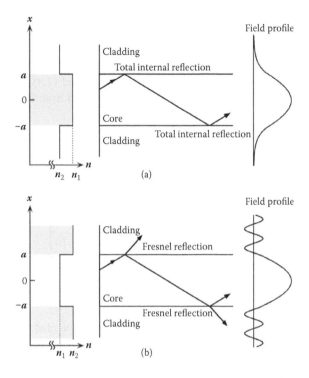

FIGURE 5.2   (a) Total internal reflection confinement-type waveguide and (b) leaky structure waveguide.

TABLE 5.1

Classification of Optical Waveguides According to the Guided Wave Principle

Waveguiding principle \ Material	Lossless dielectric material	Medium with absorption or gain
Total internal reflection ($R = 1.0$)	Normal open dielectric waveguides	Semiconductor lasers,
Fresnel reflection ($R < 1.0$)	Leaky waveguides	Electric absorption modulators, Metal cladding waveguides

5.1   GENERAL CONCEPT OF THE GUIDED MODES

5.1.1   WAVE EQUATIONS AND BOUNDARY CONDITIONS

For the Maxwell's electromagnetic equations, it is enough to consider the following two equations (the same as Equations (1.7) and (1.8)).* That is, the two equations $\nabla \cdot D = 0$ and $\nabla \cdot B = 0$ can be derived from Equations (1.7) and (1.8).[†]

$$\nabla \times E = -\frac{\partial B}{\partial t}, \tag{5.1}$$

$$\nabla \times H = J + \frac{\partial D}{\partial t}. \tag{5.2}$$

Here, when the transparent dielectric material is considered as a medium, with

$$\left.\begin{array}{ll} \mu = \mu_0 & \text{(nonmagnetic material),} \\ \sigma = 0 & \text{(insulator, therefore } J = 0) \end{array}\right\} \tag{5.3}$$

and when the dielectric constant is expressed as the refractive index in accordance to Equation (1.58), this would result in

$$\varepsilon = \varepsilon_0 n_i^2 \quad (i = 1 \text{ or } 2), \tag{5.4}$$

where $i = 1$ represents the core and $i = 2$ represents the cladding. When the time dependence of the electric field and the magnetic field is represented by

$$E = E^0(x, y, z)e^{j\omega t}, \tag{5.5}$$

$$H = H^0(x, y, z)e^{j\omega t}, \tag{5.6}$$

Equations (5.1) and (5.2) will be transformed into the following two equations:

$$\nabla \times E^0 = -j\omega\mu_0 H^0, \tag{5.7}$$

$$\nabla \times H^0 = j\omega\varepsilon_0 n_i^2 E^0. \tag{5.8}$$

Here, when the vector identical equations (refer to Appendix D)

$$\nabla \times (\nabla \times A) = \nabla(\nabla \cdot A) - \nabla^2 A, \tag{5.9}$$

$$\nabla \cdot (\psi A) = \psi(\nabla \cdot A) + (\nabla\psi)A, \tag{5.10}$$

$$\nabla \times (\psi A) = \psi(\nabla \times A) + (\nabla\psi) \times A, \tag{5.11}$$

---

* As noted in Chapter 1, the two equations $\nabla \cdot D = \rho$ and $\nabla \cdot B = 0$ were added to the Maxwell's equations for a total of four equations. Then, three auxiliary equations ($D = \varepsilon E$, $B = \mu H$, $J = \sigma E$) were added for a total of seven equations.

† When divergence is operated on both sides of Equation (1.7), $\nabla \cdot \frac{\partial B}{\partial t} = \frac{\partial(\nabla \cdot B)}{\partial t} \equiv 0$ is obtained from the vector equation $\nabla \cdot \nabla \times A \equiv 0$, which is identical to any vector A. Consequently, this would result in $\nabla \cdot B \equiv \text{const}$, which is independent of time. If $t = -\infty$ is considered physically, it would be unthinkable that a constant magnetic field existed even before the universe was born, so the constant value would be 0, that is, $\nabla \cdot B = 0$ is derived. $\nabla \cdot D = \rho$ would also be the same, except for the point where the continuity equation $\nabla \cdot J + \frac{\partial\rho}{\partial t} = 0$ was used on the electric charge. Because $\nabla \cdot D = \rho$ and $\nabla \cdot B = 0$ do not express temporal variation, it can be interpreted that they do not represent the causal laws of physics.

are used, the following *wave equations* are obtained.[*]

$$\nabla^2 E^0 + \omega^2 \varepsilon_0 \mu_0 n_i^2 E^0 = -\nabla \left(\frac{\nabla \cdot n_i^2}{n_i^2} E^0 \right) \tag{5.12}$$

$$\nabla^2 H^0 + \omega^2 \varepsilon_0 \mu_0 n_i^2 H^0 = -\left(\frac{\nabla \cdot n_i^2}{n_i^2} \right) \times (\nabla \times H^0) \tag{5.13}$$

Here, the right side of these equations is called the *gradient term of the dielectric constant*. The gradient term needs to considered for the accurate description of the propagation of the electromagnetic wave in the medium with a nonuniform refractive index. In general, however, if the difference in the refractive indices of the core and the cladding is less than a few percent, it will be small enough that the effect of this term is negligible. If this term is approximated to zero, the following wave equations will be obtained:

$$\nabla^2 E^0 + \omega^2 \varepsilon_0 \mu_0 n_i^2 E^0 = 0, \tag{5.14}$$
$$\nabla^2 H^0 + \omega^2 \varepsilon_0 \mu_0 n_i^2 H^0 = 0. \tag{5.15}$$

Here, the electromagnetic field distributions, which are dependent only on the position where the time dependency $e^{j\omega t}$ is excluded, were expressed as E^0 and H^0, but for simplicity, they will be expressed as E and H below. Consequently, in differentiation with respect to time, $j\omega$ will appear in all as the coefficient, whereas the differential operator $\frac{\partial}{\partial t}$ will not appear. However, the equation for vectors E and H appearing below will hold true even if $e^{j\omega t}$ is later multiplied to include the time dependency term.

On the other hand, the boundary conditions for the amplitude electromagnetic field at the discontinuity are expressed by the following two equations:[†]

$$(E_1 - E_2) \times n = 0 \tag{5.16}$$
$$(H_1 - H_2) \times n = 0 \tag{5.17}$$

(n is the unit normal vector for the boundary surface.)

In terms of physics, these equations mean that the normal components of E and H must be equal on both sides of the boundary surface.

---

[*] If the refractive index of the core is not uniform as expressed by $n_1 = n_1(x, y)$, then $\nabla \cdot E \neq 0$. This is because when $\nabla \cdot D = \nabla \cdot (\varepsilon E) = 0$ and Equation (5.10) are used

$$\nabla \cdot E = -\frac{\nabla \cdot \varepsilon}{\varepsilon} E.$$

Moreover, it is necessary to include the gradient term of the dielectric constant using Equation (5.11) for the wave equation of the magnetic field.

[†] Other than these two equations for the boundary conditions, there are two other equations wherein the normal components of B and D are equal at both sides of the boundary surface. These are the boundary conditions derived from $\nabla \cdot D = 0$ and $\nabla \cdot B = 0$. Then, in the case of the time-varying electromagnetic field, these two of the four Maxwell's equations can be derived from Equations (5.1) and (5.2). Similarly, the boundary conditions for the normal components of D and B will automatically be satisfied, if Equations (5.16) and (5.17) are satisfied.

If the refractive index is continuous inside the regions, the general solution of the wave equations (5.14) and (5.15) in each of the regions is obtained. For each of the solutions, if the unknown coefficient of the general solution is determined to satisfy the boundary condition equations (5.16) and (5.17) at the boundary surface of each region, then the solution for the whole system will be obtained. The wave equations (5.14) and (5.15) and the boundary condition equations (5.16) and (5.17) will contain six electromagnetic field components (three each of E and H), and so there may be six simultaneous equations. However, in fact, of the six electromagnetic components, there are only two independent components, one each from E and H. For example, in the x, y, z Cartesian coordinates, if a solution that would satisfy each of the wave equations and boundary conditions for E_z and H_z is obtained, then the remaining E_x, E_y, H_x, H_y can be expressed in terms of E_z and H_z as follows:

$$E_x = \frac{-j}{\omega^2 \varepsilon \mu - \beta^2} \left(\beta \frac{\partial E_z}{\partial x} + \omega \mu \frac{\partial H_z}{\partial y} \right), \tag{5.18}$$

$$E_y = \frac{-j}{\omega^2 \varepsilon \mu - \beta^2} \left(\beta \frac{\partial E_z}{\partial y} - \omega \mu \frac{\partial H_z}{\partial x} \right), \tag{5.19}$$

$$H_x = \frac{-j}{\omega^2 \varepsilon \mu - \beta^2} \left(-\omega \varepsilon \frac{\partial E_z}{\partial y} + \beta \frac{\partial H_z}{\partial x} \right), \tag{5.20}$$

$$H_y = \frac{-j}{\omega^2 \varepsilon \mu - \beta^2} \left(\omega \varepsilon \frac{\partial E_z}{\partial x} + \beta \frac{\partial H_z}{\partial y} \right). \tag{5.21}$$

The direction that is parallel to the axis of propagation of the waveguide is called the longitudinal direction, whereas the direction that is perpendicular to the axis of propagation of the waveguide is called the transverse direction. In this case, E_z and H_z are called the longitudinal components, whereas the remaining four components are called the transverse components. Similarly, if the wave equations and boundary conditions for E_z and H_z in the cylindrical coordinate system are satisfied, the remaining transverse electromagnetic field components E_r, E_θ, H_r, H_θ can be determined using E_z and H_z as follows:

$$E_r = \frac{-j}{\omega^2 \varepsilon \mu - \beta^2} \left(\beta \frac{\partial E_z}{\partial r} + \omega \mu \frac{1}{r} \frac{\partial H_z}{\partial \theta} \right), \tag{5.22}$$

$$E_\theta = \frac{-j}{\omega^2 \varepsilon \mu - \beta^2} \left(\beta \frac{1}{r} \frac{\partial E_z}{\partial \theta} - \omega \mu \frac{\partial H_z}{\partial r} \right), \tag{5.23}$$

$$H_r = \frac{-j}{\omega^2 \varepsilon \mu - \beta^2} \left(-\omega \varepsilon \frac{1}{r} \frac{\partial E_z}{\partial \theta} + \beta \frac{\partial H_z}{\partial r} \right), \tag{5.24}$$

$$H_\theta = \frac{-j}{\omega^2 \varepsilon \mu - \beta^2} \left(\omega \varepsilon \frac{\partial E_z}{\partial r} + \beta \frac{1}{r} \frac{\partial H_z}{\partial \theta} \right). \tag{5.25}$$

5.1.2 CLASSIFICATION OF EIGENMODES AND PROPAGATION CONSTANTS

Eigenmode has already been explained using the optical path length and the interference conditions in Section 2.1. In this section, the concept of the eigenmode will be explained using the *symmetrical three-layer slab waveguide*, of which the simplest waveguide structure is shown in Figure 5.3. For the waveguide, the core spreads infinitely in the y- and z-directions, and the boundaries of the core and cladding only exist in the x-axis. Because the refractive index of the cladding layer at the upper portion of the core is equal to that of the refractive index of the cladding layer at the bottom portion of the core, and the waveguide is symmetrical with respect to the yz plane, it is called a symmetric three-layer slab waveguide.

To simplify future notations, the spatial distributions, $E^0(x, y, z)$ and $H^0(x, y, z)$, of the electric and the magnetic fields below will be expressed simply as E and H. Moreover, the direction of propagation of light will be set as the z-direction, and with the propagation constant as β, the z-direction dependency of the electromagnetic field is assumed to be $\exp(-j\beta z)$. Consequently, the readers should multiply the $\exp[j(\omega t - \beta z)]$ term in the electromagnetic field below when reading the equation.

In the slab waveguide shown in Figure 5.3, light is confined in the x-direction, but as it is uniform in the y-direction, the differential with respect to y is zero ($\frac{\partial}{\partial y} = 0$). Then, the Maxwell's equations are written using the Cartesian coordinates x, y, z with the above condition, as summarized in Table 5.2.

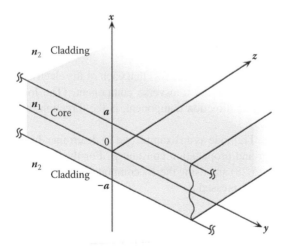

FIGURE 5.3 Symmetrical three-layer slab waveguide.

TABLE 5.2

$x\,y\,z$ Components of the Maxwell's Equations for the Slab Waveguide

	Equation (5.7)	Equation (5.8)
x	$j\beta\boxed{E_y} = -j\omega\mu_0\boxed{H_x}$ (5.26)	$j\beta\left(\!H_y\!\right) = j\omega\varepsilon_0 n_i^2\left(\!E_x\!\right)$ (5.29)
y	$-j\beta\left(\!E_x\!\right) - \dfrac{\partial\left(E_z\right)}{\partial x} = -j\omega\mu_0\left(\!H_y\!\right)$ (5.27)	$-j\beta\boxed{H_x} - \dfrac{\partial\boxed{H_z}}{\partial x} = j\omega\varepsilon_0 n_i^2\boxed{E_y}$ (5.30)
z	$\dfrac{\partial\boxed{E_y}}{\partial x} = -j\omega\mu_0\boxed{H_z}$ (5.28)	$\dfrac{\partial\left(H_y\right)}{\partial x} = j\omega\varepsilon_0 n_i^2\left(\!E_z\!\right)$ (5.31)

A careful examination of the equations in Table 5.2 reveals that, of the six electromagnetic field components, only Equations (5.26), (5.28), and (5.30) contain E_y, H_x, and H_z (enclosed in $\square$), whereas the remaining equations, Equations (5.27), (5.29), and (5.31) contain H_y, E_x, and E_z (enclosed in $\bigcirc$), and we can see that they are independent systems of simultaneous equations. Consequently, general electromagnetic fields can be expressed as a linear combination of the solutions of these two independent systems of simultaneous equations. The solutions of these two independent systems of simultaneous equations are called as

> TE mode: $E(0, E_y, 0)$, $H(H_x, 0, H_z)$,
> (transverse electric modes)
> TM mode: $E(E_x, 0, E_z)$, $H(0, H_y, 0)$.
> (transverse magnetic modes)

The names owe their origin to the fact that each of the electric fields and magnetic fields include only the transverse component. (The longitudinal component, that is, the z-direction component, is the direction of propagation of light.)

Now, with the TE mode as an example, let us determine the electromagnetic field distribution and propagation constant β. For the TE mode, from Equations (5.26) and (5.28) and using the E_y component, the two other components H_x and H_z can be expressed as

$$H_x = -\frac{\beta}{\omega\mu_0}E_y, \tag{5.32}$$

$$H_z = -\frac{j}{\omega\mu_0}\cdot\frac{\partial E_y}{\partial x}. \tag{5.33}$$

When Equations (5.32) and (5.33) are substituted to Equation (5.30), the following wave equation can be obtained:

$$\frac{\partial^2 E_y}{\partial x^2} + (\omega^2 \varepsilon_0 \mu_0 n_i^2 - \beta^2)E_y = 0 \quad (i = 1 \text{ or } 2). \quad (5.34)$$

Here, when k_0^2 is designated as

$$k_0^2 = \omega^2 \varepsilon_0 \mu_0 \quad (5.35)$$

and the wave equation (5.34) is separated for that of the core ($n = n_1$) and that of the cladding ($n = n_2$), the following two equations will be obtained:

$$\frac{\partial^2 E_y}{\partial x^2} + (k_0^2 n_1^2 - \beta^2)E_y = 0 \quad \text{(inside the core)}, \quad (5.36)$$

$$\frac{\partial^2 E_y}{\partial x^2} - (\beta^2 - k_0^2 n_2^2)E_y = 0 \quad \text{(inside the cladding)}. \quad (5.37)$$

Now, the general solution of Equations (5.36) and (5.37) can be expressed as the linear combination of sin and cos functions in the core region and can be represented as the linear combination of exponentially increasing and exponentially decreasing functions in the cladding region, respectively. The unknown coefficients of these general solutions are further determined by taking into consideration the boundary conditions. When Equations (5.16) and (5.17) are applied to the structure shown in Figure 5.3, the tangential components of the boundary surface are E_y and H_z, so this boundary condition can be expressed using the following two equations:

$$E_y(x \to \pm a_{+0}) = E_y(x \to \pm a_{-0}) \quad \text{(double sign corresponds)}, \quad (5.38)$$
$$H_z(x \to \pm a_{+0}) = H_z(x \to \pm a_{-0}) \quad \text{(double sign corresponds)}. \quad (5.39)$$

Moreover, in the cladding, the solution that has a physical meaning is the solution that satisfies

$$E(x \to \pm\infty) = \text{finite} \qquad H(x \to \pm\infty) = \text{finite}. \quad (5.40)$$

(1) Solution of the TE mode

When the solution of the wave equations (5.36) and (5.37) is determined (guided mode), first consider the case wherein the propagation constant β is within the following range:

$$k_0 n_2 \le \beta < k_0 n_1 \quad \text{(guided mode)}. \quad (5.41)$$

Assuming β to be in such a range, the coefficient term of E_y in Equations (5.36) and (5.37) will be a positive constant, so the coefficient term will be designated as follows:

$$\kappa^2 = k_0^2 n_1^2 - \beta^2, \tag{5.42}$$
$$\gamma^2 = \beta^2 - k_0^2 n_2^2. \tag{5.43}$$

Here, let us define the V *parameter* (normalized frequency) as

$$V = k_0 n_1 a \sqrt{2\Delta}, \tag{5.44}$$

where

$$\Delta = \frac{(n_1^2 - n_2^2)}{2n_1^2} \tag{5.45}$$

$$\cong \frac{(n_1 - n_2)}{n_1} \quad \text{(when } n_1 + n_2 \simeq n_1 \text{)} \tag{5.46}$$

is a fundamental parameter called the *relative index difference*. When Equations (5.42) and (5.43) are used, the following equation can be derived:

$$(\kappa a)^2 + (\gamma a)^2 = V^2. \tag{5.47}$$

If the solution of Equations (5.36) and (5.37) is determined to satisfy the conditions of Equations (5.38) and (5.40), then the following two solutions are obtained:

(a) TE even modes

$$\begin{aligned} E_y &= A_e \cdot \cos(\kappa x) & (|x| \leq a) \\ &= A_e \cdot \cos(\kappa a) \cdot e^{-\gamma(|x|-a)} & (|x| > a). \end{aligned} \tag{5.48}$$

(b) TE odd modes

$$\begin{aligned} E_y &= A_0 \cdot \sin(\kappa x) & (|x| \leq a) \\ &= \frac{x}{|x|} A_0 \cdot \sin(\kappa a) \cdot e^{-\gamma(|x|-a)} & (|x| > a). \end{aligned} \tag{5.49}$$

Moreover, in the case of the TE even modes, Equation (5.48) is substituted to Equation (5.33) and to satisfy the boundary condition of Equation (5.39), the following eigenvalue equation has to hold true:

$$\tan(\kappa a) = \frac{\gamma a}{\kappa a}. \tag{5.50}$$

The eigenvalue equation for other modes can also be obtained using the same procedure. Table 5.3 summarizes the classification of modes, their field distributions, and eigenvalue equations.

(2) Dispersion curve

As Equations (5.47) and (5.50) are simultaneous equations, when these two equations are solved, κ and γ are obtained as solutions. The propagation

TABLE 5.3

Field Distribution and Eigenvalue Equation of Each of the Guided Modes of the Symmetrical Three-Layer Slab Waveguide

	Mode Electromagnetic Field		
Modes	$\|x\| \leq a$	$\|x\| > a$	Eigenvalue Equation
			$V^2 = (\kappa a)^2 + (\gamma a)^2$
TE even	$E_y = A_e \cdot \cos(\kappa x)$	$E_y = A_e \cdot \cos(\kappa a) \cdot e^{-\gamma(\|x\|-a)}$	$\tan(\kappa a) = \dfrac{\gamma a}{\kappa a}$
TE odd	$E_y = A_0 \cdot \sin(\kappa x)$	$E_y = \dfrac{x}{\|x\|} A_0 \cdot \sin(\kappa a) \cdot e^{-\gamma(\|x\|-a)}$	$\tan(\kappa a) = -\dfrac{\kappa a}{\gamma a}$
TM even	$H_y = B_e \cdot \cos(\kappa x)$	$H_y = B_e \cdot \cos(\kappa a) \cdot e^{-\gamma(\|x\|-a)}$	$\tan(\kappa a) = \left(\dfrac{n_1}{n_2}\right)^2 \dfrac{\gamma a}{\kappa a}$
TM odd	$H_y = B_0 \cdot \sin(\kappa x)$	$H_y = \dfrac{x}{\|x\|} B_0 \cdot \sin(\kappa a) \cdot e^{-\gamma(\|x\|-a)}$	$\tan(\kappa a) = -\left(\dfrac{n_2}{n_1}\right)^2 \dfrac{\kappa a}{\gamma a}$

constant β can be determined from κ and γ using Equations (5.42) and (5.43).

$$\kappa a = X, \tag{5.51a}$$

$$\gamma a = Y. \tag{5.51b}$$

Equations (5.47) and (5.50) can be rewrittern as the following simultaneous equations:

$$X^2 + Y^2 = V^2, \tag{5.52}$$

$$Y = X \cdot \tan X. \tag{5.53a}$$

In the same way, for the odd modes, in exchange for Equation (5.53a), the following equation is obtained:

$$Y = -X \cdot \cot X. \tag{5.53b}$$

When the solution for the pair of Equations (5.52) and (5.53a) or the pair of Equations (5.52) and (5.53b) is determined using an illustration, as shown in Figure 5.4, x is the intersection of the monotonically increasing function that diverges from 0 to infinity as the interval that separates every $\pi/2$ and the circle with the radius V. The x and y coordinates of this intersection shall each correspond to κa and γa, so when these values are substituted to Equations (5.42) and (5.43), the propagation constant β is obtained.

Incidentally, the V parameter defined by Equation (5.44) is the one that is unambiguously determined once the refractive indices n_1 and n_2 of the core and cladding of the waveguide, the core half-width a, and the wavelength λ have been determined. Therefore, as previously defined in Section 2.3, the propagation constant can also be normalized, and the normalized propagation

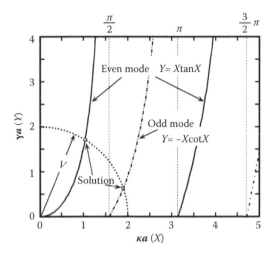

FIGURE 5.4 Graphical method for solving the mode eigenvalue equation.

$$b = \frac{(\beta/k_0)^2 - n_2^2}{n_1^2 - n_2^2} = \frac{(\gamma a)^2}{(\kappa a)^2 + (\gamma a)^2} \tag{5.54}$$

constant is used to determine the relationship of V and b. Once this relationship is obtained, there is no need to recalculate the propagation constant every time the waveguide parameter changes, so the prospects are good for the understanding of the guided properties. Figure 2.3 of Section 2.3 depicts the relationship of V and b that was determined this way. In other words, the procedure for determining the propagation constant of each mode, when the waveguide structure and the wavelength are given can be simplified with just one drawing, that is, the dispersion curve, as shown in Figure 2.4.

(3) Classification of modes

In drawing a dispersion curve, the propagation constant was assumed to be within the range designated in Equation (5.41). However, the propagation constant may be smaller than the lower limit in Equation (5.41). For example, in the dispersion curve of Figure 2.3, the upper limit value $\pi/2$ of Equation (2.16) corresponds to the point when the normalized propagation constant b of the TE first-order mode is 0 (the propagation constant β is $k_0 n_2$). When the propagation constant β is smaller than $k_0 n_2$, the field distribution of this mode is expressed by a function that oscillates inside the cladding, as is seen from Equation (5.34). So, the mode spreads throughout the whole space and is not anymore confined within the core. Such modes are called *radiation modes*.

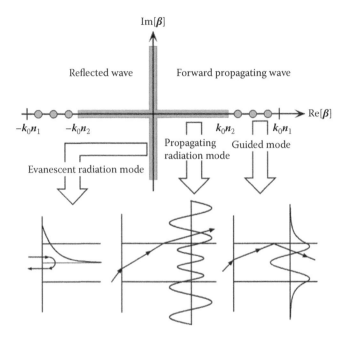

FIGURE 5.5   Classification of modes on the propagation constant axis.

Meanwhile, modes wherein the propagation constant is within the range set in Equation (5.41) and which are confined to the core are called *guided modes*. In addition, the critical point at which the guided mode changes to the radiation mode is called the cutoff of the mode, and the V value corresponding to the cuttoff is called the cutoff V *value*. The condition for the single mode of the step-index slab waveguide is that the V value of the waveguide should be smaller than the cutoff V value of the TE first-order mode.

The classification of modes is summarized as shown in Figure 5.5 by plotting on the propagation constant axis. As can be seen in Figure 5.5, the propagation constant of the guided mode (mathematically, an eigenvalue) is a discrete number, whereas the propagation constant of the radiation mode is a continuous number.

5.1.3   ELECTROMAGNETIC FIELD DISTRIBUTION, NEAR-FIELD PATTERN, AND SPOT SIZE

For the electromagnetic field distribution of the fundamental TE mode, all the electromagnetic field components can be determined by substituting Equation (5.48) for the E_y set at $\kappa a \leq \pi/2$ to Equation (5.32) and (5.33). The electric field distribution, which is expressed by the line of electric force and the line of magnetic force from these electromagnetic field components, is shown in Figure 5.6.

On the other hand, the light intensity distribution in the cross-section is expressed by the z component (actually, the real part only has a z component) of the real part of the complex Poynting vector. The light intensity distribution

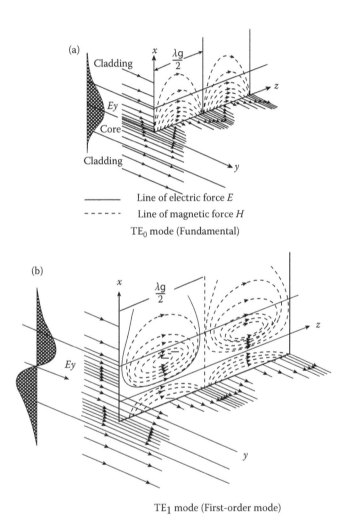

FIGURE 5.6 The electromagnetic field distribution of the TE mode of a symmetrical three-layer slab waveguide. (a) TE_0 mode (fundamental mode). (b) TE_1 mode (first-order mode).

in the cross-section is called the *near-field pattern* (which is abbreviated as NFP). Using Equation (5.32), the NFP of the TE mode is expressed by the following equation:

$$\tilde{S}_z = \frac{1}{2}(\boldsymbol{E} \times \boldsymbol{H}^*) \cdot \boldsymbol{e}_z = -\frac{1}{2}E_y H_x^*$$

$$= \frac{\beta}{2\omega\mu_0}|E_y|^2 \quad \text{(for TE modes)}. \tag{5.55}$$

Here, $*$ denotes the complex conjugate, which is needed to determine the time average by eliminating the vibrational term $\exp[j(\omega t - \beta z)]$.

As can be seen from Figure 5.6, the transverse electromagnetic field distribution of the fundamental mode is generally a monomodal gently sloping curve, which can be approximated by the *Gaussian function*

$$f(x) = \exp\left[-\left(\frac{x}{w}\right)^2\right]. \tag{5.56}$$

The parameter w involved in the Gaussian curve approximation is called the spot size, and it is one of the fundamental parameters in single-mode waveguides (in optical fibers, the length that is twice the spot size is called the mode field diameter). Several equations have been proposed to determine spot size from the transverse electromagnetic field distribution function $\psi(x)$ of the fundamental mode. By transforming the coordinate system from the cylindrical coordinate system to the x, y, z coordinate system of Petermann's formula [9], which is generally used in optical fibers, the following equation is obtained [11]:

$$w^2 = \frac{\displaystyle\int_{-\infty}^{\infty} \psi^2(x)dx}{\displaystyle\int_{-\infty}^{\infty}\left[\frac{\partial\psi(x)}{\partial x}\right]^2 dx}. \tag{5.57}$$

A more accurate equation compared to the above equation is Equation (7.49), which will be discussed in Section 7.5.

5.1.4 Mode Orthogonality and Eigenmode Expansion

The guided mode and the radiation mode that were derived mathematically in Section 5.1.2 are called the eigenmode, wherein the propagation constant is the eigenvalue, whereas the electromagnetic field distribution is the eigenfunction. As can be seen from the fact that the eigenfunctions are the orthogonal sets in mathematical eigenvalue problems, the eigenmode must also satisfy the orthogonal relationship. Although the concept of mode orthogonality has already been discussed in Section 2.4 using the scalar orthogonal function, more precisely, the *eigenmode* is a vector containing six electric field and magnetic field components. Then, the orthogonality can be expressed by the following equation:*

$$\frac{1}{2} \int_{-\infty}^{\infty} Re[E^{(\mu)} \times H^{(\nu)*}] \cdot e_z dx$$

$$= P_z^{(\mu)} \begin{cases} \delta_{\mu\nu} & : \text{Guided mode} \\ \delta(\beta^{(\mu)} - \beta^{(\nu)}) & : \text{Radiation mode.} \end{cases} \tag{5.58}$$

Here, μ and ν are the mode labels, $E^{(\mu)}$ and $H^{(\mu)}$ are the transverse components of the electric and magnetic fields belonging to the mode μ, $P_z^{(\mu)}$ is the power carried by the mode when $\mu = \nu$, and $*$ denotes the complex conjugate. (Refer to Appendix H for proof.) However, if the material of the waveguide does not have the gain and absorption and is expressed only by the real number of the refractive index, $E^{(\mu)} \times H^{(\nu)*}$ of Equation (5.58) will be a real number, and so the Re Re[ ] symbol can be omitted.

When the orthogonal relationship of this eigenmode is used, any incident electromagnetic field that is beamed to the optical waveguide can be expanded in terms of the eigenmode.†

$$\mathcal{E} = \sum_{\nu=0}^{N} C_\nu E^{(\nu)} + \sum_{\text{TE, TM}} \int_0^{\infty} C_\sigma E_\sigma d\sigma, \tag{5.59}$$

$$\mathcal{H} = \underbrace{\sum_{\nu=0}^{N} C_\nu H^{(\nu)}}_{\text{Guided mode}} + \underbrace{\sum_{\text{TE, TM}} \int_0^{\infty} C_\sigma H_\sigma d\sigma}_{\text{Radiation mode}}. \tag{5.60}$$

Here, $\mathcal{E}$ and $\mathcal{H}$ are the incident electromagnetic fields (each of which has the x, y, z coordinate components and dependent on x, y, z) and the variable of integration σ is associated with the propagation constant using

$$\sigma = \sqrt{k_0^2 n_2^2 - \beta^2} = \sqrt{-\gamma^2}. \tag{5.61}$$

Using the orthogonal relationship of Equation (5.58), the expansion coefficient of Equations (5.59) and (5.60) can be expressed as

---

* Here, it is explained with a slab waveguide as an example, so the integral over the cross-section is only a one-dimensional integral of x. In a two-dimensional cross-section, such that of optical fibers, the integral for a light that is confined in a waveguide will have to be set as $\int_{-\infty}^{\infty} \int_{-\infty}^{\infty} dx\, dy$. Because the y-direction is uniform for a slab waveguide, the mode power P_z appearing in this equation is the power that is transmitted through the cross-section spreading infinitely in the x-direction at a unit width of the y-direction (e.g., 1 m).

† Because the eigenmode is a set of electric and magnetic fields, the expansion coefficients of the electric and the magnetic fields are the same.

$$C_v = \frac{1}{2P_z^{(v)}} \int_{-\infty}^{\infty} (\mathcal{E} \times \boldsymbol{H}^{(v)*}) \cdot \boldsymbol{e}_z dx \qquad (5.62a)$$

$$= \frac{1}{2P_z^{(v)}} \int_{-\infty}^{\infty} (\boldsymbol{E}^{(v)} \times \mathcal{H}^*) \cdot \boldsymbol{e}_z dx. \qquad (5.62b)$$

As explained later in the next page, the optical power of each of the eigenmodes should be normalized for the $P_z^{(\mu)}$ in Equation (5.58) to be equal to 1.

In terms of physics, Equations (5.59) and (5.60) means that the power of the electromagnetic field distribution (optical beam) that is beamed to the optical waveguide will be separated into eigenmodes and then propagated as shown in Figure 5.7. In such a case, optical power expanded to the guided mode from the incident beam is guided to the exit end, and the portion that is expanded to the radiation mode is radiated (or reflected) from the waveguide and will not reach the exit end, so this part of the optical power is lost. This loss is called the coupling loss at the incidence end.

Moreover, the following equation can be derived from Equations (5.58) to (5.60):

$$\frac{1}{2} \int_{-\infty}^{\infty} (\mathcal{E} \times \mathcal{H}^*) \cdot \boldsymbol{e}_z dx = \sum_{v=0}^{N} |C_v|^2 P_z^{(v)} + \sum_{\text{TE, TM}} \int_0^{\infty} |C_\sigma|^2 P_z(\sigma) d\sigma. \qquad (5.63)$$

Here,

$$P_z^{(v)} = \frac{1}{2} \int_{-\infty}^{\infty} (\boldsymbol{E}^{(v)} \times \boldsymbol{H}^{(v)*}) \cdot \boldsymbol{e}_z dx \qquad (5.64)$$

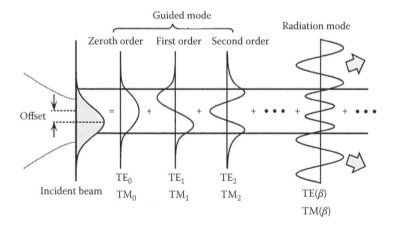

FIGURE 5.7 Eigenmode expansion of an arbitrary incident electromagnetic field.

is the mode power and is typically normalized to 1.* This equation is the same as the Perseval formula for the Fourier transform (Equation (A.28) of Appendix A), wherein the left side of the equation refers to the power of the optical beam and the right side represents the sum of the optical power that is carried by each of the modes. Consequently, this equation also reveals the physical meaning of the *eigenmode expansion* of the incident light beam.

5.1.5   FAR-FIELD PATTERN AND NUMERICAL APERTURE

The light intensity distribution in the plane that is far enough from the exit end of the optical waveguide (a region regarded as the Fraunhofer region) is called the *far-field pattern* (abbreviated as FFP). As already mentioned in Section 2.5, if light is emitted from the step-index multimode optical waveguide, each of the guide modes will be emitted as a light beam with a divergence angle (refracted at the exit end) that corresponds to the propagation angle. Consequently, if light beam is beamed to the multimode optical waveguide and the convergence angle of the light beam is greater than the maximum output angle, coupling loss will occur at the incidence end. This is equivalent to the coupling of the NA (numerical aperture).

On the other hand, as the maximum acceptance angle of the single-mode waveguide is not determined using ray optics, the near-field pattern needs to be calculated using the Fresnel-Kirchhoff integral. In such a case, because the calculation of the Fresnel-Kirchhoff integral for the mode field distribution in the core and cladding regions is not straightforward, the fundamental mode field distribution of the waveguide is approximated by the Gaussian function, and the diffraction integral is expressed by Gaussian beam diffraction. For this, it is necessary to determine the spot size from the field distribution of the fundamental mode. Equations (7.48) and (7.49), which will be discussed in Section 7.5, will be used in the determination. If the spot size of the fundamental mode is determined this way, the Gaussian beam having this spot size will exit from the exit end and spread through diffraction, and so the divergence angle and NA of the exit beam can be determined. The NA of single-mode waveguide is usually expressed as the NA of this approximated Gaussian beam.

When the light beam is beamed to the single-mode optical waveguide, the convergence angle (NA) of the incident beam should also be matched to the NA of the single-mode waveguide. Otherwise, a coupling loss will also occur at the incident end. Because there is a 1:1 correspondence with the convergence angle (NA) and the spot size at the beam waist, if the NA matches, then the spot size of the incident beam at the beam waist will coincide with the spot size of the

---

* In other words, the electric field amplitude of the mode will be set so that the power will be, for example, 1 mW.

waveguide. Consequently, the expansion coefficient (C_0 in Equation (5.59)) to the fundamental mode will approach 1.0.

5.1.6 OPTICAL CONFINEMENT FACTOR

The electromagnetic field distribution of the guided mode of the optical waveguide is not guided and confined only to the core, but rather, as can be seen from Equations (5.48) and (5.49), it also seeps into the cladding. The ratio between the optical power that exists inside the core and the total optical power that exists both in the core and the cladding is called the *confinement factor*. According to the definition of the confinement factor, the confinement factor of the symmetric three-layer slab waveguide is expressed as

$$\Gamma \triangleq \frac{\displaystyle\int_{-a}^{a} |E_y|^2 dx}{\displaystyle\int_{-\infty}^{\infty} |E_y|^2 dx}. \tag{5.65}$$

Furthermore, if Equation (5.65) is calculated for the step-index slab waveguide, it can be expressed as follows:

$$\Gamma = \frac{V + \sqrt{b}}{V + \dfrac{1}{\sqrt{b}}}. \tag{5.66}$$

Equation (5.66) is a function of V and b and as the normalized propagation constant b can be related to V by a dispersion curve, eventually Equation (5.66) is only a function of V. Consequently, the confinement factor of the TE fundamental mode can be calculated using V as shown in Figure 5.8.

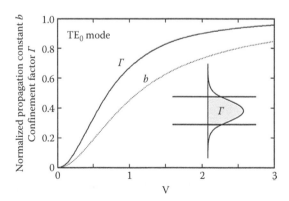

FIGURE 5.8 Confinement factor of the symmetrical three-layer slab waveguide.

A good example of using the confinement factor involves the case of waveguide with gain and absorption in the core and absorption in the cladding like that of semiconductor lasers and optical amplifiers. Here, let us derive the relationship of the optical confinement factor and the propagation constant of the waveguide in more detail. In the wave Equation (5.14), as the polarization direction of the guided wave is in a plane that is nearly perpendicular to the direction of propagation (propagation axis of the waveguide) defining the z-axis as the propagation axis, the electric field can be expressed as

$$E(r,t) = e_p \psi(x,y) e^{j(\omega t - \beta z)}. \tag{5.67}$$

Here, e_p is the unit vector representing the polarization direction, and $\psi(x,y)$ is the scalar function that represents the electric field amplitude distribution of the guided mode. Thus, the wave equation (5.14) can be rewritten as:

$$\left[\frac{\partial^2}{\partial x^2} + \frac{\partial^2}{\partial y^2} + (k_0^2 n_i^2 - \beta^2) \right] \psi(x,y) = 0. \tag{5.68}$$

The method of analyzing the electromagnetic field of a scalar function that assumes that the polarization direction is nearly perpendicular to the propagation of direction is called the scalar wave analysis, and the equation is called the scalar wave equation.* If this equation is further rewritten, it becomes

$$\mathcal{H}\psi(x,y) = \beta^2 \psi(x,y). \tag{5.69}$$

Here, $\mathcal{H}$ can also be expressed as

$$\mathcal{H} = \frac{\partial^2}{\partial x^2} + \frac{\partial^2}{\partial y^2} + k_0^2 n_i^2 = \nabla_t + k_0^2 n_i^2. \tag{5.70}$$

($\nabla_t = \frac{\partial^2}{\partial x^2} + \frac{\partial^2}{\partial y^2}$ is the transverse direction Laplacian.) Mathematically, $\mathcal{H}$ is regarded as an operator with β^2 as an *eigenvalue* and $\psi(x,y)$ as an *eigenfuncton*. If $\psi(x,y)$ is multiplied[†] from the left on both sides of Equation (5.69) and integrated on the (x,y) plane, the following equation is obtained:

$$\beta^2 = \frac{\displaystyle\int_{-\infty}^{\infty}\int_{-\infty}^{\infty} \psi(x,y)\mathcal{H}\psi(x,y)dx\,dy}{\displaystyle\int_{-\infty}^{\infty}\int_{-\infty}^{\infty} \psi^2(x,y)dx\,dy}. \tag{5.71}$$

---

* On the other hand, the method of analyzing the electromagnetic wave as a vector is called the vector wave analysis. Actually, when Equation (5.12), which contains the dielectric constant gradient term, is analyzed precisely, it is called a full vector analysis. There are also cases that include several approximations such as a semivectorial analysis, wherein one assumes a TE wave or TM wave polarization direction, and then analysis is done only on a portion of the dielectric constant gradient term.
† Because the operator $\mathcal{H}$ acts on the function that is on the right side, if one multiplies from the right side, the equality will not hold true.

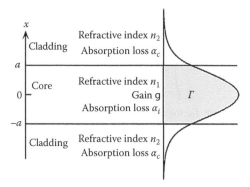

FIGURE 5.9 Waveguide with a gain and absorption loss in the core and an absorption loss in the cladding.

Such an expression is called a *variational expression*, and even if the approximation function that is close to the eigenfunction is substituted in place of $\psi(x, y)$, a significant error in the eigenvalue will not appear due to being *stationary*.[*]

Then, here, let us consider a waveguide such as that shown in Figure 5.9, which has a gain and absorption loss inside the core and an absorption loss in the cladding. As discussed in Section 1.6, the gain and absorption loss is represented by the imaginary part of the refractive index, and as the imaginary part is small compared to the real part (except in the case of absorption by metals), the refractive index distribution is expressed as

$$n^2(x, y) = (n_{r1} + jn_{i1})^2 \simeq n_{r1}^2 + 2jn_{r1}n_{i1} \quad \text{(inside the core)}, \qquad (5.72)$$

$$= (n_{r2} + jn_{i2})^2 \simeq n_{r2}^2 + 2jn_{r2}n_{i2} \quad \text{(inside the cladding)}. \quad (5.73)$$

If the refractive index distribution is substituted to Equation (5.71) and the eigenfunction is approximated with the eigenfunction $\psi^{(0)}$ in the absence of gain and absorption loss (that is, if $n_{i1} = 0$ and $n_{i2} = 0$ in Equations (5.72) and (5.73)), the following equation will be obtained:

$$\beta^2 = \beta^{(0)2} + k_0^2 \frac{\int_{-\infty}^{\infty} \int_{-\infty}^{\infty} n_r(x, y)n_i(x, y)\psi^{(0)2}(x, y)dx\,dy}{\int_{-\infty}^{\infty} \int_{-\infty}^{\infty} \psi^{(0)2}(x, y)dx\,dy}$$

$$= \beta^{(0)2} + j2k_0^2 n_{r1}n_{i1}\Gamma + j2k_0^2 n_{r2}n_{i2}(1 - \Gamma). \qquad (5.74)$$

---

[*] The characteristic wherein a significant error does not result even when substituting a function that is slightly deviating from the eigenfunction is called *stationary characteristic*. It is possible to prove that Equation (5.71) satisfies the stationary condition; however, we will omit it here to save space.

Here, $\beta^{(0)}$ is the eigenvalue of the eigenfunction $\psi^{(0)}(x, y)$ and Γ

$$\Gamma = \frac{\iint_{\text{core}} \psi^{(0)2}(x, y)dx\,dy}{\int_{-\infty}^{\infty}\int_{-\infty}^{\infty} \psi^{(0)2}(x, y)dx\,dy} \tag{5.75}$$

is the core confinement factor that is defined by the two-dimensional cross-section. Here, if the difference between the real part of the refractive index in the core and the cladding is small, the approximation of $\beta^{(0)} \simeq k_0 n_{r1} \simeq k_0 n_{r2}$ can be used, and Equation (5.74) will become

$$\beta = \beta^{(0)} + jk_0 n_{i1}\Gamma + jk_0 n_{i2}(1 - \Gamma). \tag{5.76}$$

Because the relationship between the imaginary part of the refractive index and the gain or loss is given by Equation (1.139) of Section 1.6, the optical gain constant (mode gain) G of the waveguide mode can be approximately expressed in terms of Γ by

$$G = (g - \alpha_i)\Gamma - \alpha_c(1 - \Gamma) \quad [\text{Neper/cm}]. \tag{5.77}$$

Here,

$$g = \text{gain constant of the core medium}$$
$$\alpha_i = \text{absorption coefficient of the core medium}$$
$$\alpha_c = \text{absorption coefficient of the cladding medium}$$
$$(\text{free electron absorption, etc.}).$$

PROBLEM 5-1

Show that the confinement factor Γ in a step-index slab waveguide or a step-index cylindrical fiber can be represented by

$$\Gamma = \frac{1}{2}\left[b + \frac{d(Vb)}{dV}\right] = b + \frac{1}{2}V\frac{db}{dV}. \tag{5.78}$$

SOLUTION 5-1

Let us partially integrate the transverse Laplacian of the numerator of Equation (5.71) and rewrite it as follows:

$$\int_{-\infty}^{\infty}\int_{-\infty}^{\infty} \psi(x, y)\left[\frac{\partial^2 \psi(x, y)}{\partial x^2} + \frac{\partial^2 \psi(x, y)}{\partial y^2}\right]dx\,dy$$
$$= \int_{-\infty}^{\infty}\left[\psi(x, y)\frac{\partial \psi(x, y)}{\partial x}\right]_{-\infty}^{\infty}dy + \int_{-\infty}^{\infty}\left[\psi(x, y)\frac{\partial \psi(x, y)}{\partial y}\right]_{-\infty}^{\infty}dx$$
$$- \int_{-\infty}^{\infty}\int_{-\infty}^{\infty}\left[\left(\frac{\partial \psi(x, y)}{\partial x}\right)^2 + \left(\frac{\partial \psi(x, y)}{\partial y}\right)^2\right]dx\,dy. \tag{5.79}$$

When Equation (5.79) is substituted to Equation (5.71), Equation (5.71) can be transformed into the following equation:

$$\beta^2 = \frac{\int_{-\infty}^{\infty}\int_{-\infty}^{\infty}\left[k_0^2 n_i^2 \psi(x,y)^2 - \left(\frac{\partial \psi(x,y)}{\partial x}\right)^2 - \left(\frac{\partial \psi(x,y)}{\partial y}\right)^2\right]dx\,dy}{\int_{-\infty}^{\infty}\int_{-\infty}^{\infty}\psi^2(x,y)dx\,dy}.$$

(5.80)

Now, if Equation (5.80) is differentiated with respect to k_0, and the wavelength dependency of the refractive index can be ignored, then this would result in

$$\beta\frac{d\beta}{dk_0} = k_0 n_1^2 \frac{\iint_{core}\psi^2(x,y)dx\,dy}{\int_{-\infty}^{\infty}\int_{-\infty}^{\infty}\psi^2(x,y)dx\,dy}$$

$$+ k_0 n_2^2 \frac{\iint_{cladding}\psi^2(x,y)dx\,dy}{\int_{-\infty}^{\infty}\int_{-\infty}^{\infty}\psi^2(x,y)dx\,dy}$$

$$= k_0[n_1^2\Gamma + n_2^2(1-\Gamma)] - \frac{1}{2}\frac{\partial}{\partial k_0}\left(\frac{1}{\omega_x^2} + \frac{1}{\omega_y^2}\right). \quad (5.81)$$

(If the wavelength dependency of the refractive index is taken into consideration, then the equation can be derived by Equation (8.34) of Section 8.4.)

On the other hand, let us compare when the same $\frac{d\beta}{dk_0}$ is derived using a totally different equation. The starting point for this is Equation (5.54). That is, using

$$b = \frac{(\beta/k_0)^2 - n_2^2}{n_1^2 - n_2^2} \quad (5.82)$$

and replacing β with b and k_0 with V, the following equation will be obtained:

$$\frac{d\beta}{dk_0} = \frac{\left[n_2^2 + (n_1^2 - n_2^2)\left(b + \frac{1}{2}V\frac{db}{dV}\right)\right]}{\left[n_2^2 + (n_1^2 - n_2^2)b\right]^{\frac{1}{2}}}. \quad (5.83)$$

When k_0 is multiplied to the denominator on the right side of this equation, it becomes β, as can be seen in Equation (5.82). Consequently, if k_0 is multiplied to the numerator and denominator at the right side of Equation (5.83), and the β of the denominator is further multiplied with both sides of the equation, the following will be obtained:

$$\beta\frac{d\beta}{dk_0} = k_0\left[n_2^2 + (n_1^2 - n_2^2)\left(b + \frac{1}{2}V\frac{db}{dV}\right)\right]. \quad (5.84)$$

Equation (5.78) can be derived when this equation is compared to Equation (5.81) ignoring the second term at the right side of Equation (5.81).

As can be seen from the derivation process, Equation (5.78) can be applied to all optical waveguide structures with a refractive index of n_1 for the core and a refractive index of n_2 for the cladding and is independent of the shape of the boundary of the core and the cladding. That is, it will also hold true for a rectangular or a ridge waveguide. Here, the wavelength dependency of the refractive index in Equations (5.81) and (5.83) was ignored, but in Sections 6.2.2 and 8.4, an exact derivation that includes the wavelength dependency of the refractive index will be done.   **[End of Solution]**

5.1.7 SINGLE-MODE CONDITION AND MODE NUMBER

As seen in Figure 2.3 of Section 2.3, if the V parameter in the symmetric three-layer slab waveguide is smaller than $\pi/2$, only the fundamental mode will be propagated in the waveguide. Such a waveguide is called a single-mode waveguide (called a single-mode fiber for a circular cross-section), and the condition for a waveguide to be a single-mode waveguide is called the *single-mode condition*. The single-mode condition is equivalent to the condition that the V parameter of the waveguide is less than the cutoff V value V_c of the first-order mode. For example, the single-mode condition for the symmetric three-layer slab waveguide is expressed by Equation (2.16). The V parameter, as can be seen from the definition of Equation (2.13), is determined by the wavelength λ of the light that is propagated in a waveguide with refractive indices n_1 and n_2, respectively, for the core and the cladding and a core half-width a, so if the wavelength λ of the incident light satisfied the condition

$$\lambda \geq \frac{2\pi n_1 a \sqrt{2\Delta}}{V_c}, \tag{5.85}$$

then the waveguide will become single mode.

Now, the single-mode condition for symmetric three-layer slab waveguide is given by Equation (2.16), but when the refractive index in the core has a distribution, the conditional expression will be different. In general, the refractive index distribution is expressed as

$$\begin{aligned} n^2(x) &= n_1^2[1 - 2\Delta f(x)] & (|x| \leq a) \\ &= n_2^2 & (|x| > a), \end{aligned} \tag{5.86}$$

where $f(x)$ is assumed to be symmetrical with respect to $x = 0$. The cutoff V value V_c of the TE$_1$ mode of this symmetric three layer is given by the following equation [16] (the detailed calculation is omitted because of space limitations):

$$V_c \cong \left[2 \int_0^1 \{1 - f(\eta)\} \left\{ \int_0^\eta \xi^2 \{1 - f(\xi)\} d\xi \right\} d\eta \right]^{-\frac{1}{4}}. \tag{5.87}$$

Here, the coordinates ξ and η of Equation (5.87) have been normalized by the core half-width a. The error of the above approximation equation is less than 2%.

Now, in the symmetric three-layer slab waveguide with a uniform refractive index in the core, the number of modes increases by 1 with the increase of V by $\pi/2$, so if V is sufficiently larger than $\pi/2$, then the mode number M can be approximated by the following equation:

$$M \cong \frac{2V}{\pi}. \tag{5.88}$$

On the other hand, if the refractive index is given by Equation (5.86), the mode number M can be determined using the Wentzel Kramerrs Brillouin (WKB) method (derived in detail in Section 6.1.8 and Appendix J) to be

$$M \cong \frac{1}{\pi} \int_0^a k_0^2(n^2(x) - n_2^2)dx. \tag{5.89}$$

5.2   FUNDAMENTAL STRUCTURE AND MODE OF THE OPTICAL WAVEGUIDE

The concept and the basic characteristics of the waveguide mode were described in Section 5.1, with a symmetric three-layer slab waveguide, which has the same refractive index in the cover and the lower cladding layers that sandwich the core as an example. In this section, the structure of various optical waveguides and the characteristics of their guided modes will be discussed.

5.2.1   TWO-DIMENSIONAL SLAB WAVEGUIDE

The general two-dimensional slab waveguide, as shown in Figure 5.10, does not necessarily have equal refractive indices in the cladding layers at the top and bottom of the core, and the refractive index inside the core is not always uniform. Moreover, it may also happen that the core and cladding layers will be multilayered. Some typical examples of such a waveguide will be discussed later, with an overview of the mode characteristics and analytical methods.

(1)   Asymmetric three-layer slab waveguide

This waveguide is one wherein the refractive indices at the top and bottom of the core are not equal. As the structure is not symmetric with respect to the yz plane, it is not necessary to position the origin of the x coordinate in the middle of the core as defined by the coordinate system. Here, as shown in Figure 5.10, the coordinate system is defined with the boundary surface of the core and the upper cladding layer as the yz plane, with the thickness of the core

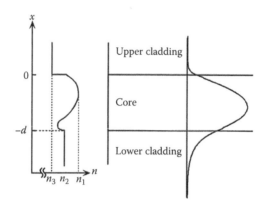

FIGURE 5.10 Two-dimensional slab waveguides with a general refractive index distribution.

designated as d. As such, the refractive index distribution is expressed as the following equation:

$$
\begin{aligned}
n(x) &= n_3 \quad (x \geq 0) \\
&= n_1 \quad (-d \leq x < 0) \\
&= n_2 \quad (x < -d).
\end{aligned}
\tag{5.90}
$$

Here, $n_3 \leq n_2 < n_1$. The solution of the wave equation (5.34) for the TE mode is given by the following equation for a guided mode:

$$
\begin{aligned}
E_y &= A e^{-\delta x} & (x > 0) \\
&= A \cos \kappa x + B \sin \kappa x & (-d \leq x < 0) \\
&= (A \cos \kappa d - B \sin \kappa d) e^{\gamma(x+d)} & (x < -d).
\end{aligned}
\tag{5.91}
$$

Here, both κ and γ are defined by Equations (5.42) and (5.43), respectively, the same as those for a symmetrical three-layer slab waveguide. In addition, δ is defined as

$$
\delta^2 = \beta^2 - k_0^2 n_3^2
\tag{5.92}
$$

and is an attenuation constant of the upper cladding boundary like that of γ. As can be seen in Equation (5.91), the field distribution is not symmetrical with respect to the origin $x = 0$. As the refractive index distribution is not symmetrical, the solution cannot be divided into an even mode and an odd mode.

The two other electromagnetic field components H_x and H_z of the TE mode can be determined by substituting Equation (5.91) to Equations (5.32) and (5.33), respectively. Thus, the following eigenvalue equation can be obtained from the boundary condition that the tangential component at the boundary surface of the core and the cladding will be continuous:

$$\tan \kappa d = \frac{\kappa(\gamma + \delta)}{(\kappa^2 - \gamma\delta)}. \tag{5.93}$$

For the TM mode, the following eigenvalue equation can be derived starting from H_y:

$$\tan \kappa d = \frac{n_1^2 \kappa (n_3^2 \gamma + n_2^2 \delta)}{(n_2^2 n_3^2 \kappa^2 - n_1^4 \gamma\delta)}. \tag{5.94}$$

Now, in an asymmetrical three-layer slab waveguide, the thickness of the core is expressed as d, so the V parameter is expressed as

$$V' = k_0 n_1 d \sqrt{2\Delta}. \tag{5.95}$$

When $n_3 = n_2$, this waveguide becomes a symmetrical three-layer slab waveguide, and as in such a case $d = 2a$, there is a need to note that when this is compared to Equation (2.13), $V' = 2V$. Moreover, the definition of the relative index difference Δ is the same as Equation (5.46) and is given by

$$\Delta = \frac{(n_1^2 - n_2^2)}{2n_1^2}. \tag{5.96}$$

If the relationship between V' and b (the same as the definition of Equation (5.54)) is determined, then the dispersion relationship will also be obtained; however, as in the case of Equation (2.14) for symmetrical three-layer slab waveguide, the analytical derivation of the equation that can determine V' from b is difficult. The equation for the inverse relation is obtained from the eigenvalue equations (5.93) and (5.94) as [1]

$$V' = \frac{1}{\sqrt{1-b}} \left[\tan^{-1} \chi_2 \sqrt{\frac{b}{1-b}} + \tan^{-1} \chi_3 \sqrt{\frac{b+a'}{1-b}} + N\pi \right] \tag{5.97}$$

Here, a' in Equation (5.97) is the parameter that expresses the asymmetry of the refractive index distribution and is defined as

$$a' = \frac{n_2^2 - n_3^2}{n_1^2 - n_2^2}, \tag{5.98}$$

and χ_i ($i = 2$ or 3) is defined as follows for the TE and the TM modes, respectively:

$$\chi_i = \begin{cases} 1 & : \text{TE mode} \\ \left(\dfrac{n_1}{n_i}\right)^2 & : \text{TM mode.} \end{cases} \tag{5.99}$$

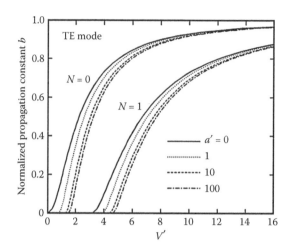

FIGURE 5.11 Dispersion curve of an asymmetrical three-layer slab waveguide.

Here, when the symmetrical condition of $n_3 = n_2$ holds true, $a' = 0$. Using Equation (5.99), and drawing the dispersion curve with a' as the parameter, would result in Figure 5.11 [1].

The cutoff V value of the zero-order mode and the first-order mode can be obtained by substituting $b = 0$, when $N = 0$ or $N = 1$ in Equation (5.97). Then, the single-mode condition for the asymmetric three-layer slab waveguide is given by

$$\tan^{-1} \chi_3 \sqrt{a'} < V' \leq \tan^{-1} \chi_3 \sqrt{a'} + \pi. \qquad (5.100)$$

From Equation (5.100), cutoff is also seen to exist in the fundamental mode in the case of asymmetric slab waveguide. In addition, in contrast to the symmetric three-layer slab waveguide where the cutoff V value V_c of the higher order mode is in agreement in the TE and the TM modes, it is important to note that the cutoff value V in an asymmetric three-layer slab waveguide is different for the TE and the TM modes.

When the mode of the asymmetric three-layer slab waveguide is classified in terms of the propagation constant, there will be two kinds of radiation modes as shown in Figure 5.12, and we can see that there is a mode that radiates only to the substrate side (to the lower cladding side with a refractive index of n_2). This mode is sometimes referred to as the substrate radiation mode.

(2) Distributed refractive index slab waveguide

Slab waveguides with a uniform refractive index are discussed in detail in Section 5.1 and number (1). However, for a core in which refractive index is

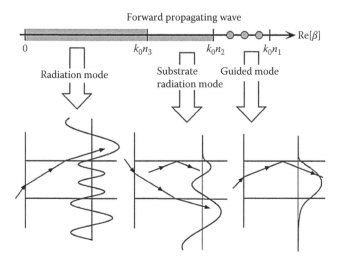

FIGURE 5.12 Classification of the asymmetrical three-layer slab waveguide modes.

generally given by Equation (5.86), it is difficult to express analytically the eigenvalue equation. The analytical method for such a case will be described in (3); however, for a distribution wherein the refractive index distribution function $f(x)$ decreases in proportion to x^2, the analytical solution can be obtained under certain approximations. For this, the refractive index distribution is expressed as

$$n^2(x) = n^2(0)[1 - (gx)^2] \quad (|x| \leq a)$$
$$= n_2^2 \quad (|x| > a). \tag{5.101}$$

Substituting the above distribution to the wave equation (5.34), the following equation is obtained:

$$\frac{d^2 E_y}{dx^2} + [k_0^2 n^2(x) - \beta^2] E_y = 0. \tag{5.102}$$

If Equation (5.102) is solved correctly, the distribution inside the core and the cladding must be considered separately. However, as the exact solution will not be easily obtained, in the usual case of the electromagnetic field of light being well confined within the core (i.e., for multimode waveguides), the following approximation is used to determine the propagation constant and field distribution of the mode:

Assumption (1) Assume that the squared distribution is continuous from $x \rightarrow \pm\infty$. With this assumption, the existence of the cladding will be negligible, and the refractive index will be $-\infty$, when $x \rightarrow \pm\infty$. In terms of physics, it is not possible for a refractive index to be less than 1; however, when the field distribution is well-confined inside the core, the light will only feel the refractive index inside the core, and a good approximation can be obtained.

Assumption (2) $E_y(x \to \pm\infty) = 0$

That is, this is the same as Equation (5.40), and a physically meaningful solution can be determined.

When the solution for Equation (5.102) is determined by the above assumptions, it can be represented by the following *Hermite–Gaussian function*:

$$E_y^{(p)}(x) = \frac{1}{[2^p p! w_0 \sqrt{\frac{\pi}{2}}]^{1/2}} H_p\left(\sqrt{2}\frac{x}{w_0}\right) e^{-\left(\frac{x}{w_0}\right)^2}. \tag{5.103}$$

Here, $H_p(x)$ is an Hermite polynomial, with the lower order functions shown in Table 5.4. For more details, please refer to reference books on special functions and Appendix G.

Here, w_0 of Equation (5.103) is the same spot size (characteristic spot size) as that of Equation (5.56) and can be related to the refractive index distribution by*

$$w_0 = \sqrt{\frac{2}{k_0 n(0) g}} \tag{5.105}$$

TABLE 5.4
Hermite Polynomials

p	$H_p(x)$
0	1
1	$2x$
2	$4x^2 - 2$
3	$8x^3 - 12x$

---

* As mentioned in the footnote of Section 3.3, there are also some textbooks that use the Hermite–Gaussian function, wherein a portion of the Gaussian function is expressed as

$$\exp\left[-\frac{1}{2}\left(\frac{x}{w}\right)^2\right]. \tag{5.104}$$

In the calculation of diffraction integration, this format is convenient because it does not change the form after a Fourier transform. However, the definitions of spot size in the International Standards are "the distance between the point where the amplitude distribution is $\frac{1}{e}$ of that of the center and the center of the beam" and "the distance between the point where the light intensity is $\frac{1}{e^2}$ and the center of the beam," so one has to take note of the definition of spot size.

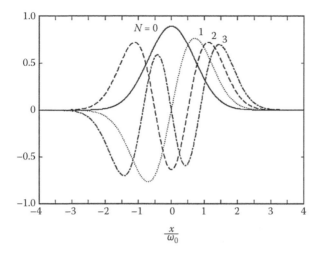

FIGURE 5.13 Field distribution of the Hermite–Gaussian mode.

Some of the mode functions of Equation (5.103) are shown in Figure 5.13. The eigenvalue is given by the following equation:

$$\beta_p^2 = k_0^2 n(0)^2 - k_0 g(2p + 1). \tag{5.106}$$

(3) Multilayer slab waveguide

The symmetric three-layer slab waveguide and the asymmetric three-layer slab waveguide that have been mentioned so far have the simplest structure among the optical waveguides, so the eigenvalue equation and the mode function can be analytically derived. Thus, they are good examples that could be used in understanding the general properties of guided modes. However, in the case of a distributed index waveguide with a refractive index distribution in the core that is given by Equation (5.86) or a multilayer structure that is configured with multiple layers of different refractive indices, the derivation of an analytical solution is often difficult. For such a case, if the distributed refractive index waveguide is approximated to have a stair-like refractive index distribution in the core as shown in Figure 5.14, the eigenvalue wave equation can be derived using the matrix form, which is suitable for numerical calculations [17]–[19]. This section deals with the *matrix method* as an analytical method for optical waveguides with a multilayer structured core layer as shown in Figure 5.14.

First, as the boundary surface between each of the layers of the optical waveguide is parallel to the yz plane, the components that must satisfy Equations (5.16) and (5.17) are E_y and H_z, pair for the TE mode, and H_y and E_z,

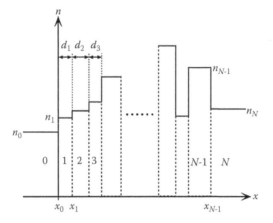

FIGURE 5.14 Refractive index distribution of the multilayer structure slab waveguide and the coordinate definition.

pair for the TM mode. To describe these two polarization modes in a unified manner, the electromagnetic field distribution is expressed as follows:

$$\begin{bmatrix} \psi_y(x) \\ \psi_z(x) \end{bmatrix} = \begin{bmatrix} E_y(x) \\ jZ_0H_z(x) \end{bmatrix} \quad \text{(TE mode)} \tag{5.107a}$$

$$= \begin{bmatrix} Z_0H_y(x) \\ -jE_z(x) \end{bmatrix} \quad \text{(TM mode)}. \tag{5.107b}$$

Here, Z_0 is the impedance of vacuum, which is the same as that in Equation (1.36) in Section 1.3.2 and is defined as

$$Z_0 = \sqrt{\frac{\mu_0}{\varepsilon_0}}. \tag{5.108}$$

Here, let us consider point x in the ith layer. When

$$x_{i-1} \leq x \leq x_i \tag{5.109}$$

using the field distribution at the right edge x_i of the layer, the field distribution at position x can be expressed by the following equation:

$$\begin{bmatrix} \psi_y(x) \\ \psi_z(x) \end{bmatrix} = [F_i(x - x_i)] \cdot \begin{bmatrix} \psi_y(x_i) \\ \psi_z(x_i) \end{bmatrix}. \tag{5.110}$$

Here, if $\beta < k_0 n_i$ in the i layer,

$$[F_i(x)] = \begin{bmatrix} \cos \kappa_i x, & \dfrac{k_0 \zeta_i}{\kappa_i} \sin \kappa_i x \\ -\dfrac{\kappa_i}{k_0 \zeta_i} \sin \kappa_i x, & \cos \kappa_i x \end{bmatrix}, \tag{5.111}$$

$$\kappa_i = \sqrt{k_0^2 n_i^2 - \beta^2}, \tag{5.112}$$

$$\zeta_i = \begin{cases} 1 & : \text{TE mode} \\ n_i^2 & : \text{TM mode}. \end{cases} \tag{5.113}$$

On the other hand, if $\beta > k_0 n_i$ in the i layer,

$$[F_i(x)] = \begin{bmatrix} \cosh \gamma_i x, & \dfrac{k_0 \zeta_i}{\gamma_i} \sinh \gamma_i x \\ \dfrac{\gamma_i}{k_0 \zeta_i} \sinh \gamma_i x, & \cosh \gamma_i x \end{bmatrix} \tag{5.114}$$

$$\gamma_i = \sqrt{\beta^2 - k_0^2 n_i^2} \tag{5.115}$$

are used.

Now, as ψ_y and ψ_z are continuous at the boundary, the matrix $[F_i(x)]$ can be connected in sequence. Consequently, region 0 and region N can be related using the following equation:

$$\begin{bmatrix} \psi_y(x_0) \\ \psi_z(x_0) \end{bmatrix} = [T] \cdot \begin{bmatrix} \psi_y(x_{N-1}) \\ \psi_z(x_{N-1}) \end{bmatrix}, \tag{5.116}$$

where

$$[T] = \begin{bmatrix} A & B \\ C & D \end{bmatrix} = \prod_{i=1}^{N-1} [F_i(-d_i)]. \tag{5.117}$$

From the condition designating $e^{\gamma_0 x}$ for region 0 and $e^{-\gamma_N x}$ for region N, a physically meaningful solution will result in the following eigenvalue equation:

$$\frac{k_0 \zeta_N}{\gamma_N} A - B - \frac{k_0^2 \zeta_0 \zeta_N}{\gamma_0 \gamma_N} C + \frac{k_0 \zeta_0}{\gamma_0} D = 0. \tag{5.118}$$

In the analysis based on the numerical calculation, the propagation constant is first assumed, then the matrix is calculated and its elements are substituted to Equation (5.118). The assumed propagation constant will be outputted as the solution, if the value on the left side approaches 0 (if less than the criterion value ε that was set up initially). If the value on the left side does not approach 0, the propagation constant will be corrected, and the calculation will be repeated until the left hand of Equation (5.118) approaches 0.

This analytical method can be characterized as follows:

- If the refractive index in the core has a distribution, it can be applied to any refractive index distribution by approximating the distribution by staircase index profile.
- The field distribution will be known once the propagation constant β is determined. So, there is no need to initially specify the range of the core.
- Because it is presumed that the field distribution attenuates at infinity, it is difficult to apply to a leaky waveguide.

(4)   ARROW and leaky waveguide

The preceding paragraph (3) described the analytical method for waveguides with multilayer structured cores. (Here, the confinement of light is not limited to all layers of the multilayer structure, it is more accurate to say that the core and the cladding have multilayers.) Conversely, for a leaky waveguide having a multilayer cladding with a reflectance that is less than 1 (i.e., there is no total internal reflection), it is necessary to think of the concept of mode in another way.

As shown in Figure 5.2(b), optical power gradually radiates to the cladding along with the propagation of a leaky waveguide, so a radiation loss occurs. Consequently, it is impossible to use this over long distance transmission lines, such as optical fibers. However, by some ingenuity, it is sufficiently possible to use this on an optical integrated circuit with a length of several centimeters. One such waveguide is the *ARROW waveguide*, which has been newly developed by Duguay et al. [20] and Baba and Kokubun [21]. As shown in Figure 5.15, this waveguide is an optical waveguide that was fabricated on a semiconductor substrate with a high-refractive index, wherein a pair of two layers of interference-type antireflection films are sandwiched between the core and the high-refractive index substrate as the cladding. The cladding at the top of the core may be the same as the interference-type antireflection cladding, or an ordinary low-refractive index cladding can also be used. The cladding layer, which is close to the core, is called the first cladding layer, and the other one is called the second cladding layer. The first cladding layer will either have a very large or a very small refractive index compared to the core, whereas the second cladding layer will have a refractive index that is the same as that of the core. The structure wherein the refractive index of the first cladding layer is much larger than that of the core is called the ARROW, whereas the waveguide wherein the refractive index of the first cladding layer is smaller than that of the core is called the ARROW-B.

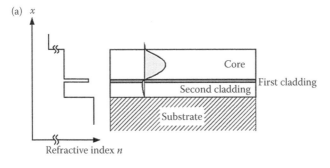

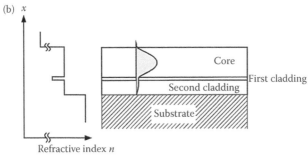

FIGURE 5.15 ARROW-type waveguide. (a) ARROW [20], [21]. (b) ARROW-B [21].

Light is guided by repeating the total internal reflection at the boundary of the core and the upper cladding and the interference reflection from the interference cladding. Because the interference-type antireflection is not a total internal reflection, the reflectance is slightly less than 1.0. However, as it is very close to 1.0 (normally more than 99.9%), the radiation loss to the substrate is less than 0.1 dB/cm, which is practically considered as a low loss. Moreover, compared to conventional waveguides (i.e., those that use total internal reflection), ARROW waveguides have the following advantages suited to integrated optics:

- Layers with a large difference in the refractive indices are combined, so there is no need to precisely control the refractive index, this results in ease of fabrication.
- Because the cladding layer may be thin, it is easy to deposit.
- The optical waveguide has its own light control functions, such as polarization filtering and wavelength filtering.
- The upper surface can be flat even if it is made into a channel waveguide (structure where the light can also be confined to a direction parallel to the surface of the substrate), so it is suitable for stacked integration.

- When a directional coupler is configured, the coupling coefficient for the conventional types decreases in proportion to the exponential function of the waveguide interval, whereas it becomes a periodic function of the waveguide interval in the ARROW, so remote coupling is possible.

Because the ARROW is a leaky waveguide, the boundary conditions, such as in Equation (5.40), wherein the electromagnetic field in the cladding converges to 0 at infinity cannot be applied. Consequently, the analytical method that has been discussed so far is not applicable. Below, we will present a summary of the *interference matrix method* that was developed by the authors in their laboratory to analyze leaky waveguides, wherein, as shown in Figure 5.16, the refractive index in the core is lower than the refractive index in the cladding.

For this analytical method, first assume a layer that can be considered as the core (where light is confined). In a leaky waveguide, total internal reflection does not occur at the boundary surface of the core and the cladding, and only a simple Fresnel reflection takes place. Consequently, as shown in Figure 5.16, light is radiated from the core to the cladding by refraction, every time light is reflected at the boundary surface, and the portion that is ratiated is called the radiation loss. However, as reflection also occurs, the light is guided. (Such a waveguide, wherein reflection is very close to 1, loss is low, and features are new, is the ARROW.) For such a light that is guided through this zigzag optical path, if the sum of the phase change due to two reflections and a single optical path length of the zigzag optical path is not an integer multiple of 2π, the mode cannot be guided while maintaining a nearly constant electromagnetic field distribution in the transverse direction. Here, when the angle of propagation of the mode is designated as θ_v, and the phase change due to the reflection

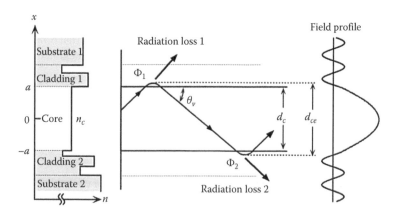

FIGURE 5.16 Leaky waveguide.

that occurs at the top and the bottom of the boundary surface is designated as Φ_1 and Φ_2, the condition (phase condition) that would support the mode is given by the following equation:

$$2k_0 n_c d_c \sin\theta_v + \Phi_1 + \Phi_2 = 2\pi v \quad (v : \text{mode number}). \tag{5.119}$$

The radiation loss can be calculated from the percentage of light, which is radiated from the core at a single radiation, and the number of reflections, which are repeated during propagation at a unit length. Here, when the power reflectance at the upper and the lower boundary surfaces is designated as R_1, R_2, and v, the radiation loss α_v of the vth-order mode is expressed by the following equation:

$$\alpha_v = 2.17(2 - R_1 - R_2)\frac{\tan\theta_v}{d_{ce}} \quad [\text{dB/m}]. \tag{5.120}$$

Here, d_{ce} is the equivalent thickness taking into consideration the spread of the electric field distribution outside the core and can be approximated as

$$d_{ce} = d_c + \frac{2\pi - \Phi_1 - \Phi_2}{2k_0 n_c \sin\theta_v}. \tag{5.121}$$

Using complex reflectances r_1 and r_2 of the electromagnetic field described in Equation (1.109) for TE polarization and in Equation (1.110) for TM polarization, the phase changes Φ_1 and Φ_2 and the power reflectivity R_1 and R_2 can be expressed using the following equations:

$$\Phi_\ell = -\arg(r_\ell) \quad (\ell = 1 \text{ or } 2), \tag{5.122}$$

$$R_\ell = |r_\ell|^2 \quad (\ell = 1 \text{ or } 2). \tag{5.123}$$

Here, if the cladding layer consists of a multilayer film, using the interference matrix in Section 4.5, r_1 and r_2 can be calculated using the following equation:

$$r_\ell = \frac{m_{11} - Y_c^{-1} Y_s m_{22} + Y_s m_{12} - Y_c^{-1} m_{21}}{m_{11} + Y_c^{-1} Y_s m_{22} + Y_s m_{12} + Y_c^{-1} m_{21}}. \tag{5.124}$$

In Equation (5.124), Y_c and Y_s are the admittances of the core and the substrate, respectively. m_{11} to m_{22} are the components of the interference matrix M and are defined by the following equation:

$$M = \begin{bmatrix} m_{11} & m_{12} \\ m_{21} & m_{22} \end{bmatrix} = \prod_i \begin{bmatrix} \cos\phi_i & jY_i^{-1}\sin\phi_i \\ jY_i\sin\phi_i & \cos\phi_i \end{bmatrix}. \tag{5.125}$$

Here, ϕ_i represents the phase change and absorption (i.e., a complex number) of light in the i layer, and Y_i is the admittance in the i layer. Using the thickness of each of the layers d_i, the refractive index n_i, angle θ_i of the ray with respect to the boundary surface when the ray is passing through the layers, and the wavelength λ_0, they are given by each of the following equations:

$$\phi_i = k_0 n_i d_i \sin\theta_i, \tag{5.126}$$

$$Y_i = \sqrt{\frac{\epsilon_0}{\mu_0}} \begin{cases} -n_i \sin\theta_i & : \text{TE mode} \\ \dfrac{n_i}{\sin\theta_i} & : \text{TM mode.} \end{cases} \tag{5.127}$$

Here, θ_i is obtained from θ_v using Snell's Law $n_i \cos\theta_i = n_c \cos\theta_v$. If there is light absorption in the i layer and total internal reflection at the boundary with the next $(i+1)$ layer, it will become a complex number. Figure 5.17 shows the steps of this analytical method.

In addition, for the ARROW that is shown in Figure 5.15a, the optimum value of the thickness of the first cladding that would result in a minimum radiation loss of the TE mode is given by

$$\frac{d_1}{\lambda} \simeq \frac{1}{4n_1}\left[1-\left(\frac{n_c}{n_1}\right)^2+\left(\frac{\lambda}{2n_1 d_{ce}}\right)^2\right]^{-1/2} \cdot (2N+1) \quad (N=0,1,2,\ldots), \tag{5.128}$$

whereas the optimum value for the second cladding is almost half of the core layer ($d_2 = \frac{d_{ce}}{2}$). Here, d_{ce} is the equivalent core thickness, taking into consideration the spread of the optical electromagnetic field to the upper cladding layer and is expressed by

$$d_{ce} \simeq d_c + \frac{\lambda}{2\pi\sqrt{n_c^2 - n_0^2}}. \tag{5.129}$$

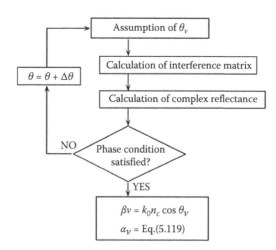

FIGURE 5.17 Steps in calculation for the interference matrix method.

The radiation loss of the TE mode normalized by the wavelength is expressed by

$$\alpha\lambda \simeq 0.543 \left(\frac{\lambda}{d_c}\right)^5 \frac{(v+1)^4}{n_c(n_1^2 - n_c^2)\sqrt{n_s^2 - n_c^2}} \quad (\text{dB}\cdot\lambda/\text{m}) \quad (5.130)$$

with v as a mode number [21]. As seen from this equation, the radiation loss of the higher order ARROW mode is large, so the single-mode waveguide can be substantially realized. Analytical expressions of the optimum film thickness and radiation loss of the TM mode and those of ARROW-B have been derived, but these are omitted because of space limitations.

PROBLEMS

1. Let us consider an *asymmetric four-layer slab waveguide* as in Figure 5.18. Assume that the refractive indices are $n_3 \leq n_4 < n_2 \leq n_1$. Derive the dispersion relation of the waveguide using a format similar to Equation (5.97).

   Solution
   When the parameters are defined as

   $$V_1 = k_0 a \sqrt{n_1^2 - n_4^2}, \quad b = \frac{(\beta/k_0)^2 - n_4^2}{n_1^2 - n_4^2}$$

   $$V_2 = k_0 d \sqrt{n_2^2 - n_4^2}, \quad g = \frac{n_2^2 - n_4^2}{n_1^2 - n_4^2}$$

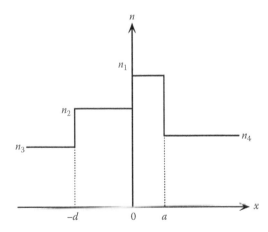

FIGURE 5.18 Refractive index distribution and definition of coordinates of an asymmetrical four-layer slab waveguide.

$$
a' = \frac{n_4^2 - n_3^2}{n_1^2 - n_4^2}, \quad \eta_{ij} = \begin{cases} 1.0 & : \text{TE mode} \\ \frac{n_i^2}{n_j^2} & : \text{TM mode.} \end{cases} \tag{5.131}
$$

And, if $k_0 n_4 \leq \beta \leq k_0 n_2$, then

$$
V_1 = \frac{1}{\sqrt{1-b}} \left[N\pi + \tan^{-1}\left(\eta_{14}\sqrt{\frac{b}{1-b}} \right) \right.
$$

$$
+ \tan^{-1} \left\{ \eta_{12}\sqrt{\frac{g-b}{1-b}} \tan\left(\tan^{-1}\left(\eta_{23}\sqrt{\frac{a'+b}{g-b}} \right) \right. \right.
$$

$$
\left. \left. \left. - \frac{V_2}{\sqrt{g}}\sqrt{g-b} \right) \right\} \right].
$$

On the other hand, $k_0 n_2 \leq \beta < k_0 n_1$, then

$$
V_1 = \frac{1}{\sqrt{1-b}} \left[N\pi + \tan^{-1}\left(\eta_{14}\sqrt{\frac{b}{1-b}} \right) \right.
$$

$$
+ \tan^{-1} \left\{ \eta_{12}\sqrt{\frac{b-g}{1-b}} \coth\left(\tanh^{-1}\left(\frac{1}{\eta_{23}}\sqrt{\frac{b-g}{a'+b}} \right) \right. \right.
$$

$$
\left. \left. \left. + \frac{V_2}{\sqrt{g}}\sqrt{b-g} \right) \right\} \right].
$$

2. Let us consider a symmetric five-layer slab waveguide as shown in Figure 5.19. Assume that the refractive indices are $n_3 < n_2 \leq n_1$. Derive the dispersion relation of the waveguide using a format similar to Equation (5.97).

3. How can radiation loss to a high-refractive index substrate be calculated?

4. Show that the confinement factor of an asymmetric three-layer slab waveguide is expressed by the following equations:

$$
\text{TE}: \Gamma_{\text{core}} = \frac{V' + \sqrt{b} + \frac{\sqrt{b+a'}}{1+a'}}{V' + \frac{1}{\sqrt{b+a'}} + \frac{1}{\sqrt{b}}},
$$

$$
\Gamma_{\text{cover}} = \frac{(1-b)\sqrt{b+a'}}{\sqrt{b}\,(b+a')V' + \sqrt{b} + \sqrt{b+a'}},
$$

$$
\Gamma_{\text{upper}} = \frac{\frac{(1-b)\sqrt{b}}{(1+a')}}{\sqrt{b}\,(b+a')V' + \sqrt{b} + \sqrt{b+a'}}.
$$

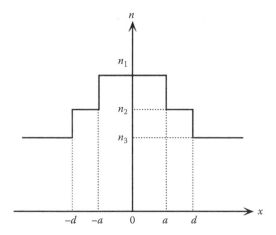

FIGURE 5.19 Refractive index distribution and definition of coordinates of a symmetrical five-layer slab waveguide.

5.2.2 THREE-DIMENSIONAL WAVEGUIDES

An optical waveguide that confines light to a cross-sectional structure, not just in the x-direction but also in the y-direction is called a *three-dimensional waveguide* (or a *channel waveguide*). In the actual application of waveguide-type devices, a three-dimensional waveguide configuration would be required, wherein light is also confined in the lateral direction. As shown in Figure 5.1, there are several configurations for the three-dimensional waveguide structure. Here, we will describe the approximate analysis method for the representative rectangular cross-section configuration in Figure 5.1b and the configuration, wherein the core film thickness is distributed in the lateral direction, as shown in Figure 5.1c.

(1) Solution for the rectangular waveguide by the Marcatili's method

The numerical analysis method is required for a rigorous analysis of a three-dimensional waveguide structure with a rectangular core cross-section, as shown in Figure 5.20. Here, the method used to find an approximate analytical solution is the *Marcatili's method* [22]. In this method, the electromagnetic field distribution in the hatched line portion of Figure 5.20 is ignored, and only the boundary conditions at the four sides of the core are taken into consideration. Consequently, a good approximation can be done if the light is well-confined to the core; however, the approximation becomes poor in a single-mode waveguide where light spreads out to the cladding. The main points of the assumptions and analysis used in this analytical method are as follows:

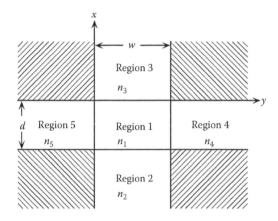

FIGURE 5.20 Region when considering the field distribution of rectangular waveguide in the analysis by Marcatili's method [22].

Point: (1) Ignore the hatched line portion of Figure 5.20.

$\rightarrow$ The error in single-mode waveguide is large

Point: (2) Assume that the transverse electromagnetic field distribution can be separated owing to a separation of variables such that

$$F(x, y) = f(x) \cdot g(y)$$

Point: (3) Assume a TE- or TM-like polarization.

Under the above assumptions, the solutions that satisfy the wave equation and boundary conditions are obtained as follows:

For example: TM-like mode that is polarized mainly in the x-direction electromagnetic field of region 1:

$$E_z = A \cos \kappa_x (x + \xi) \cos \kappa_y (y + \eta),$$

$$H_z = -A \left(\frac{\varepsilon_0}{\mu_0} \right)^{1/2} n_1^2 \left(\frac{\kappa_y}{\kappa_x} \right) \left(\frac{k_0}{\beta} \right) \sin \kappa_x (x + \xi) \sin \kappa_y (y + \eta),$$

$$E_x = \left(\frac{jA}{\kappa_x \beta} \right) (k_0^2 n_1^2 - \kappa_x^2) \sin \kappa_x (x + \xi) \cos \kappa_y (y + \eta),$$

$$E_y = -jA \left(\frac{\kappa_y}{\beta} \right) \cos \kappa_x (x + \xi) \sin \kappa_y (y + \eta),$$

$$H_x = 0,$$

$$H_y = jA \left(\frac{\varepsilon_0}{\mu_0} \right)^{1/2} n_1^2 \left(\frac{k_0}{\kappa_x} \right) \sin \kappa_x (x + \xi) \cos \kappa_y (y + \eta).$$

Electromagnetic field of region 2:

$$E_z = A \cos \kappa_x(\xi - d) \cos \kappa_y(y + \eta) \exp[\gamma_2(x + d)],$$

$$H_z = -A \left(\frac{\varepsilon_0}{\mu_0}\right)^{1/2} n_2^2 \left(\frac{\kappa_y}{\gamma_2}\right) \left(\frac{k_0}{\beta}\right) \cos \kappa_x(\xi - d) \sin \kappa_y(y + \eta)$$
$$\times \exp[\gamma_2(x + d)],$$

$$E_x = jA \left(\frac{\gamma_2^2 + k_0^2 n_2^2}{\gamma_2 \beta}\right) \cos \kappa_x(\xi - d) \cos \kappa_y(y + \eta) \exp[\gamma_2(x + d)],$$

$$E_y \cong 0,$$

$$H_x = 0,$$

$$H_y = jA \left(\frac{\varepsilon_0}{\mu_0}\right)^{1/2} n_2^2 \left(\frac{k_0}{\gamma_2}\right) \cos \kappa_x(\xi - d) \cos \kappa_y(y + \eta) \exp[\gamma_2(x + d)].$$

Electromagnetic field of region 3:

$$E_z = A \cos \kappa_x \xi \cos \kappa_y(y + \eta) \exp(-\gamma_3 x),$$

$$H_z = -A \left(\frac{\varepsilon_0}{\mu_0}\right)^{1/2} n_3^2 \left(\frac{\kappa_y}{\gamma_3}\right) \left(\frac{k_0}{\beta}\right) \cos \kappa_x \xi \sin \kappa_y(y + \eta) \exp(-\gamma_3 x),$$

$$E_x = jA \left(\frac{\gamma_3^2 + k_0^2 n_3^2}{\gamma_3 \beta}\right) \cos \kappa_x \xi \cos \kappa_y(y + \eta) \exp(-\gamma_3 x),$$

$$E_y \cong 0,$$

$$H_x = 0,$$

$$H_y = jA \left(\frac{\varepsilon_0}{\mu_0}\right)^{1/2} n_3^2 \left(\frac{k_0}{\gamma_3}\right) \cos \kappa_x \xi \cos \kappa_y(y + \eta) \exp(-\gamma_3 x).$$

Electromagnetic field of region 4:

$$E_z = A \left(\frac{n_1^2}{n_4^2}\right) \cos \kappa_x(x + \xi) \cos \kappa_y(w + \eta) \exp[-\gamma_4(y - w)],$$

$$H_z = -A \left(\frac{\varepsilon_0}{\mu_0}\right)^{1/2} n_1^2 \left(\frac{\gamma_4}{\kappa_x}\right) \left(\frac{k_0}{\beta}\right) \sin \kappa_x(x + \xi) \cos \kappa_y(w + \eta)$$
$$\times \exp[-\gamma_4(y - w)],$$

$$E_x = jA \left(\frac{n_1^2}{n_4^2}\right) \left(\frac{k_0^2 n_4^2 - \kappa_x^2}{\kappa_x \beta}\right) \sin \kappa_x(x + \xi) \cos \kappa_y(w + \eta) \exp[-\gamma_4(y - w)],$$

$$E_y \cong 0,$$

$$H_x = 0,$$

$$H_y = jA \left(\frac{\varepsilon_0}{\mu_0}\right)^{1/2} n_1^2 \left(\frac{k_0}{\kappa_x}\right) \sin \kappa_x(x + \xi) \cos \kappa_y(w + \eta) \exp[-\gamma_4(y - w)],$$

Electromagnetic field of region 5:

$$E_z = A\left(\frac{n_1^2}{n_5^2}\right)\cos \kappa_x(x+\xi)\cos \kappa_y \eta \exp(\gamma_5 y),$$

$$H_z = A\left(\frac{\varepsilon_0}{\mu_0}\right)^{1/2} n_1^2 \left(\frac{\gamma_5}{\kappa_x}\right)\left(\frac{k_0}{\beta}\right)\sin \kappa_x(x+\xi)\cos \kappa_y \eta \exp(\gamma_5 y),$$

$$E_x = jA\left(\frac{n_1^2}{n_5^2}\right)\left(\frac{k_0^2 n_5^2 - \kappa_x^2}{\kappa_x \beta}\right)\sin \kappa_x(x+\xi)\cos \kappa_y \eta \exp(\gamma_5 y),$$

$$E_y \cong 0,$$

$$H_x = 0,$$

$$H_y = jA\left(\frac{\varepsilon_0}{\mu_0}\right)^{1/2} n_1^2 \left(\frac{k_0}{\kappa_x}\right)\sin \kappa_x(x+\xi)\cos \kappa_y \eta \exp(\gamma_5 y).$$

Here, the boundary surfaces are the surface that is parallel to the yz plane and the surface that is parallel to the xz plane, so the eigenvalue equation will be the following two simultaneous equations:

$$\tan \kappa_x d = \frac{n_1^2 \kappa_x (n_3^2 \gamma_2 + n_2^2 \gamma_3)}{n_3^2 n_2^2 \kappa_x^2 - n_1^4 \gamma_2 \gamma_3}, \tag{5.132}$$

$$\tan \kappa_y w = \frac{\kappa_y(\gamma_4 + \gamma_5)}{\kappa_y^2 - \gamma_4 \gamma_5}. \tag{5.133}$$

Here, phases ξ and η of the transverse direction are determined by the following equations:

$$\tan \kappa_x \xi = -\left(\frac{n_3}{n_1}\right)^2 \frac{\kappa_x}{\gamma_3}, \tag{5.134}$$

$$\tan \kappa_y \eta = -\frac{\gamma_5}{\kappa_y}. \tag{5.135}$$

In addition, the following is the relationship between κ_x, κ_y, γ_2, γ_3, γ_4, and γ_5:

$$k_0^2 n_1^2 - \beta^2 = \kappa_x^2 + \kappa_y^2, \tag{5.136}$$

$$k_0^2 n_2^2 - \beta^2 = \kappa_y^2 - \gamma_2^2, \tag{5.137}$$

$$k_0^2 n_3^2 - \beta^2 = \kappa_y^2 - \gamma_3^2, \tag{5.138}$$

$$k_0^2 n_4^2 - \beta^2 = \kappa_x^2 - \gamma_4^2, \tag{5.139}$$

$$k_0^2 n_5^2 - \beta^2 = \kappa_x^2 - \gamma_5^2. \tag{5.140}$$

(2) Equivalent index method

For a ridge waveguide as shown in Figure 5.1c, the structural changes in the lateral direction are smooth, so if we approximate that the change of the field distribution in the lateral direction is smoother compared with that of the longitudinal direction, the approximate solution shown below can be used.

First, let us consider the structure where the core film thickness is distributed to the lateral direction (y-direction), as shown in Figure 5.21. The analysis for this kind of waveguide involves the reduction of the two-dimensional cross-section to a one-dimensional one by replacing the thickness distribution in the x-direction with the y-direction equivalent index distribution. The main points of the assumptions in this analytical method are as follows:

Point (1): Assume that the polarization is a TE or TM mode.

Point (2): When calculating the electromagnetic field distribution and propagation constant, the film thickness is first set as $d(y_0)$ at $y = y_0$, then the propagation constant of the slab waveguide, when the film thickness is assumed to be uniform in the y-direction, is determined.

Point (3): Next, assume that the change in the y-direction of the electromagnetic distribution is sufficiently gradual. Depending on each of the values of the film thickness distribution $d(y)$ placed at different points on the y-axis,

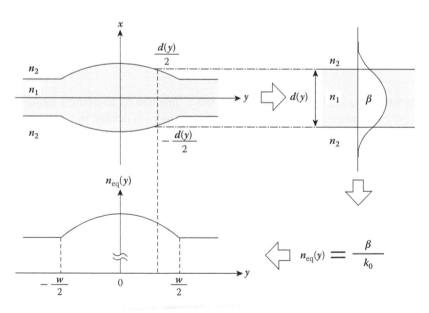

FIGURE 5.21 Equivalent index method, where the thickness distribution is replaced with the equivalent index distribution in the y-direction.

the y-direction distribution of the propagation constant is determined based on the assumption in step (2).

Point (4): The y-direction distribution of the propagation constant, which was determined in step (3), has an equivalent refractive index distribution in the y-direction, and as this is equal to the slab waveguide with a uniform film thickness w in the x-direction, the propagation constant of the equivalent slab waveguide can be determined.

With the TE mode as an example, based on the assumption in step (2), the field distribution is expressed using the following equation:

$$E_y(x, y) = \psi(y)\cos[\kappa(y)x]. \tag{5.141}$$

Here, $\kappa(y)$ is obtained from the following eigenvalue equation:

$$\tan\left(\frac{\kappa(y)d(y)}{2}\right) = \frac{\sqrt{k_0^2(n_1^2 - n_2^2) - \kappa^2(y)}}{\kappa(y)}. \tag{5.142}$$

Here, based on the assumption in step (3), Equation (5.141) is substituted to the wave equation (5.7) to obtain the following wave equation:

$$\frac{d^2\psi(y)}{dy^2} + (k_0^2 n_1^2 - \kappa^2(y) - \beta^2)\psi(y) \cong 0. \tag{5.143}$$

Equation (5.143) is equivalent to the wave equation of the slab waveguide that spreads infinitely in the x-direction and has an equivalent refractive index distribution in the y-direction, which is

$$n_{eq}^2 = n_1^2 - \frac{\kappa^2(y)}{k_0^2} \tag{5.144}$$

in which the thickness distribution in the y-direction is determined through $\kappa(y)$ by Equation (5.142). Here, the equivalent index is defined by

$$n_{eq} = \frac{\beta}{k_0} \tag{5.145}$$

and

$$n_2 \leq n_{eq} < n_1 \tag{5.146}$$

always holds true for the guided mode of the slab waveguide as shown in Figure 5.3. As Equation (5.144) is equal to the equivalent index, this analytical method is called the *equivalent index method* [23].

In this method, the three-dimensional configuration is replaced with an equivalent two-dimensional slab waveguide (from two- to one-dimension for a cross-sectional dimension), and so the numerical analysis method or the approximate analytical method is used in each case to solve the resulting wave equation (5.143).

(3) Numerical analysis method for a three-dimensional waveguide

Although Marcatili's method and the equivalent index method provide good out-look on the solution owing to numerical formulas, the accuracy of both methods is limited because they are approximate solution methods. To exactly determine the propagation constant and the electromagnetic field of a three-dimensional waveguide of an arbitrary cross-sectional configuration and refractive index, one has to absolutely rely on the numerical analysis method. Examples of the numerical analysis method include the finite element method and the boundary element method, along with the vector finite element method and the scalar finite element method, which depends on whether it contains a dielectric constant gradient term in the wave equation, which is the basis for the analysis. These numerical analysis methods will not be discussed here because of space limitations. When a reader needs to analyze optical waveguides with complex cross-sectional configuration and when high precision analysis is required, analytical tools for optical waveguide that have been made available commercially in the market in recent years can be used. These tool sets contain the beam propagation method and/or finite difference time domain method that can simulate the light propagation in optical waveguides of which cross-sectional structure changes in the longitudinal direction and numerical analysis program for three-dimensional waveguides of arbitrary cross-sectional configuration.

6 Optical Fibers

This chapter describes the optical propagation and signal propagation inside an *optical fiber*. Optical fibers are classified on the basis of their materials and structure. In terms of materials, optical fibers can be classified as silica fibers* and plastic fibers, whereas in terms of structure, they can be categorized as multimode fibers and single-mode fibers. However, plastic fibers only consist of multimode fibers. (Single-mode fibers are possible, but they have no advantage[†] over single-mode silica fibers, so they are not produced.) Moreover, silica fibers have for long been installed in many optical communication systems starting with long-distance trunk lines, and the diameters of the silica glass portion (portion that also includes the cladding), the buffer layer, and the jacket layer are standardized internationally as shown in Figure 6.1. Moreover, as shown in Table 6.1, silica fibers can also be classified in terms of the refractive index profile in the core. Single-mode fibers have core diameters of about 8–10 μm, with a relative refractive index difference of about 0.3% between the core and the cladding (actually, the standard is determined by the spot size,[‡] which will be discussed in subsection 6.1.5), and the diameter of the glass portion is 125 μm. More than 90% of the optical fibers currently being used in optical communications are single-mode silica fibers. The single-mode silica fiber is the best communication transmission channel owing to its ultralow propagation loss and ultrahigh transmission capacity.

---

[*] Multicomponent glass fibers existed at the time when optical fibers were initially developed; however, silica fibers were dramatically developed due to the ultralow propagation loss and ultrahigh transmission capacity. Therefore, multicomponent glass fibers have not been manufactured ever since.

[†] The advantage of plastic fibers is that a plastic fiber can be bent even if its diameter is several times thicker than that of multimode silica fibers, and it is easy to connect owing to its thick diameter, resulting in reduced costs. The bandwidth of about 1 GHz·km is also possible by the distributed-index profile, as will be discussed in subsection 6.3.2. Moreover, as the propagation loss is not so low compared with silica fibers, it can be installed in short-range optical communications of less than several hundred meters.

[‡] The spot size is the parameter that represents the beam radius when a light beam is approximated using a Gaussian function. However, as the beam size is expressed in diameters in optical fibers by tradition, it is called the *mode field diameter*, to avoid confusion.

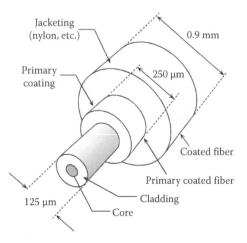

FIGURE 6.1   Silica fiber configuration.

TABLE 6.1

Classification of Silica Optical Fibers According to the Refractive Index Distribution in the Core

	Classification of optical fiber		
	Step-index multimode fiber	**Distributed-index multimode fiber**	**Single mode fiber**
Structure	$n(r)$ $\Delta\sim 1\%$ $2a = 50\ \mu m$ 125 μm	$n(r)$ $\Delta\sim 1\%$ $\propto r^2$ $2a = 50\ \mu m$ 125 μm	$n(r)$ $\Delta\sim 0.3\%$ $2a = 8\text{–}10\ \mu m$ 125 μm
Bandwidth	A few tens **MHz · km**	~ 1 **GHz · km** (at optimum wavelength)	>1 **THz** $\cdot \sqrt[4]{\text{km}}$ (1.3 μm zero-dispersion wavelength) ~ 25 **GHz** $\cdot \sqrt{\text{km}}$ (at 1.55 μm wavelength region)

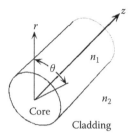

FIGURE 6.2 Cylindrical coordinate system for round optical fibers.

6.1 OPTICAL FIBER MODES

For optical fibers with a circular cross-sectional configuration, it is more convenient to use the cylindrical coordinate system instead of the x, y, z coordinate system. The mode will be determined according to the coordinate system defined by Figure 6.2.

6.1.1 EIGENVALUE EQUATIONS OF OPTICAL FIBERS

Let us assume that the z-direction dependency of the field is $e^{-j\beta z}$. The z components of the wave equations (5.14) and (5.15) are written in terms of the cylindrical coordinates as follows:

$$\frac{\partial^2 E_z}{\partial r^2} + \frac{1}{r}\frac{\partial E_z}{\partial r} + \frac{1}{r^2}\frac{\partial^2 E_z}{\partial \theta^2} + (k_0^2 n^2 - \beta^2)E_z = 0, \qquad (6.1)$$

$$\frac{\partial^2 H_z}{\partial r^2} + \frac{1}{r}\frac{\partial H_z}{\partial r} + \frac{1}{r^2}\frac{\partial^2 H_z}{\partial \theta^2} + (k_0^2 n^2 - \beta^2)H_z = 0. \qquad (6.2)$$

When a *step-index round optical fiber* with a uniform refractive index profile inside the core is taken into consideration, the n^2 in Equations (6.1) and (6.2) can be expressed separately in the core and in the cladding as follows:

$$n^2(r) = \begin{cases} n_1^2 & (r \le a) \\ n_2^2 = n_1^2[1 - 2\Delta] & (r > a). \end{cases} \qquad (6.3)$$

When Equation (6.3) is substituted to Equations (6.1) and (6.2) and the solutions for the guided modes of E_z and H_z are determined, using the method of separation of variables, the dependency of the azimuth angle θ is expressed as a trigonometric function, whereas the dependency of the radial r direction will become the oscillatory solution in the core and will be expressed as $J_\nu(x)$, the Bessel function of the first kind. From the condition that light will converge to zero at infinity, the solution in the cladding is expressed as $K_\nu(x)$, the modified Bessel function of the second kind [refer to (Appendix F)]. Therefore, the solutions in the core and the cladding are expressed as follows (here, the time and z-dependency term $e^{j(\omega t - \beta z)}$ is omitted):

In the core ($r \leq a$):

$$E_z = A_\ell J_\ell(\kappa r)\cos(\ell\theta + \phi_\ell), \tag{6.4}$$

$$H_z = B_\ell J_\ell(\kappa r)\cos(\ell\theta + \psi_\ell). \tag{6.5}$$

In the cladding ($r > a$):

$$E_z = A_\ell \frac{J_\ell(\kappa a)}{K_\ell(\gamma a)} K_\ell(\gamma r)\cos(\ell\theta + \phi_\ell), \tag{6.6}$$

$$H_z = B_\ell \frac{J_\ell(\kappa a)}{K_\ell(\gamma a)} K_\ell(\gamma r)\cos(\ell\theta + \psi_\ell). \tag{6.7}$$

Here, κ and γ are defined by Equations (5.42) and (5.43) in the same way as slab waveguides. $J_\ell(x)$ and $K_\ell(x)$ are the Bessel function (cylindrical function of the first kind) and modified Bessel function of the second kind, respectively. The outlines of these functions are shown in Figure 6.3a and b. As can be seen from the figure, the Bessel function corresponds to the cos function for a one-dimensional vibration problem, so as an example, the vibration of guitar strings is expressed as a cos function as it is a one-dimensional problem, whereas the vibration of drums and round speakers is expressed by the Bessel function. On the other hand, the modified Bessel function of the second kind corresponds to $e^{-\gamma x}$ for a one-dimensional problem and expresses the solution for attenuation in a two-dimensional problem. ℓ is the mode number (mode label) of the direction of the azimuth angle θ, and ϕ_ℓ is the phase term, which is introduced to avoid the field distribution containing the sin function being identical to zero, when $\ell = 0$.

Now, all the electromagnetic components can be determined by substituting Equations (6.4) to (6.7) into Equations (5.22) to (5.25). Here, the boundary condition Equations (5.16) and (5.17), wherein the tangential components of the electromagnetic field become continuous at the boundary of the core and the cladding, will have to be satisfied. In Equations (6.4) through (6.7), E_z and H_z are the tangential components with respect to the boundary surface (cylindrical surface) of the core and the cladding, and they were already determined so that the coefficients for the solutions will become continuous at $r = a$. The remaining tangential components are E_θ and H_θ, and the condition that they will be continuous at the boundary of the core and the cladding is expressed by the following equations:

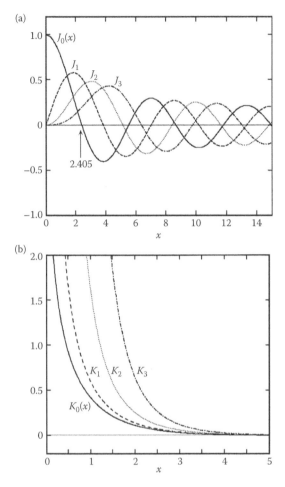

FIGURE 6.3 Cylinder functions: (a) Bessel function $J_\ell(x)$ and (b) modified Bessel function of the second kind $K_\ell(x)$.

$$E_\theta(r \rightarrow a_{+0}) = E_\theta(r \rightarrow a_{-0}), \tag{6.8}$$

$$H_\theta(r \rightarrow a_{+0}) = H_\theta(r \rightarrow a_{-0}). \tag{6.9}$$

E_θ and H_θ can be obtained by substituting Equations (6.4) through (6.7) to Equations (5.23) and (5.25), respectively. Two simultaneous homogeneous equations with respect to the two amplitude coefficients A_ℓ and B_ℓ can be obtained from the conditions where the above components satisfy Equations (6.8) and (6.9). For these homogeneous equations to have identical nonzero solutions, it is necessary for the determinants of the coefficient to be zero, so the following equation is obtained [24]:

$$
\frac{k_0^2 \left[\frac{J_\ell'(\kappa a)}{\kappa a J_\ell(\kappa a)} + \frac{K_\ell'(\gamma a)}{\gamma a K_\ell(\gamma a)} \right] \left[n_1^2 \frac{J_\ell'(\kappa a)}{\kappa a J_\ell(\kappa a)} + n_2^2 \frac{K_\ell'(\gamma a)}{\gamma a K_\ell(\gamma a)} \right]}{\ell^2 \beta^2 \left(\frac{1}{(\kappa a)^2} + \frac{1}{(\gamma a)^2} \right)^2}
$$
$$
= -\frac{\sin(\ell\theta + \phi_\ell)\sin(\ell\theta + \psi_\ell)}{\cos(\ell\theta + \phi_\ell)\cos(\ell\theta + \psi_\ell)}. \tag{6.10}
$$

Although Equation (6.10) represents a boundary condition, this boundary condition has to hold true everywhere at the boundary ($r = a$) between the core and the cladding. Therefore, the right side of Equation (6.10) must be independent of θ. Thus, it is necessary for the following relationship between ϕ_ℓ and ψ_ℓ to hold true:

$$
\phi_\ell - \psi_\ell = \pm\frac{\pi}{2}. \tag{6.11}
$$

Equation (6.11) means that the dependency of E_z and H_z on the azimuth angle θ only deviates by $\frac{\pi}{2}$ and that if one side is $\cos \ell\theta$, then the other side will be expressed as $\pm \sin \ell\theta$.* When this condition is substituted to Equation (6.10), the right side of Equation (6.10) will become $+1$, so the boundary condition will result in the following transcendental equation, which contains β:

$$
\left[\frac{J_\ell'(\kappa a)}{\kappa a J_\ell(\kappa a)} + \frac{K_\ell'(\gamma a)}{\gamma a K_\ell(\gamma a)} \right] \left[\frac{J_\ell'(\kappa a)}{\kappa a J_\ell(\kappa a)} + (1 - 2\Delta) \frac{K_\ell'(\gamma a)}{\gamma a K_\ell(\gamma a)} \right]
$$
$$
= \left(\frac{\ell\beta}{k_0 n_1} \right)^2 \left(\frac{1}{(\kappa a)^2} + \frac{1}{(\gamma a)^2} \right)^2. \tag{6.12}
$$

This is an eigenvalue equation for a step-index round optical fiber. This eigenvalue equation is more complex compared to a symmetric three-layer slab waveguide; however, as will be discussed in Section 6.1.2, if $\Delta \ll 1$, then it will become somewhat easier. The V parameter and the normalized propagation constant b are defined by Equations (2.13) and (2.10) in the same way as the symmetric three-layer slab waveguide. The κ and γ are defined by Equations (5.42) and (5.43) in the same way. Consequently, the following relationship will hold true:

$$
\kappa a = V\sqrt{1-b}, \tag{6.13}
$$
$$
\gamma a = V\sqrt{b}, \tag{6.14}
$$
$$
(\kappa a)^2 + (\gamma a)^2 = V^2. \tag{6.15}
$$

---

* That is, this means that the electric field and the magnetic field are orthogonal to one another. If the right side of Equation (6.11) is $\frac{\pi}{2}$, then it will correspond to the HE mode, and if it is $-\frac{\pi}{2}$, it will correspond to the EH mode. This will be discussed in Section 6.1.2.

6.1.2 WEAKLY GUIDING APPROXIMATION

If

$$\Delta = \frac{n_1^2 - n_2^2}{2n_1^2} \cong \frac{n_1 - n_2}{n_1} \ll 1 \tag{6.16}$$

holds true in the eigenvalue equation (6.12) and β can be approximated as

$$\beta \cong k_0 n_1. \tag{6.17}$$

Then, the eigenvalue equation can be simplified as follows:

$$\left[\frac{J'_\ell(\kappa a)}{\kappa a J_\ell(\kappa a)} + \frac{K'_\ell(\gamma a)}{\gamma a K_\ell(\gamma a)}\right] = \chi \ell \left(\frac{1}{(\kappa a)^2} + \frac{1}{(\gamma a)^2}\right), \tag{6.18}$$

(here, $\chi = +1$ or -1.)

Such an approximation is generally referred to as the *weakly guiding approximation* [25].

If the Bessel function formulas (F.5), (F.6), (F.10), and (F.11) in Appendix F are used, Equation (6.18) can be rewritten for $\chi = +1$ or $\chi = -1$ and $\ell \neq 0$ as follows:

* $\chi = -1$ (HE mode)

$$\frac{J_{\ell-1}(\kappa a)}{\kappa a J_\ell(\kappa a)} - \frac{K_{\ell-1}(\gamma a)}{\gamma a K_\ell(\gamma a)} = 0. \tag{6.19}$$

* $\chi = +1$ (EH mode)

$$\frac{J_{\ell+1}(\kappa a)}{\kappa a J_\ell(\kappa a)} + \frac{K_{\ell+1}(\gamma a)}{\gamma a K_\ell(\gamma a)} = 0. \tag{6.20}$$

These eigenvalue equations are considerably simplified compared to Equation (6.12). Next, let us take a look at how the aforementioned modes are classified by means of these eigenvalue equations.

6.1.3 CLASSIFICATION OF MODES

In the case of three-layer slab waveguides, as the six simultaneous equations are divided into two pairs of three simultaneous equations when Maxwell's equations are divided into each of the coordinate components, the modes are decomposed into the TE and the TM modes. However, in the process of deriving the eigenvalue equation (6.12) of optical fibers, all the six electromagnetic components are coupled by Equations (5.22) through (5.25) and the boundary condition Equations (6.8) and (6.9), and as a general solution, they cannot be decomposed into the TE and the TM modes. (Of course, the TE and the TM modes can be derived by first assuming that $E_z = 0$ and $H_z = 0$; however, these are singular solutions and not general solutions.) Consequently, an optical fiber mode generally has all six electromagnetic components, and such a mode is commonly referred to as the *hybrid mode*.

(a) *HE mode:* When $\chi = -1$ and $\ell \geq 1$, the mode number is designated anew as

$$\nu = \ell - 1. \tag{6.21}$$

Then, Equations (6.13) and (6.14) are used to express κa and γa in terms of V and b, and using the Bessel function formulas (F.2) and (F.7) of Appendix F, the eigenvalue equation, Equation (6.19), can be transformed into the following equation:

$$\frac{J_{\nu-1}(\sqrt{1-b}V)}{J_\nu(\sqrt{1-b}V)} \cdot \frac{K_\nu(\sqrt{b}V)}{K_{\nu-1}(\sqrt{b}V)} = -\sqrt{\frac{b}{1-b}}. \tag{6.22}$$

If the eigenvalue b (or the propagation constant β) obtained from this eigenvalue equation is numbered as $m = 1, 2, \ldots$ arranged starting from the largest eigenvalue, then this number will be the mode number of the radial r direction. This mode is generally referred to as the $HE_{\ell,m}$ mode. This differs with the slab waveguide as the cross-section is two-dimensional and would require two mode numbers, that is, the mode number ℓ of the azimuth angle θ direction and the mode number m of the radial r direction.

(b) *TE and TM modes:* If $\ell = 0$, then the following equation will be obtained from Equation (6.20) or (6.19):

$$\frac{J_0(\sqrt{1-b}V)}{J_1(\sqrt{1-b}V)} \cdot \frac{K_1(\sqrt{b}V)}{K_0(\sqrt{b}V)} = -\sqrt{\frac{b}{1-b}}. \tag{6.23}$$

(Equation (6.23) can also be derived by substituting $\ell = 0$ and $\Delta \ll 1$ in the wave equation (6.12) and then using the formula $J_0'(x) = -J_1(x)$ and $K_0'(x) = -K_1(x)$.) Because $\ell = 0$, designating $\phi_\ell = \frac{\pi}{2}$ in Equations (6.4) through (6.7) of the z field distribution will result in $E_z = 0$, so the TE mode can be derived. Meanwhile, designating $\phi_\ell = 0$ will result in $H_z = 0$, so the TM mode can be likewise derived.

(c) *EH mode:* When $\chi = +1$ and $\ell \geq 1$, the mode number is designated anew as

$$\nu = \ell + 1. \tag{6.24}$$

Then, in the same way as that of the HE mode, the following equation will be obtained:

$$\frac{J_{\nu-1}(\sqrt{1-b}V)}{J_\nu(\sqrt{1-b}V)} \cdot \frac{K_\nu(\sqrt{b}V)}{K_{\nu-1}(\sqrt{b}V)} = -\sqrt{\frac{b}{1-b}}. \tag{6.25}$$

6.1.4 LP Mode and Dispersion Curves

Now, if the eigenvalue equations (6.22), (6.23), and (6.25) are compared with each other, we can see that all these have the same pattern. That is, when the mode number ℓ of the azimuth angle direction is converted to

$$
\nu = \begin{cases} \ell - 1 & : \mathrm{HE}_{\ell,m} \text{ mode} \\ \ell + 1 & : \mathrm{TE}_{0,m}, \mathrm{TM}_{0,m}, \mathrm{EH}_{\ell,m} \text{ mode} \end{cases} \tag{6.26}
$$

the $\mathrm{HE}_{\nu+1,m}$ mode and the $\mathrm{EH}_{\nu-1,m}$, with the same value for ν, will have equal propagation constants. Here, m is the mode number of the radial r direction. Such a case, wherein different eigenvalues of eigenfunctions become equal, is called *degeneration*. Eigenfunctions under degeneration can form a new set of eigenfunctions by the linear combination (i.e., they can be orthogonalized). Therefore, a linearly polarized mode, which is polarized in one direction inside the cross-section, can be formed by the linear combination of these eigenmodes. Such a mode is called a *LP (linearly polarized) mode* [25] and is expressed as $\mathrm{LP}_{\nu,m}$ with mode numbers. The degenerate relationship between the LP mode and the HE, EH, TE, and TM modes are summarized in Table 6.2.

By solving the eigenvalue equation with mode $\mathrm{LP}_{\nu,m}$, the dispersion curve as shown in Figure 6.4 is obtained.

Note that the mode number (mode label) of optical fibers starts from 0 for ν such as $\nu = 0, 1, 2, \ldots$ whereas the mode number m in the radial r direction starts from 1. The mode number (mode order or mode label) for a slab waveguide starts from 0, and this corresponds to the number of zeros in the field profile (number of points where the field profile becomes zero, and the number of zeros

---

TABLE 6.2
LP Mode and Strict Mode Correspondence

LP Mode Approximation	Strict Mode	Cutoff V Value
$\mathrm{LP}_{0,m}$	$\mathrm{HE}_{1,m}$	$V_c = 0 \ (m = 1)$ $J_1(V_1) = 0,\ m - 1\text{th}^{*}\text{root} \ (m \geq 2)$
$\mathrm{LP}_{1,m}$	$\mathrm{HE}_{2,m}$	$V_c = 2.4048 \ (m = 1)$ $J_0(V_c) = 0,\ m\text{th root} \ (m \geq 2)$
	$\mathrm{TE}_{0,m}$ $\mathrm{TM}_{0,m}$	
$\mathrm{LP}_{\nu,m}$ $(\nu \geq 2)$	$\mathrm{HE}_{\nu+1,m}$ $\mathrm{EH}_{\nu-1,m}$	$J_{\nu-1}(V_c) = 0,\ m\text{th}^{*}\text{root}$

* When the root of $J_\nu(x) = 0 \ (\nu \geq 1)$ is counted in the ascending order, $x = 0$ is normally not included.

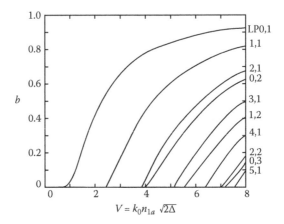

FIGURE 6.4 Dispersion curve of the LP mode of a step-index optical fiber.

in a fundamental mode is 0 since the field profile of fundamental mode does not become zero up to infinity). However, as the mode number of the radial r direction in optical fibers traditionally starts from 1, m corresponds to the number of peaks in the radial direction of the light intensity distribution. On the other hand, another mode number ν corresponds to half the number of zeros (nodes) in the azimuth angle θ direction.

Now, the reason why degeneration in the LP mode occurs is that the weakly guiding approximation $\Delta \ll 1$ was introduced to the eigenvalue equation (6.12), and to be exact, the propagation constants of $HE_{\nu+1,m}$ mode and the $EH_{\nu-1,m}$ mode only slightly differs from that of the LP mode. However, as this difference is small, if a linearly polarized light is beamed at the incident end of the optical fiber, the LP mode can propagate with the linearly polarized light remaining almost unchanged. Moreover, as the fundamental mode $LP_{0,1}$ is configured only from the $HE_{1,1}$ mode, there is theoretically no change in the polarization state with propagation. (Actually, the polarization state can change due to stress resulting from bending or various other reasons, and if the polarization state should be maintained, a polarization-maintaining fiber, which will be discussed in Section 6.1.6, can be used.)

Now, as can be seen from Figure 6.4, even in a pair of LP modes with different mode numbers, there are some combinations of LP modes with nearly close dispersion curves (e.g., $LP_{2,1}$ and $LP_{0,2}$). Although the detailed theoretical background is being omitted, such combination of LP modes is given by a mode number of

$$N = 2m + \nu - 2 \tag{6.27}$$

and is called the *principal mode number*. The rough method of comprehending such a mode number is used to analyze the transmission characteristics of multimode fiber with a large number of modes (from thousands to more than ten thousand). Using the principal mode number, the approximate equation of the dispersion relation of the step-index round fiber can be derived as in the following equation:

$$V = \frac{1}{\sqrt{1-b}}\left[\tan^{-1}\sqrt{\frac{b}{1-b}} + \frac{2N+1}{4}\pi \right].\qquad (6.28)$$

When this equation is compared with Equation (2.14) of Section 2.3, we can see that the approximate dispersion curve of the step-index round fiber can be obtained by shifting the dispersion curve of the slab waveguide by $\frac{\pi}{4}$ to the horizontal axis V direction. This approximation is derived when the V value is big and $b \simeq 1.0$. However, somehow the degree of approximation is good even in a single-mode region with a small V value and can be used in approximating the dispersion curve of single-mode fibers as will be discussed in Section 6.2.2.

6.1.5 FUNDAMENTAL MODE AND SINGLE-MODE FIBERS

As seen in Figure 6.4, the $LP_{0,1}$ mode (identical to the $HE_{1,1}$ mode) is the lowest order mode, that is, it is the fundamental mode. If v in Equation (6.22) is set as $v = 0$, it will become

$$\frac{J_1(\sqrt{1-b}V)}{J_0(\sqrt{1-b}V)} \cdot \frac{K_0(\sqrt{b}V)}{K_1(\sqrt{b}V)} = \sqrt{\frac{b}{1-b}}.\qquad (6.29)$$

The maximum solution among the solutions of this equation gives the normalized propagation constant b of the fundamental mode. Because κa and γa can be determined from V and b, these can be used to obtain the field distribution of the fundamental mode from Equations (6.4) through (6.7) and Equations (5.22) through (5.25). The electromagnetic field of the fundamental mode is shown in Figure 6.5.

Although the light intensity distribution is not well understood from this figure, because the light intensity distribution of the fundamental mode closely resembles the Gaussian-type intensity distribution as in the case of slab waveguides, it is convenient to express the size of the electromagnetic distribution by the spot size (in optical fibers, a length that is twice the length of the spot size is called the mode field diameter) of the Gaussian-type profile. The equation that is frequently used as an approximate equation of the spot size is the Petermann's formula [9]. (More precisely, there are two propositions in

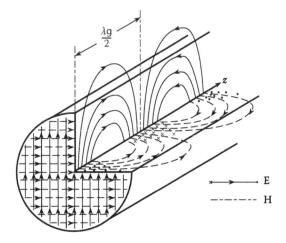

FIGURE 6.5   Field distribution of the fundamental mode $LP_{0,1}$ of the step-index fiber.

the Petermann's formula, and the frequently used one is the equation with the second proposition.) This expression is written as follows (refer to Appendix K for the derivation):

$$
w = \left[\frac{2 \displaystyle\int_0^\infty \psi^2(r) r\, dr}{\displaystyle\int_0^\infty \left(\frac{d\psi}{dr} \right)^2 r\, dr} \right]^{\frac{1}{2}}.
\tag{6.30}
$$

The approximate formula 5.57 for the spot size of slab waveguide is derived by converting the coordinates of Equation (6.30).

Here, as will be discussed in detail in Section 7.5, Equation (6.30) is derived from the stationary representation of the wave equation, so its validity as a formula for approximating the propagation constant from field profile is recognized. However, as this is not an equation wherein the field distribution $\psi(r)$ is approximated by the least squares fit to the Gaussian function, this approximation is said to be mathematically incomplete. It is necessary to use the expression focusing on the size of the field distribution in the approximation of the axial misalignment loss characteristics described in Section 8.1.2. Equations 7.51 and 7.52 are highly accurate equations that approximate the field distribution by the method of least squares using the Gaussian function.

Because the $LP_{1,1}$ mode is the next higher mode order after the fundamental mode, the single-mode condition of optical fibers is given by the cutoff V value of the $LP_{1,1}$ mode. As mentioned in Section 6.1.3, the $LP_{1,1}$ mode is expressed by the superposition of the $TE_{0,1}$, $TM_{0,1}$, and $HE_{2,1}$ modes,

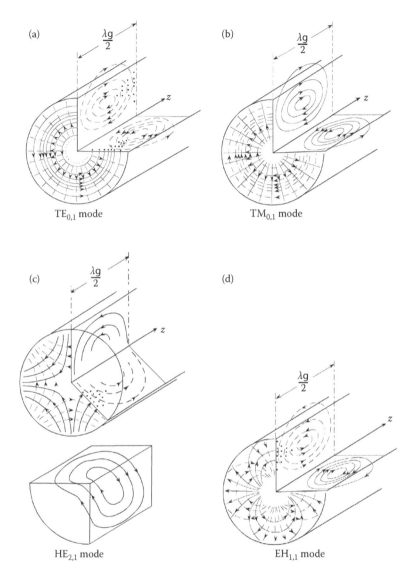

FIGURE 6.6 Field distribution of the higher-order modes of step-index fibers: (a) $TE_{0,1}$ mode, (b) $TM_{0,1}$ mode, (c) $HE_{2,1}$ mode, and (d) $EH_{1,1}$ mode.

and the electromagnetic field distributions of these three modes are shown in Figure 6.6. The $LP_{1,1}$ mode for this superimposed linear polarized light (there are four degenerate modes resulting from the method of superimposition, and two cases are shown here) is shown in Figure 6.7.

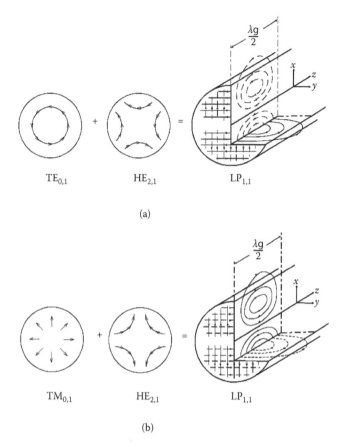

FIGURE 6.7 Field distribution of the $LP_{1,1}$ mode of step-index fibers: (a) Superposition of the $TE_{0,1}$ mode and $HE_{2,1}$ mode and (b) superposition of the $TM_{0,1}$ mode and $HE_{2,1}$ mode.

Now, if the solution of the cutoff V value of the $LP_{1,1}$ mode is determined by Equation (6.23) with $b = 0$, it will be

$$\text{The first solution for } J_0(V_c) = 0 \quad V_c = 2.405. \tag{6.31}$$

That is, the single-mode condition for the step-index round optical fiber is expressed as

$$V \leq 2.405. \tag{6.32}$$

Moreover, if the core has a refractive index distribution, and the refractive index distribution function is expressed in the same way as Equation (5.86) by

$$n^2(r) = \begin{cases} n_1^2[1 - 2\Delta f(r)] & (r \le a) \\ n_2^2 & (r > a), \end{cases} \tag{6.33}$$

where $f(r)$ is an arbitrary function, the cutoff V value of the $LP_{1,1}$ mode can be calculated by the following equation [16], [26]:

$$V_c = \left[\frac{1}{2} \int_0^1 \frac{1}{\eta} [1 - f(\eta)] \left\{ \int_0^\eta \xi^3 [1 - f(\xi)] d\xi \right\} d\eta \right]^{-\frac{1}{4}}. \tag{6.34}$$

Here, the variables have been normalized as $\eta = \frac{r}{a}$ and $\xi = \frac{r}{a}$, and the approximate margin for error of this equation is less than 2%.

Moreover, if the dispersion relation is approximated by Equation (6.28), the confinement factor of the step-index round fiber can be determined from the dispersion curve using Equation (5.66) for a slab waveguide.

6.1.6 POLARIZATION PROPERTIES OF SINGLE-MODE FIBERS AND POLARIZATION-MAINTAINING FIBER

The $LP_{0,1}$ mode ($HE_{1,1}$ mode), which is the fundamental mode of the single-mode fiber, is a linearly polarized mode. If the core of the optical fiber is a completely perfect circle and the propagation axis is straight without any external force or bending, the shape is completely symmetrical with respect to the central axis, and, therefore, the propagation constants of the x and the y polarizations are equal (i.e., they are degenerate). However, it is difficult to actually make an optical fiber core that is a completely perfect circle and then use it without adding any external force or bending. Thus, if this results in a deviation from the axial symmetry, a principal polarization axis (the principal birefringent axis mentioned in Section 1.5) will be formed in a direction inside the cross-section and the direction that is orthogonal to such direction. In particular, the phenomenon wherein birefringence occurs because of an external force is called the *photoelastic effect*. In general, the impact of the photoelastic effect because of an external force is bigger than that resulting from the elliptical deformation of the core. Let us set the principal polarization axis as the x- and y-axes as shown in Figure 6.8.

The two $LP_{0,1}$ modes that are polarized in the primary axis direction are called the $LP_{0,1}^{even}$ mode and the $LP_{0,1}^{odd}$ mode, respectively (even and odd mean the even function and the odd function, respectively; the former is because of the proportionality of E_r to $\cos \ell\theta$, whereas the latter is because of the proportionality of E_r to $\sin \ell\theta$). Thus, if the degeneracy is removed, the propagation constant of these polarization modes will become

$$\beta_x \ne \beta_y. \tag{6.35}$$

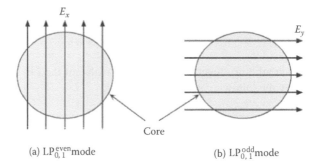

(a) $LP_{0,1}^{even}$ mode   (b) $LP_{0,1}^{odd}$ mode

FIGURE 6.8 Two principal polarization modes of a single-mode fiber with elliptical deformation or under an external force in one direction: (a) $LP_{0,1}^{even}$ mode and (b) $LP_{0,1}^{odd}$ mode.

This resembles the propagation of the plane wave in a birefringent medium as mentioned in Section 1.5, which means that the polarization state of the fundamental mode in the optical fiber will change according to the propagation distance. That is, if the linearly polarized light inclined from the main polarization axes (refer to the x- and y-axes here) of the $LP_{0,1}$ mode is incident on the optical fiber, the polarization state will change along with the propagation distance as shown in Figure 6.9. This polarization state will be periodic if the birefringence of the optical fiber is uniform in the direction of propagation, with the period L_B given by

$$L_B = \frac{2\pi}{|\beta_x - \beta_y|}. \tag{6.36}$$

This L_B is called the beat length.

Because the strength and the direction of the bending and external force change along the propagation direction in an actual optical fiber, the direction of the principal birefringent axis and also the birefringence are not uniform in the propagation direction. In addition, the birefringence itself changes with time because of fluctuations in temperature. Consequently, the polarization state of the output light from the single-mode optical fiber after a long-range propagation changes slowly with time, so the direction of the primary axis is also not constant. Such instability of the polarization state is irrelevant in optical communications using the On–Off keying modulation in which only the power of the output light from optical fibers is detected. However, in the heterodyne detection (refer to Exercise 1 in Section 4.1), or in an interferometer, which is configured using an optical fiber, if the polarization states of the signal light (output light of the optical fiber) and the reference light are not matched,

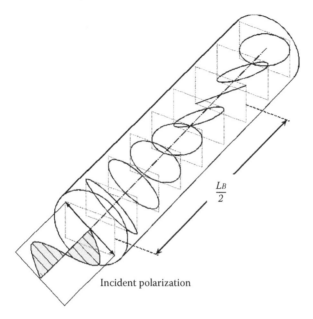

FIGURE 6.9 Change in the polarization state when the linearly polarized light that is inclined from the principal polarization axis is incident on the optical fiber.

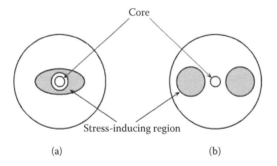

(a)   (b)

FIGURE 6.10 Polarization-maintaining fibers: (a) elliptical intermediate cladding type. (b) PANDA fiber.

interference will not occur. To solve this problem, *polarization-maintaining fibers* have been developed to keep the linear polarization of incident light to the output end of the optical fiber. Figure 6.10 shows the cross-sectional configurations of such fibers. The principle of these polarization-maintaining fibers involves the deliberate retention of internal stress to the fiber to create a principal polarization axis, with the birefringence being made larger than that induced by an external force, so that any external influence is eliminated. Let us look into the PANDA fiber of Figure 6.10b as a means of creating an internal residual stress. Here, at the state of the preform rod before drawing* the fiber, a pair of holes is made using a glass lathe in the cladding on both sides of the core, and then inserted into the hole is a SiO_2 glass rod added with B_2O_3 and other components. This SiO_2 glass added with B_2O_3 and other components has a lower melting point and a different linear expansion coefficient compared to pure SiO_2 glass, so during the rapid cooling stage in a subsequent fiber drawing process, an internal residual stress is applied from one direction in the core because of the difference in the linear expansion coefficient. As a result, a large birefringence that is unaffected by external force occurs. To correctly use the polarization-maintaining fiber, it is necessary to correctly find out the direction of the principal polarization axis at the incident end and to adjust the direction of linearly polarized light to the principal polarization axis of the polarization-maintaining fiber.

6.1.7   DISTRIBUTED INDEX SINGLE-MODE FIBERS

In the derivation of the guided modes from Section 6.1.1 to Section 6.1.5, we assumed that the refractive index profile in the core is uniform at n_1. However, in recent years, fibers with refractive index profiles in the core like that in Equation (6.33) are sometimes used even for single-mode fibers. The classification of the modes of the distributed index fibers is the same as that described so far; however, as the refractive index is not uniform inside the core, the electromagnetic field distribution inside the core cannot be expressed by the Bessel function. Consequently, the dispersion curve will be different from that shown in Figure 6.4. It is necessary to use the numerical analysis method to accurately analyze the electromagnetic field and propagation constant of the distributed index fibers, and the *matrix method* [27] (the basic principle is the same as the matrix method described in Section 5.2.1 (3)), *finite element method* [28], *variational method* [29] (Rayleigh–Ritz method), and the *direct integration method* [30] have been developed.

---

* The manufacturing process of silica-based optical fibers involves the production of a glass rod with a thickness of several centimeters that has a core of high-refractive index (added with GeO_2 with a refractive index of about 1.60) that is covered by a material of slightly lower refractive index (e.g., pure SiO_2). This glass rod is called the preform rod. The tip of this preform rod is gradually lowered from the top downwards into an electric furnace with a temperature of about 2,000°C to stretch it into thin fiber. This is called the drawing process. During the drawing process, the optical fiber that comes down from the electric furnace is wound on drums with a diameter of about 30 cm. The winding speed is controlled and the diameter of the optical fiber is adjusted to 125 μm.

6.1.8 Distributed Index Multimode Fibers

It was shown in Section 5.2.1 that if the eigenmode of the slab waveguide with a parabolic refractive index profile is far from the cutoff (b is not near 0), it can be expressed as a Hermite–Gaussian function. If an optical fiber has a refractive index distribution that is proportional to the square of the radius r, the index distribution has parabolic index profiles in both the x- and the y-directions due to the relation $x^2 + y^2 = r^2$, and the separation of variables for the wave equation will be possible. Consquently, the x- and the y-directions will have eigenmode functions that are each expressed as Hermite–Gaussian functions. However, a more common method of expressing the refractive index profile is beneficial for optimal design. Thus, the refractive index profile in the core is represented as

$$n^2(r) = \begin{cases} n_1^2 \left[1 - 2\Delta \left(\dfrac{r}{a} \right)^\alpha \right] & (r \le a) \\ n_2^2 & (r > a), \end{cases} \tag{6.37}$$

where the profile is expressed as proportional to r^α. A fiber with this kind of refractive index distribution is called an α-*power profile fiber*. The α-power index profile inside the core for various values of α is shown in Figure 6.11. If $\alpha = 2$, it corresponds to a parabolic index fiber, whereas if $\alpha = \infty$, it corresponds to a step-index profile. Let us derive the dispersion relation equation

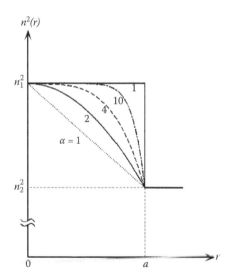

FIGURE 6.11   The refractive index profile of the α-power profile multimode fiber.

of this α-power profile fiber, when the mode number is very large using the approximation method that is called the WKB method [37].

First, assume an LP mode and consider a guided mode that is polarized in the x-direction. For the small refractive index difference to which weakly guiding approximation can be applied, the electric field and the magnetic field exist almost in the transverse direction, so we will take into consideration the x component E_x of the electric field. This kind of approximation, which takes into account only the electromagnetic field in the transverse direction, is called the transverse electromagnetic wave (TEM wave) approximation. The wave equation in cylindrical coordinates is as follows:

$$\frac{\partial^2 E_x}{\partial r^2} + \frac{1}{r}\frac{\partial E_x}{\partial r} + \frac{1}{r^2}\frac{\partial^2 E_x}{\partial \theta^2} + (k_0^2 n^2(r) - \beta^2)E_x = 0. \qquad (6.38)$$

Here, when the separation of variables is used and $E_x(r,\theta)$ is designated as

$$E_x(r,\theta) = R(r)\Theta(\theta), \qquad (6.39)$$

the solution (the θ direction mode number of the LP mode is ν) for the azimuth angle θ direction can be easily written as

$$\Theta(\theta) = \cos(\nu\theta + \varphi_\nu), \qquad (6.40)$$

(φ_ν is a constant.) On the other hand, the equation for variable $R(r)$ of the radial r direction is the following equation:

$$\frac{d^2 R(r)}{dr^2} + \frac{1}{r}\frac{d R(r)}{dr} + \left(k_0^2 n^2(r) - \beta^2 - \frac{\nu^2}{r^2}\right) R(r) = 0. \qquad (6.41)$$

How to solve this equation for an arbitrary refractive index profile $n(r)$ is a problem. To make the differential equation easier, further change of variables is performed resulting in

$$R(r) = \frac{1}{\sqrt{r}}F(r), \qquad (6.42)$$

and when Equation (6.41) is transformed into a differential equation for $F(r)$, the following equation is obtained:

$$\frac{d^2 F(r)}{dr^2} + U^2(r)F(r) = 0. \qquad (6.43)$$

Here, $U^2(r)$ is given by the following equation (the use of $U^2(r)$ and not $U(r)$ does not mean that it is a positive real number, so care must be taken; actually, a negative number is also possible for $U^2(r)$):

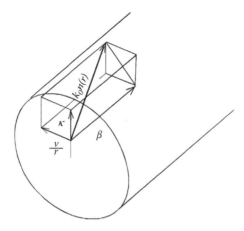

FIGURE 6.12   The wave vector relationship for each of the directions in the α-power profile multimode fiber.

$$U^2(r) = k_0^2 n^2(r) - \beta^2 - \frac{v^2 - \frac{1}{4}}{r^2}. \qquad (6.44)$$

If $U^2(r)$ is a constant that is independent of r, the solution for this differential equation is the vibration solution expressed by $\cos(Ur)$ or $\sin(Ur)$ for $U^2 > 0$ and is the attenuation solution expressed by $e^{|U|r}$ or $e^{-|U|r}$ for $U^2 < 0$. Similarly, if $U^2(r)$ is dependent on r, then it would result in the following solutions:

$$U^2(r) > 0 : \quad \text{vibration solution,}$$
$$U^2(r) < 0 : \quad \text{attenuation solution.}$$

Here, let us consider the physical meaning of $U^2(r)$. As we are taking into consideration the case wherein the mode number is very large, the term $\frac{1}{4}$ in Equation (6.44) can be ignored. As shown in Figure 6.12, there is a wave vector with magnitude $k_0 n(r)$ in the direction perpendicular to the wave front of the plane wave configuring the mode. As its z component is β, and the θ component is $\frac{v}{r}$, then $U(r)$ is the r component. Then, at the range wherein $k_0^2 n^2(r) - \beta^2$ is bigger than $\frac{v^2}{r^2}$, $U^2(r)$ will become positive and have a vibration solution, so the region wherein $U^2(r) > 0$ for any arbitrary β is graphically represented as shown in Figure 6.13.

In the region between r_1 and r_2 of this figure, $U^2(r) > 0$ and the solutions are vibration solutions. For the solution to be attenuated outside this region, it is necessary to form a standing wave in the region of the vibration solution (refer to Appendix J for the method to determine the solution of $F(r)$ using

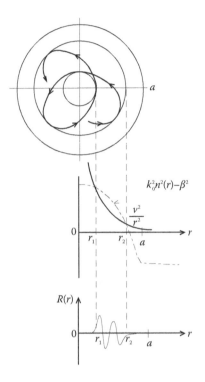

FIGURE 6.13 Region wherein the radial direction propagation constant $U^2(r)$ is positive and its solution.

the WKB method). If the number of standing waves included in this region is set to m,

$$\int_{r_1}^{r_2} U(r)\,dr = \int_{r_1}^{r_2} \left[k_0^2 n^2(r) - \beta^2 - \frac{\nu^2}{r^2} \right]^{\frac{1}{2}} dr = m\pi \qquad (6.45)$$

should hold true. This is the generalized eigenvalue equation of the multimode fiber with any arbitrary refractive index profile, with ν as the mode number of the azimuth angle θ direction, and m as the mode number of the radial r direction.

Now, the mode number pair (ν, m) that satisfied Equation (6.45) is expressed as a point in the two-dimensional plane shown in Figure 6.14, where ν is the horizontal axis and m is the vertical axis. Here, the possible number of m for a given number of ν will become larger when $U(r)$ takes its biggest value in the range $r_1 < r < r_2$. Consequently, the maximum number of m is given when $\beta = k_0 n_2$, that is, it is the cutoff value. On the other hand, the possible maximum value of ν is given when $m = 0$ in Equation (6.45).

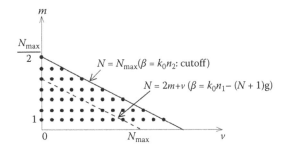

FIGURE 6.14 Number of possible combinations of the mode number (v, m) and the maximum mode number.

From these considerations, when v is given, the maximum value of m, lies on the solid line in Figure 6.14. That is, as the (v, m) number pair in the triangle enclosed by this solid line and the vertical and the horizontal axes corresponds to the total number of modes, the area of the triangle has to be calculated to determine the total number of modes. Moreover, we need to further consider here that the x polarization was taken into account in Equation (6.38); however, exactly the same number of modes also exists for the y polarization. In addition, even if Equation (6.40) has the same mode number v, different modes having different electromagnetic field distributions (rotated $90°$) exist corresponding to $\varphi_v = 0$ and $\varphi_v = \frac{\pi}{2}$ (i.e., $\cos(v\theta)$ and $\sin(v\theta)$). That is, the mode number is doubly degenerate in the polarization direction, with the mode number v in the θ direction also doubly degenerate, totaling to a four-fold degeneration. Thus, four times the area of the triangle is the total number of modes. From this argument, the total number of modes M is given by the following equation:

$$M = 4 \int_0^{v_{\max}} m(v)|_{\beta=k_0 n_2} dv$$

$$= \frac{4}{\pi} \int_0^{v_{\max}} \left[\int_{r_1}^{r_2} \left\{ k_0^2 (n^2(r) - n_2^2) - \frac{v^2}{r^2} \right\}^{\frac{1}{2}} dr \right] dv. \qquad (6.46)$$

Here, the integration region of the double integral is shown in Figure 6.15, and if the order of the integration is switched, it will become the following equation:

$$M = \frac{4}{\pi} \int_0^a \int_0^{k_0 r \sqrt{n^2(r) - n_2^2}} \left[k_0^2 (n^2(r) - n_2^2) - \frac{v^2}{r^2} \right]^{\frac{1}{2}} dv dr$$

$$= k_0^2 \int_0^a (n^2(r) - n_2^2) r \, dr. \qquad (6.47)$$

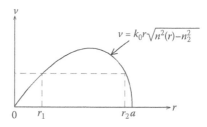

FIGURE 6.15 The integration range of the double integral when the total number of modes is calculated.

That is, the total number of modes is equal to the number resulting from multiplying by k_0^2; the volume enclosed by the curved surface in three-dimension representing the refractive index distribution $n^2(r) - n_2^2$ (to be exact, the coordinate system is r, θ) is drawn in three-dimensions.

In the derivation process of Equation (6.47) for the total number of modes, the lower limit of β was assumed to be $\beta = k_0 n_2$, and the total number of modes was calculated when the propagation constant is within the $k_0 n_2 \leq \beta < k_0 n_1$ range. Then, designating a value that is bigger than $k_0 n_2$ as β', how many number of modes $\mu(\beta')$ are there within the range of $\beta' \leq \beta < k_0 n_1$. First, refer to the ν and m pair (ν, m) that satisfies Equation (6.45) when $\beta = \beta'$, which is on the dashed line in Figure 6.14. To solve this problem, just determine the area enclosed by this dashed line, the vertical axis, and the horizontal axis. The result is given by the following equation:

$$\mu(\beta') = \int_0^{r_0} (k_0^2 n^2(r) - \beta'^2) r \, dr. \tag{6.48}$$

Here, r_0 is the solution for r of the following equation:

$$k_0 r \sqrt{n^2(r) - \left(\frac{\beta'}{k_0}\right)^2} = 0. \tag{6.49}$$

This mode number is the number of modes that has the propagation constant within the $\beta' \leq \beta < k_0 n_1$ range. That is, when the mode is numbered in the descending order of propagation constant from the fundamental mode as $1, 2, 3$, the μth mode has the propagation constant β'. Consequently, if we think of μ as a new mode number,[*] then we can consider that the propagation constant of the mode number μ is given by the solution of Equation (6.48).

---

[*] However, this is not the principal mode number. This is because, for the principal mode number, the mode number pairs (ν, m) that give almost the same propagation constant β are all counted as one principal mode number. But here, different (ν, m) pairs are also counted.

Now, finally, let us substitute the specific α-power index profile Equation (6.37) through Equations (6.47) and (6.48). The results are as follows:

$$M = \frac{1}{2}\frac{\alpha}{\alpha+2}V^2, \tag{6.50}$$

$$\mu(\beta') = \frac{1}{2}\frac{\alpha}{\alpha+2}V^2\left[\frac{n_1^2 - \left(\frac{\beta'}{k_0}\right)^2}{n_1^2 - n_2^2}\right]. \tag{6.51}$$

Moreover, when conversely solving for β' in Equation (6.51), the following equation will be obtained:

$$\beta' = k_0n_1\left[1 - 2\Delta\left(\frac{\mu}{M}\right)^{\frac{\alpha}{\alpha+2}}\right]^{\frac{1}{2}}, \tag{6.52}$$

$$b = \frac{\left(\frac{\beta'}{k_0}\right)^2 - n_2^2}{n_1^2 - n_2^2} = 1 - \left(\frac{\mu}{M}\right)^{\frac{\alpha}{\alpha+2}}. \tag{6.53}$$

These equations express directly the propagation constant and the normalized propagation constants, when given the mode number μ.

6.2   SIGNAL PROPAGATION IN OPTICAL FIBER

6.2.1   GROUP DELAY AND DISPERSION

The concept of group velocity has already been discussed in Section 2.6.1. In this section, let us determine the changes in the waveform associated with transmission for general pulse waveforms. In this section, an exact calculation is done using the Fourier transform of waveforms to derive the pulse width broadening. For readers who only want to know the concept of why the pulse width broadens, it is better to begin reading from Section 6.2.2.

As shown in Figure 6.16, when a pulse of arbitrary waveform is incident on a single-mode fiber, the amplitude of the waveform that is emitted from the exit end will be generally attenuated because of loss inside the fiber. Moreover, the waveform will be distorted (generally, it broadens).

When the angular frequency of the carrier wave is set to ω_0, the waveform of the electric field amplitude at the incident end $z = 0$ is expressed as

$$E(x, y, 0, ; t) = A(x, y, 0; t)e^{j\omega_0t}. \tag{6.54}$$

Here, the field distribution of the fundamental mode of the single-mode optical fiber is the same (to be precise, they are similar figures as the amplitude will be attenuated because of loss) at the incident end and the exit end, if the optical fiber is uniform along the length, so the dependency on the coordinates

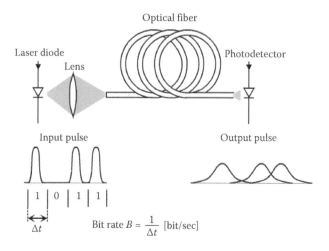

FIGURE 6.16 Incident waveform and exit waveform of the single-mode optical fiber.

x, y is ignored, and the electric field is expressed as $E(z; t)$, whereas the amplitude function is represented as $A(z; t)$. When the Fourier transform of the amplitude function $A(0; t)$ at $z = 0$ is expressed as $A_F(0; \omega)$, the Fourier transform $E_F(0; \omega)$ of Equation (6.54) is expressed by

$$E_F(0; \omega) = A_F(0; \omega - \omega_0), \qquad (6.55)$$

using Equation (A.8) of Appendix A.

Next, when the pulse waveform propagates by a distance z, a phase change $\exp[-j\beta z]$ that is due to propagation will be applied, and so the Fourier spectrum at the exit end will be expressed as

$$A_F(z; \omega) = A_F(0; \omega - \omega_0) \exp[-j\beta z]. \qquad (6.56)$$

(Because the waveform shape is taken into consideration here, the attenuation of the amplitude due to loss is ignored.) That is, $\exp[-j\beta z]$ is the transfer function of the optical fiber. Consequently, the time waveform at the exit end is expressed by the inverse Fourier transform of Equation (6.56), so $E(z; t)$ can be written as

$$E(z; t) = \frac{1}{2\pi} \int_{-\infty}^{\infty} A_F(0; \omega - \omega_0) \exp[-j\beta z] e^{j\omega t} d\omega. \qquad (6.57)$$

Moreover, transforming the variable as $\omega - \omega_0 = u$

$$E(z; t) = \frac{1}{2\pi} e^{j\omega_0 t} \int_{-\infty}^{\infty} A_F(0; u) \exp[j(ut - \beta(u)z)] du \qquad (6.58)$$

is obtained.

Here, as the propagation constant β in the optical fiber is dependent on the angular frequency ω, it is written as $\beta(\omega)$. However, the frequency of the modulated signal is much smaller compared to the angular frequency ω_0 of the light (carrier), and so $\beta(\omega)$ at the vicinity of ω_0 is expressed in terms of the Taylor series expansion as follows:

$$\beta(\omega) = \beta(\omega_0) + \left.\frac{d\beta}{d\omega}\right|_{\omega_0} (\omega - \omega_0) + \frac{1}{2}\left.\frac{d^2\beta}{d\omega^2}\right|_{\omega_0} (\omega - \omega_0)^2 \cdots . \quad (6.59)$$

When Equation (6.59) is substituted to Equation (6.58), and then rearranged (the $\omega - \omega_0 = u$ is also taken into consideration)

$$E(z; t) = \frac{1}{2\pi} e^{j[\omega_0 t - \beta(\omega_0)z]} \int_{-\infty}^{\infty} A_F(0; u) \exp\left[ju\left(t - \frac{z}{v_g}\right) - j\frac{\beta''}{2} z u^2 \right] du \quad (6.60)$$

is obtained. Here, the following are defined as

$$\frac{1}{v_g} = \left.\frac{d\beta}{d\omega}\right|_{\omega_0}, \quad (6.61)$$

$$\beta'' = \left.\frac{d^2\beta}{d\omega^2}\right|_{\omega_0} . \quad (6.62)$$

Here, v_g is the group velocity that was already discussed in Section 2.6.1, and β'' is a quantity that is related to the *dispersion**, which will now be explained.

Because $\exp[-j\beta z]$ is the transfer function $H(\omega)$ of the optical fiber, from Equations (6.59), (6.61), and (6.62), $H(\omega)$ can be written as

$$H(\omega) = \exp\left[-j\left\{ \beta(\omega_0)z + \frac{z}{v_g}(\omega - \omega_0) + \frac{1}{2}\beta'' z(\omega - \omega_0)^2 \right\} \right]. \quad (6.63)$$

To be able to understand the meaning of Equation (6.60), first let us set $\beta'' = 0$. Then, Equation (6.60) will become

$$E(z; t) = \frac{1}{2\pi} e^{j[\omega_0 t - \beta(\omega_0)z]} \int_{-\infty}^{\infty} A_F(0; u) \exp\left[ju\left(t - \frac{z}{v_g}\right) \right] du$$

$$= A\left(0; t - \frac{z}{v_g}\right) e^{j[\omega_0 t - \beta(\omega_0)z]}. \quad (6.64)$$

So, the pulse waveform (if the light is received as a light intensity, it is proportional to $|E(z; t)|^2$) that progresses by z at time t is equal to the waveform at

---

* Dispersion generally refers to a physical property that is dependent on the wavelength. Therefore, it should be accurately called the *group velocity dispersion* (abbreviated as GVD).

$z = 0$ at $t - \frac{z}{v_g}$. In other words, the waveform, keeping its shape as a similar figure, will progress at a velocity of v_g, so Equation (6.61) represents the group velocity. Then, the time it takes for the propagation of the pulse

$$\tau \equiv \frac{z}{v_g} = z\frac{d\beta}{d\omega} \qquad (6.65)$$

is called the *group delay*.

On the other hand, as further transformation of the equation would be difficult for a generalized waveform to look into the effect of dispersion, let us consider the Gaussian pulse waveform expressed by the following equation as the amplitude waveform $A(0; t)$ of Equation (6.54)*:

$$A(0; t) = f(z = 0; t) = A_0 \exp\left[-\left(\frac{t}{\tau_0}\right)^2\right]. \qquad (6.66)$$

Based on Equation (A.20) of Appendix A, the Fourier transform $F(\omega)$ of this Gaussian pulse waveform is given by

$$F(\omega) = A_0\sqrt{\pi}\,\tau_0 \exp\left[-\frac{\tau_0^2\omega^2}{4}\right]. \qquad (6.67)$$

Here, let us think about the width of the pulse waveform. In a Gaussian beam, the half-width at which the electric field amplitude becomes $1/e$ was defined as the spot size of the beam, but in a pulse waveform, the waveform width is usually expressed as the FWHM (discussed in Section 4.2 as the spectral width of a resonator). Therefore, when the FWHM of the pulse intensity $|f(t)|^2$ is set as T_0, the following relationship is obtained:

$$T_0 = \tau_0\sqrt{2\ln 2}. \qquad (6.68)$$

On the other hand, when the FWHM of the angular frequency spectral intensity $|F(\omega)|^2$ is set as $\Delta\omega$, then from Equation (6.67), $\Delta\omega$ can be expressed as

$$\Delta\omega = \frac{2\sqrt{2\ln 2}}{\tau_0} = \frac{4\ln 2}{T_0}. \qquad (6.69)$$

Consequently, the FWHM T_0 of the time waveform and the FWHM $\Delta\nu(= \frac{\Delta\omega}{2\pi})$ of the spectral intensity will have the following relationship:

$$T_0\Delta\nu = \frac{4\ln 2}{2\pi} \simeq 0.441. \qquad (6.70)$$

---

* Because the Gaussian function does not become zero, when t is not $t = \pm\infty$, it possesses a value from the infinite past, and if infinite time has not elapsed, it will remain a pulse that would not become zero. Therefore, in reality, such a pulse waveform does not exist. However, we adopt this functional form here for ease of calculation.

This relationship holds true for the Gaussian waveform that did not give rise to the chirping (see Equation (6.78)) due to the dispersion, and the waveform that satisfies such a relationship is called the *transform limited pulse*.

Now, substituting Equation (6.67) with $F(\omega - \omega_0) = F(u)$ to Equation (6.60), and using Equation (A.21) of Appendix A, the following equation will determine the inverse Fourier transform of the Gaussian function (Equation (A.20) also holds true for a complex number of α):

$$E(z; t) = A_0 \frac{\tau_0}{\sqrt{\tau_0^2 + j2\beta''z}} \exp\left[-\frac{\left(t - \frac{z}{v_g}\right)^2}{\tau_0^2 + j2\beta''z}\right] e^{j[\omega_0 t - \beta(\omega_0)z]}, \quad (6.71)$$

$$= A\left(z; t - \frac{z}{v_g}\right) e^{j[\omega_0 t - \beta(\omega_0)z]}. \quad (6.72)$$

When the absolute value and the phase of $A(z; t) = |A(z; t)|e^{j\Phi(z;t)}$ in Equation (6.72) is determined (the center of the pulse $t = \frac{z}{v_z}$ at the exit end is set as $t = 0$ at the new time origin), they are expressed by the following equations:

$$|A(z; t)| = A_0 \frac{\tau_0}{[\tau_0^4 + (2\beta''z)^2]^{1/4}} \exp\left[-\frac{t^2}{\tau_0^2 + \frac{(2\beta''z)^2}{\tau_0^2}}\right], \quad (6.73)$$

$$\Phi(z; t) = \frac{2\beta''z}{\tau_0^4 + (2\beta''z)^2} t^2 - \frac{1}{2}\tan^{-1}\left[\frac{2\beta''z}{\tau_0^2}\right]. \quad (6.74)$$

From Equations (6.73) and (6.74), we can see how dispersion affects the deformation of the pulse waveform. That is, from Equation (6.73), the width of the Gaussian pulse broadens with increasing z along with the propagation. When the pulse broadening is expressed by the half-width $T_{1/e}$, wherein the pulse amplitude is $1/e$, it will become

$$T_{1/e} = \left[\tau_0^2 + \frac{(2\beta''z)^2}{\tau_0^2}\right]^{\frac{1}{2}} = \tau_0\left[1 + \left(\frac{2\beta''z}{\tau_0^2}\right)^2\right]^{\frac{1}{2}}, \quad (6.75)$$

and we can see that the spot size of the Gaussian beam being represented in Equation (7.10) of Section 7.1 has the same broadening manner. Moreover, the width of the pulse waveform is often expressed as a FWHM, so when the FWHM T_{FWHM} is determined from Equation (6.73),

$$T_{\text{FWHM}} = T_{1/e}\sqrt{2\ln 2} = T_0\left[1 + \left(4(\ln 2)\frac{\beta''z}{T_0^2}\right)^2\right]^{\frac{1}{2}} \quad (6.76)$$

is obtained. Here, T_0 is the FWHM of the incident pulse intensity waveform given by Equation (6.68). When the distance at which the pulse width broadens by $\sqrt{2}$ times, the T_0 is set as L_D, the L_D can be expressed as

$$L_D = \frac{T_0^2}{4(\ln 2)|\beta''|}. \tag{6.77}$$

The pulse propagation inside the dispersive medium can be expressed by the distance normalized by this L_D.

On the other hand, for the phase in Equation (6.74), as can be seen when Equation (6.74) is substituted to Equation (6.72), the total phase $\Phi_T(t)$ of the amplitude $E(z; t)$ is $\Phi(z; t) + \omega_0 t - \beta(\omega_0)z$, so the instantaneous angular frequency $\omega(t)$ obtained from the time differentiation of this phase is

$$\omega(t) = \frac{\partial \Phi_T(t)}{\partial t} = \omega_0 + \frac{4\beta''z}{\tau_0^4 + (2\beta''z)^2}t. \tag{6.78}$$

This equation means that the instantaneous frequency gradually changes even in the pulse waveform. The phenomenon wherein the carrier frequency (i.e., the optical frequency) inside the pulse waveform changes is called *chirping*. As shown in Figure 6.17, in the pulse waveform of which the width broadens due to dispersion as compared to that of an incident waveform, the chirping with a different instantaneous frequency before and after the center of the waveform occurs.[*] Now, let us consider once again the broadening of the FWHM of the pulse waveform.

If the incident waveform is a transform-limited pulse (i.e., a pulse where chirping does not occur), the FWHM of an optical pulse wave that is determined by Equation (6.76) broadens in a manner close to a z-squared curve (parabola) when the propagation distance is short, and the longer the propagation distance the more its width, roughly in proportion to z. This dependence of pulse width on the propagation distance is shown by the solid lines in Figure 6.18. Incidentally, if the incident pulse is made to undergo chirping from the beginning, and set up to

$$A(z = 0; t) = A_1 \exp\left[-(1 + jC_p)\frac{t^2}{\tau_0^2}\right] \tag{6.79}$$

and made to propagate, the full width at half minimum (FWHM) of the pulse will become

$$T_{\text{FWHM}} = T_0 \left[\left(1 - 4(\ln 2)\frac{\beta''C_p z}{T_0^2}\right)^2 + \left(4(\ln 2)\frac{\beta''z}{T_0^2}\right)^2\right]^{\frac{1}{2}}, \tag{6.80}$$

---

[*] $\beta'' = -\frac{\lambda^2}{2\pi c}\sigma_T$, so the signs for β'' and σ_T are reversed and require care.

(a)

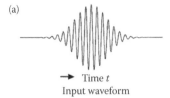

Time t
Input waveform

(b)

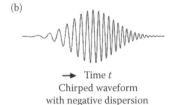

Time t
Chirped waveform
with negative dispersion

(c)

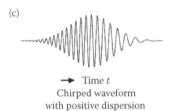

Time t
Chirped waveform
with positive dispersion

FIGURE 6.17 Pulse waveform distortion due to dispersion in a single-mode optical fiber: (a) incident waveform, (b) chirped waveform due to negative dispersion, and (c) chirped waveform due to positive dispersion.

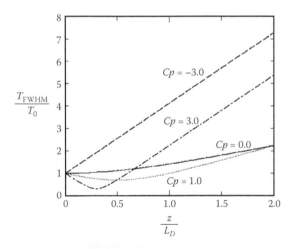

FIGURE 6.18 Change in full width at half minimum of optical pulse due to dispersion in a single-mode optical fiber.

and if $C_p > 0$, the pulse width will initially narrow down along with the propagation, and with a minimum value at $z = \frac{T_0^2 C_p}{4(\ln 2)\beta''(1+C_p^2)}$, T_{min} will be

$$T_{min} = \frac{1}{1+C_p^2} T_0, \qquad (6.81)$$

and from this minimum, it will return to the original FWHM T_0 at a distance of $z = 2\frac{T_0^2 C_p}{4(\ln 2)\beta''(1+C_p^2)}$, after which the broadening of the pulse width will begin. This phenomenon is illustrated by the dotted line in Figure 6.18. By using this phenomenon, in some cases, the transmission distance can be extended even with the same transmission capacity (modulated bandwidth) (but within the limit optical fiber loss) and such a technique is called the prechirping technique. However, if the absolute value of the prechirping is large, the pulse width will rapidly broaden along with the transmission distance, as indicated by the dashed line of Figure 6.18.

Next, let us consider the region wherein the pulse width broadens in proportion to the propagation distance. If $4\ln 2\frac{\beta'' z}{T_0^2} \gg 1$ is satisfied in Equation (6.76), using the relationship in Equation (6.69), the pulse width can be expressed as

$$T_{FWHM} \simeq \frac{4(\ln 2)\beta'' z}{T_0} = \beta'' z \Delta\omega = 2\pi\beta'' z \Delta\nu \qquad (6.82)$$

and is proportional to the product of the dispersion β'', the propagation distance z, and the full width at half minimum $\Delta\omega$ ($|F(\omega)|^2$ of FWHM) of spectral intensity. Because this broadening of the pulse width can be interpreted as the dispersion of the group delay τ, it is rewritten as $\delta\tau$. Then, replacing β'' by $\frac{d^2\beta}{d\omega^2}\Big|_{\omega_0}$ using Equation (6.62) and introducing the ratio of the spectral angular frequency width and the carrier angular frequency (or the ratio of the spectral wavelength broadening due to modulation and the wavelength of the light source), Equation (6.82) can be rewritten as follows:

$$\delta\tau = L\frac{\Delta\omega}{\omega_0}\left(\omega\frac{d^2\beta}{d\omega^2}\right)_{\omega=\omega_0} = -\frac{L}{c}\frac{\delta\lambda}{\lambda_0}\left(k\frac{d^2\beta}{dk^2}\right)_{\lambda=\lambda_0}. \qquad (6.83)$$

Here, the propagation distance is expressed as $z = L$, λ_0 is the wavelength of the light source in vacuum, $\delta\lambda$ is the spectral wavelength broadening of the modulated signal (or the spectral width of the light source, if the wavelength broadening of the source is wide), and $k = \frac{2\pi}{\lambda}$. In the case of optical fibers, this group delay dispersion can be further classified into two parts, which is dependent on the material and the structure, and so in Section 6.2.2, we will consider the specific group delay dispersion of optical fibers.

6.2.2 DISPERSION IN SINGLE-MODE OPTICAL FIBERS

If the group delay of Equation (6.65) is expressed in terms of the Taylor series expansion, the group delay dispersion in Equation (6.83) can be written as

$$\tau(\omega) = L\frac{d\beta}{d\omega} = L\left[\left.\frac{d\beta}{d\omega}\right|_{\omega_0} + \left.\frac{d^2\beta}{d\omega^2}\right|_{\omega_0}(\omega - \omega_0) + \cdots\right], \tag{6.84}$$

and so we can still see in the same way that the second term inside the [ ] corresponds to the pulse width broadening $\delta\tau$. Here, let us determine the $k\frac{d^2\beta}{dk^2}$ that appears in Equation (6.83). β is transformed using the normalized propagation constant b, and k^* can be transformed using the normalized V parameter, so let us first express $\frac{d\beta}{dk}$ as b and V. First, using β and k, the group delay will be written as

$$\tau = \frac{L}{v_g} = L\frac{d\beta}{d\omega} = \frac{L}{c}\frac{d\beta}{dk}. \tag{6.85}$$

Here, L is the propagation distance and c is the speed of light. When β is expressed as

$$\beta = k[n_1^2 b + n_2^2(1-b)]^{\frac{1}{2}} \tag{6.86}$$

based on Equation (2.10), τ is obtained as follows:

$$\tau = \frac{L}{c}\frac{d\beta}{dk} = \frac{L}{c}\frac{\left[n_2 N_2 + (n_1 N_1 - n_2 N_2)\left(b + \frac{1}{2}V\frac{db}{dV}\right)\right]}{\left[n_2^2 + (n_1^2 - n_2^2)b\right]^{\frac{1}{2}}}. \tag{6.87}$$

Here,

$$N_i = n_i + k\left.\frac{dn_i}{dk}\right|_{\omega=\omega_0} = n_i - \lambda\left.\frac{dn_i}{d\lambda}\right|_{\lambda=\lambda_0} \quad (i = 1 \text{ or } 2) \tag{6.88}$$

is called the *group index*, which is an index that includes the wavelength dependency of the refractive index (this is the same as the effective index

---

* In Section 6.2.1, the wave number in vacuum was written as k_0, whereas the wave number in a medium of refractive index n was written as $k = k_0 n$. However, in this section, to distinguish between k_0, which is the constant, and k, which is the variable corresponding to ω_0, we write k as $\frac{2\pi}{\lambda_0}$.

that was defined in Equation (4.20) in Section 4.2). Moreover, if the weakly guiding approximation ($\Delta \ll 1$) can be applied to Equation (6.87), τ can also be written as

$$\tau \simeq \frac{L}{c} \left[N_2 + (N_1 - N_2) \frac{d(Vb)}{dV} \right]. \tag{6.89}$$

Next, if $\frac{d\beta}{dk}$ is once again differentiated with respect to k to determine $\delta\tau$, then the following equation is obtained*:

$$\delta\tau = L\delta\lambda \left[\underbrace{ \left(-\frac{1}{c\lambda_0} \right) \left\{ k\frac{dN_2}{dk} + \left(k\frac{dN_1}{dk} - k\frac{dN_2}{dk} \right) \left(b + \frac{1}{2}V\frac{db}{dV} \right) \right\} }_{\sigma_m \,:\, \text{Material dispersion}} \right.$$

$$\left. + \underbrace{ \left(-\frac{1}{c\lambda_0} \right) \frac{1}{2} \frac{(n_1 N_1 - n_2 N_2)^2}{n_{eq}(n_1^2 - n_2^2)} V\frac{d^2(Vb)}{dV^2} }_{\sigma_w \,:\, \text{Waveguide dispersion}} \right]. \tag{6.90}$$

This differential group delay $\delta\tau$ means that the time is deviating from the center of the pulse depending on the frequency component (or the wavelength component) in the exit pulse waveform. Consequently, carrier frequency is slightly different at the front and the rear of the pulse waveform (chirping that was discussed in Section 6.2.1). This concept is illustrated in Figure 6.19.

(1)   Material dispersion

The first term inside the [ ] at the right side of Equation (6.90)

$$\sigma_m = -\frac{1}{c} \left[\frac{k}{\lambda}\frac{dN_2}{dk} + \left(\frac{k}{\lambda}\frac{dN_1}{dk} - \frac{k}{\lambda}\frac{dN_2}{dk} \right) \left(b + \frac{1}{2}V\frac{db}{dV} \right) \right]_{\lambda=\lambda_0} \tag{6.91}$$

is called the *material dispersion*. Here,

$$\frac{k}{\lambda}\frac{dN_i}{dk} = \lambda\frac{d^2 n_i}{d\lambda^2} \quad (i = 1 \text{ or } 2) \tag{6.92}$$

represents the dispersion of the material itself of the core or the cladding layer. As derived in Equation (5.78) in Section 5.1.6, because the confinement factor in the step-index fiber (same as for the symmetrical three-layer slab waveguide) is expressed by

---

* This calculation is enormous to derive it by the computation on paper. However, recently, formula manipulation programs can be used to do the calculation. The equation that is written here has undergone slight approximation to simplify the coefficients.

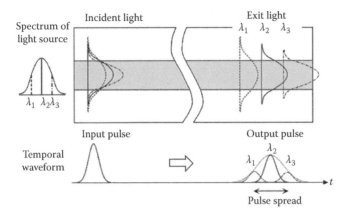

FIGURE 6.19 Concept of the pulse width broadening due to dispersion in a single-mode optical fiber.

$$\Gamma = b + \frac{1}{2} V \frac{db}{dV} = \frac{1}{2} \left[b + \frac{d(Vb)}{dV} \right], \qquad (6.93)$$

Equation (6.91) can be rewritten as the following equation:

$$\sigma_m = -\frac{1}{c} \left[\Gamma \lambda \frac{d^2 n_1}{d\lambda^2} + (1 - \Gamma) \lambda \frac{d^2 n_2}{d\lambda^2} \right]_{\lambda = \lambda_0}. \qquad (6.94)$$

The data on the wavelength dependence of the refractive index of the core and the cladding materials are required when calculating the material dispersion using Equation (6.94). Fortunately, for silica (SiO_2 glass) optical fibers, the refractive index of SiO_2 is precisely measured as a function of the wavelength and so is given by the following equation [31]:

$$n^2 = 1 + \sum_{i=1}^{3} \frac{A_i \lambda^2}{\lambda^2 - (l_i)^2}. \qquad (6.95)$$

This equation is called the *Sellmeier's formula*. In an actual optical fiber, GeO_2 is doped to increase the refractive index of the core to cause a slight difference in the refractive index of the core and the cladding. In this case, each of the coefficients in Equation (6.95) is slightly different depending on the concentration of the dopant. Detailed measurements are given in various references [32], [33], and one example is given in Table 6.3. Here, to simplify, let us calculate the material dispersion using the value of pure SiO_2.

When Equation (6.95) is used, $\frac{dn}{d\lambda}$ and $-\frac{1}{c}\lambda\frac{d^2 n}{d\lambda^2}$ can be calculated, and they are plotted in Figure 6.20 (as $\frac{dn}{d\lambda}$ is a negative value, it is denoted in Figure 6.20

TABLE 6.3
Coefficients of Sellmeier's Formula

Composition		Sellmeier Coefficients					
SiO_2 [31]		A_1	A_2	A_3	λ_1	λ_2	λ_3
100.0		0.6961663	0.4079426	0.8974794	0.0684043	0.1162414	9.896161
SiO_2	GeO_2 [33]	A_1	A_2	A_3	λ_1^2	λ_2^2	λ_3^2
97.3	6.3	0.7083952	0.4203993	0.8663412	0.007290464	0.01050294	97.93428
91.3	8.7	0.7133103	0.4250904	0.8631980	0.006910297	0.01165674	97.93434
88.8	11.2	0.7186243	0.4301997	0.8543265	0.004026394	0.01632475	97.93440
85.0	15.0	0.7249180	0.4381220	0.8221368	0.007596374	0.01162396	97.93472
80.7	19.3	0.7347008	0.4461191	0.8081698	0.005847345	0.01552717	97.93484
GeO_2 [34]		A_1	A_2	A_3	λ_1	λ_2	λ_3
100.0		0.80686642	0.71815848	0.85416831	0.068972606	0.15396605	11.841931

* The unit for dopant concentration is mol%.

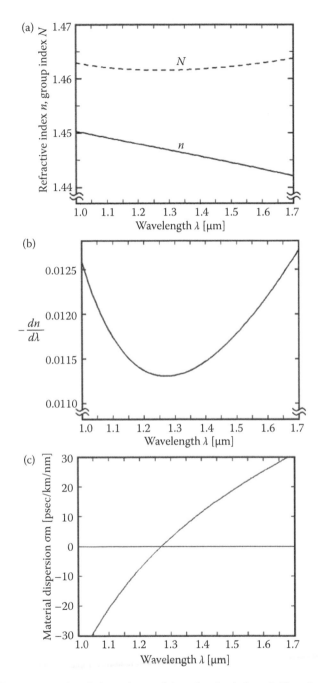

FIGURE 6.20 Wavelength dependence of the refractive index of silica glass and the material dispersion: (a) Refractive index and group index, (b) first derivative of the refractive index, $-\frac{dn}{d\lambda}$, and (c) material dispersion, $-\frac{1}{c}\lambda\frac{d^2n}{d\lambda^2}$.

as $-\frac{dn}{d\lambda}$). In Figure 6.20, we can see that the wavelength wherein the material dispersion is 0 is slightly shorter than 1.3 μm.

(2) Waveguide dispersion

The second term inside the [ ] at the right side of Equation (6.90)

$$\sigma_w = -\frac{1}{c\lambda}\frac{1}{2}\frac{(n_1 N_1 - n_2 N_2)^2}{n_{\text{eq}}(n_1^2 - n_2^2)}V\frac{d^2(Vb)}{dV^2} \tag{6.96}$$

is called the *waveguide dispersion*. Here, if the weakly guiding approximation ($\Delta \ll 1$) can be applied, σ_w can be approximated from Equation (6.89) as

$$\sigma_w \simeq -\frac{1}{c\lambda}(N_1 - N_2)V\frac{d^2(Vb)}{dV^2}. \tag{6.97}$$

Here, the $\frac{d(Vb)}{dV}$ appearing in Equation (6.89) and the $V\frac{d^2(Vb)}{dV^2}$ appearing in Equation (6.97) are calculated and shown by the solid lines in Figure 6.21 together with the dispersion curve [35].*

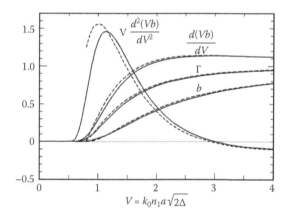

FIGURE 6.21 The waveguide dispersion $V\frac{d^2(Vb)}{dV^2}$, b, and the confinement factor Γ of a step-index cylindrical fiber.

---

* The figure in the first reference [35] that calculated $V\frac{d^2(Vb)}{dV^2}$ has calculation errors (these are perhaps numerical calculation errors), and a lot of textbooks have since cited the erroneous figure in this reference. Therefore, the reader should note that a figure wherein the maximum value of $V\frac{d^2(Vb)}{dV^2}$ is about 1.0 is incorrect, and figures with a maximum value of about 1.5 is correct.

As discussed in the last part of Section 6.1.4, the dispersion relation of the step-index cylindrical fiber can be approximated by Equation (6.28). Consequently, $\frac{d(Vb)}{dV}$ and $V\frac{d^2(Vb)}{dV^2}$ can be analytically expressed in terms of V and b, to obtain the following equations:

$$\frac{d(Vb)}{dV} \simeq \frac{V(2-b)+\sqrt{b}}{V+\frac{1}{\sqrt{b}}}, \tag{6.98}$$

$$V\frac{d^2(Vb)}{dV^2} \simeq \frac{2V^2(1-b)}{(V\sqrt{b}+1)^3}\left[1-b-3b(V\sqrt{b}+1)+2\frac{\sqrt{b}}{V}(V\sqrt{b}+1)^2\right]. \tag{6.99}$$

When these approximation equations are used to calculate $\frac{d(Vb)}{dV}$ and $V\frac{d^2(Vb)}{dV^2}$, the results are plotted as the dashed line in Figure 6.21, where we can see a certain level of accuracy in the single-mode region.

Incidentally, the horizontal axis of Figure 6.21 is the V parameter, which is inversely proportional to the wavelength ($V = \frac{2\pi}{\lambda}a\sqrt{n_1^2-n_2^2}$). Then, expressing the total dispersion σ_T by

$$\sigma_T = \sigma_m + \sigma_w, \tag{6.100}$$

drawing it on the same wavelength axis in Figure 6.20c, Figure 6.22 is obtained. From the figure, we can see that when the relative index difference Δ is small, the waveguide dispersion σ_w is smaller compared to the material dispersion σ_m,

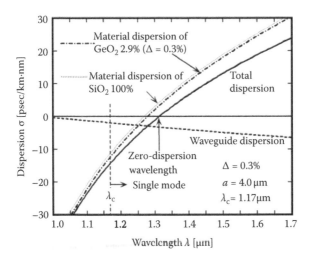

FIGURE 6.22 Total dispersion of a single-mode silica optical fiber.

and the *zero-dispersion wavelength* is roughly around a wavelength of 1.3 μm. In addition, the total dispersion at the 1.55 μm wavelength of an optical fiber, with a zero-dispersion wavelength that is in the 1.3 μm wavelength range, is about 16 ps/km · nm.

6.2.3 Transmission Bandwidth of Single-Mode Fibers

Up to now, we have discussed the pulse width broadening because of dispersion of single-mode fibers. However, pulse width broadening describes the transmission characteristics of the fiber in the time domain, and so, for the discussion of the transmission bandwidth, it is necessary to describe the transmission characteristics in the frequency domain.

The frequency characteristic is determined from the transfer function of the transmission line, so if the Fourier transform Equation (6.67) of the Gaussian incident pulse expressed in Equation (6.66) and the Fourier transform of Equation (6.73) of the exit pulse amplitude are compared, the transfer function of the modulated signal can be determined from the following equation[*]:

$$|H(\omega)|^2 = \frac{1}{\tau_0} \left[\tau_0^2 + \frac{(2\beta''z)^2}{\tau_0^2} \right]^{\frac{1}{2}} \exp \left[-2 \left(\frac{2\beta''z}{\tau_0} \right)^2 \omega^2 \right]. \qquad (6.101)$$

Here, the portion of the coefficient excluding the ω^2 of the Gaussian function at the right side of Equation (6.101) represents the pulse width, which corresponds to the fraction of broadening in comparison with τ_0^2, the denominator of the Gaussian function of Equation (6.73). Then, when this portion is related to $\delta\tau$ (here, when this portion is related to T_{FWHM} of Equation (6.82)), it will be expressed as follows:

$$|H(\omega)|^2 = \frac{1}{\tau_0} \left[\tau_0^2 + \frac{(2\beta''z)^2}{\tau_0^2} \right]^{\frac{1}{2}} \exp \left[-\frac{\delta\tau^2}{\ln 2} \omega^2 \right]. \qquad (6.102)$$

The angular frequency $\Delta\omega$ wherein this transfer function decreases by -3 dB (the half value) from the value of $\omega = 0$ can be determined from $-\frac{\delta\tau^2}{\ln 2}\Delta\omega^2 = -\ln 2$, so the *3 dB bandwidth B* is expressed as

$$B = \Delta\nu = \frac{\Delta\omega}{2\pi} = \frac{\ln 2}{2\pi |\delta\tau|}. \qquad (6.103)$$

It should be noted that the coefficient of this equation is the value in the Gaussian pulse. From this equation, the reader may think that B can be calculated

---

[*] When the optical pulse is directly detected using a photodetector, the optical power of the pulse becomes an electric signal, and the transfer function, as the electric signal, can be expressed as $|H(\omega)|^2$, when the transfer function of the optical electric field is set as $H(\omega)$.

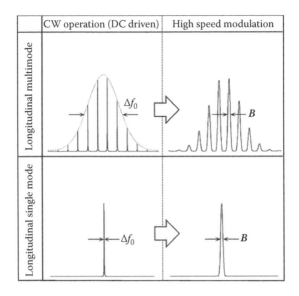

FIGURE 6.23 Oscillation spectrum of the longitudinal single-mode laser and the longitudinal multimode laser.

by expressing $\delta\tau$ by $\beta''z$ and the incident pulse width τ_0. However, on second thought, the incident pulse width τ_0 is actually related to the modulated band in Equation (6.69), so when calculating Equation (6.103), two cases should be considered. One is when the frequency of the light source is stable. This corresponds to the case wherein the frequency width Δf_0 of the light source is sufficiently narrow compared to the modulated bandwidth B of signal (or the same as $\delta\nu$ in Equation (6.70)). The other case is the opposite wherein the frequency bandwidth of the light source is wider than the modulated bandwidth B.

These two cases are based on the difference in light sources, one being a laser spectrum such as that from a DFB laser (refer to Section 4.2) and the other is a spectrum from an FP laser such as the one shown in Figure 4.6. Figure 6.23 shows the difference between the two light sources when comparing the steady state (continuous oscillation or the continuous wave, which is abbreviated as CW and is also called the CW oscillation) and the modulated case.

(1) Case where the frequency bandwidth of the light source is narrower than the signal bandwidth

In this case, $\Delta f_0 \ll B$, and as $\delta\tau$ is proportional to $\Delta\nu = B$ by Equation (6.82) ($\delta\tau$ of Equation (6.83) is the same as T_{FWHM} of Equation (6.82)), the $B\sqrt{L}$ is given by

$$B\sqrt{L} = \frac{\sqrt{\ln 2}}{2\pi} \frac{1}{\sqrt{|\beta''|}}. \tag{6.104}$$

Moreover, from Equations (6.82) and (6.90), β'' can be expressed as $\beta'' = -\frac{\lambda^2}{2\pi c}\sigma_T$, so Equation (6.104) will be written as

$$B\sqrt{L} = \sqrt{\frac{\ln 2}{2\pi}} \frac{\sqrt{c}}{\lambda\sqrt{|\sigma_T|}} = \frac{181.9}{\lambda\sqrt{|\sigma_T|}} \quad [\text{GHz}\sqrt{\text{km}}]. \tag{6.105}$$

Because the unit of the frequency λ that is assigned to the above formula is micrometer, the unit for the dispersion σ is ps/(km·nm). Equation (6.105) shows the outstanding feature of the single-mode optical fiber as a signal transmission line. In other words, the transmission bandwidth is not inversely proportional to the distance L, but rather is inversely proportional to the square root of L. Therefore, the transmission bandwidth of distance 100 km is reduced to only one-tenth of the transmission bandwidth of distance 1 km. When the characteristic of coaxial cable is expressed by $\sqrt{B}L = \text{const.}$, it is a contrastive characteristic.

On the other hand, the transmission bandwidth of a single-mode fiber with a zero-dispersion wavelength close to 1.3 µm, as shown in Figure 6.22, is calculated with respect to the wavelength, and the result is shown in Figure 6.24. According to Equation (6.105), at the *zero-dispersion wavelength* where the total dispersion $\sigma_T = 0$, the transmission bandwidth becomes infinite, but zero dispersion is a condition wherein the $\frac{d^2\beta}{d\omega^2}$ term in Equation (6.84) is zero, and as higher order terms are not zero, the transmission bandwidth will not

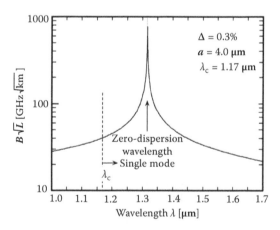

FIGURE 6.24 Transmission band of the 1.3 µm wavelength zero-dispersion single-mode fiber.

become infinite. The bandwidth in the zero-dispersion wavelength will take the form of $B^4\sqrt{L}$ = const. and will have a value of approximately 2–3 $THz^4 \cdot \sqrt{km}$ [36].

(2) Case where the frequency bandwidth of the light source is wider than the signal bandwidth

In this case, $\Delta f_0 > B$, and in Equation (6.90), which shows the relationship of $\delta\tau$ and $\delta\lambda$, the $\delta\lambda$ is not a spectral width that is determined based on the modulated frequency of the signal, so it should be replaced by the spectral width $\Delta\lambda_0$ of the light source. Consequently, the following equation is obtained from Equation (6.103):

$$BL = \frac{\ln 2}{2\pi\,\Delta\lambda_0|\sigma_T|} = \frac{110.3}{\Delta\lambda_0|\sigma_T|} \quad [\text{GHz}\cdot\text{km}]. \qquad (6.106)$$

Here, the unit of dispersion σ is ps/(km·nm) and the unit of the spectral width $\Delta\lambda_0$ is nm. This case is different from the case in (1), and the transmission bandwidth is inversely proportional to the distance and will deteriorate. Consequently, in the long-haul high-capacity optical communications, it is necessary for the spectral width of the light source to be sufficiently narrower than the transmission band, which is the case in (1).

PROBLEMS

1. Determine the material dispersion and the waveguide dispersion of a step-index single-mode silica fiber that was doped with GeO_2 with a relative index difference $\Delta = 0.9\%$, and a core radius $a = 2.0$ μm, with a $\lambda = 1.0$–1.7 μm range as the wavelength function.

    [Comment]: A single-mode fiber like this with a large refractive index difference and a small core diameter will have zero dispersion at a wavelength of around $\lambda = 1.55$ μm.

2. In the fiber in Question 1, which pulse width broadening at a wavelength of 1.3 μm as shown in Figure 6.17b and c in Section 6.2.1 will correspond to chirping?

3. In the fiber in Question 1, when light from a DFB semiconductor laser with a 1 MHz light source spectral linewidth and a wavelength of 1.55 μm is propagated, calculate the value of the $B\sqrt{L}$ product when it is modulated by an external modulator.

4. In the fiber in Question 1, when light from an FP semiconductor laser with a 3 nm spectral linewidth and a wavelength of 1.3 μm is propagated, calculate the value of the BL product when direct modulation (i.e., current flowing to the laser is modulated) is applied.

5. Prove that $T_0 \cdot \Delta\nu = 0.315$ for a $sech^2$ pulse.

6.2.4 DISPERSION-SHIFTED FIBER AND DISPERSION COMPENSATION

In a single-mode fiber with a small relative index difference Δ, the zero dispersion wavelength is close to 1.3 μm, as shown in Figure 6.22. Consequently, a high-capacity transmission is possible within this wavelength range. However, as shown in Figure 1.2, as the minimum loss wavelength region in silica fibers is in the range of 1.55 μm to realize a high-capacity long-range transmission, it is necessary to match the minimum loss wavelength and the zero-dispersion wavelength. However, the minimum loss wavelength is determined by the properties of the silica glass material and will not change significantly even if the structure of the optical fiber is devised. On the other hand, as the zero-dispersion wavelength is determined by balancing the material dispersion and the waveguide dispersion, it can be controlled by devising the structure of the optical fiber. In particular, as the waveguide dispersion is roughly proportional to the refractive index difference according to Equation (6.97), the zero-dispersion wavelength can be shifted to a longer wavelength than 1.3 μm by increasing the refractive index difference. Moreover, in a normal single-mode fiber, the V value in the operating wavelength is set to be slightly less than the single-mode condition $V_c = 2.405$. However, as can be seen in Figure 6.21, $V \frac{d^2 (Vb)}{dV^2}$ that is proportional to the waveguide dispersion takes its maximum value at around $V = 1.1$, which is much smaller than 2.405. Therefore, the refractive index difference between the core and the cladding is increased from the normal 0.3% to about 1%, and the core radius a is decreased to about 2.0 μm (diameter is 4.0 μm) to reduce the V value at the operational wavelength, so that the zero-dispersion wavelength can be shifted close to the minimum loss wavelength of 1.55 μm. A fiber wherein the zero-dispersion wavelength is shifted to a wavelength that is longer than 1.3 μm is called a *dispersion-shifted fiber*.

A dispersion-shifted fiber of which the relative index difference is merely increased and the core diameter is reduced has a small core diameter and a small V parameter, so the spot size is small and the confinement of the light in the core is weak. Therefore, the connection loss because of axial misalignment and bending loss will increase. Thus, several structures have been proposed to control the spot size and the bending loss resulting from the shift of the zero-dispersion wavelength to about the same level as conventional fibers. Examples of these structures and their corresponding dispersions are shown in Figure 6.25.

Moreover, as shown in Figure 6.17 of Section 6.2.1, the pulse with pulse broadening due to dispersion gives rise to chirping. In other words, the pulse broadened by dispersion can go back to its original pulse width if it is passed through a device with an inverse dispersion. In addition, because of the development of optical fiber amplifiers, attenuation can be compensated. So, by inserting an optical fiber amplifier every several tens of kilometers to long-haul communication systems, such as submarine cables, the incident optical

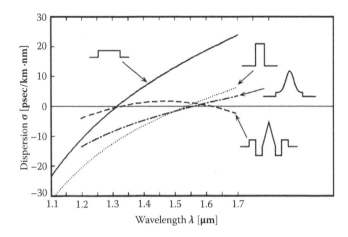

FIGURE 6.25 Various dispersion-shifted fibers and their corresponding dispersions.

pulse will not be converted into an electrical signal midway, but rather the light would be propagated for several thousands of kilometers and reach the receiver as is. Consequently, as the propagation distance of the optical pulse spans several thousands of kilometers, the technology that can compensate for the pulse broadening due to dispersion by inverse dispersion becomes very important. Even if dispersion-shifted fibers are used in long-haul optical fiber transmission lines, a little dispersion (such as the sign being distributed as + or −) remains due to manufacturing errors. Therefore, such remaining dispersion will be measured for all the fibers to ascertain the cumulative value (integral value) over the length, and a dispersion-compensating fiber (a fiber wherein dispersion is deliberately increased) is used to cancel out this cumulative value to negate the dispersion and restore the pulse width broadening. This technology is called *dispersion management*.

6.3 TRANSMISSION CHARACTERISTICS OF DISTRIBUTED INDEX MULTIMODE FIBERS

The transmission capacity of the step-index multimode waveguide was derived in Section 2.6.2. Actually, even for the transmission capacity of the step-index multimode fiber, the difference in the propagation time of the fastest guided mode and the slowest guided mode is expressed by the difference in the refractive indices of the core and the cladding, and it will derive the same results as in Section 2.6.2. However, as the wavelength dependence of the refractive index was not considered in Section 2.6.2, here, let us think about the effect of the wavelength dependence of the refractive index on the transmission characteristics of the multimode fiber.

6.3.1 GROUP DELAY OF MULTIMODE OPTICAL FIBERS

In single-mode fibers, the frequency dependence of group delay of fundamental mode causes pulse broadening. However, in multimode fibers, the intermode group delay difference is bigger than the intramode pulse broadening; hence, as shown in Figure 6.26, the pulse broadening is determined by the *intermode group delay difference*. Because the group delay is expressed in Equation (6.89) based on the weakly guiding approximation, the pulse broadening is determined based on the difference in the $\frac{d(Vb)}{dV}$ value of each mode. Then, the wavelength dependence of the refractive index is included in the group index N_1 and N_2. The $\frac{d(Vb)}{dV}$ of the fundamental mode is shown in Figure 6.21; however, the value for the higher order modes in the region, where the V value is large, is different depending on the refractive index profile in the core.

Here, for the α-power profile fiber in Section 6.1.8, some numerically calculated results of the intermode group delay difference for several values of α is shown in Figure 6.27 [30]. As can be seen from Equation (6.89) and Figure 6.27, the $\frac{d(Vb)}{dV}$ value in step-index fiber changes in the range from -1 to $+1$. If a mode in a certain V that is near the cutoff is excluded, this value will be distributed in the range from 0 to $+1$. Thus, the intermode group delay difference is expressed as

$$\Delta\tau = \frac{L}{c}(N_1 - N_2) \simeq \frac{L}{c}N_1\frac{N_1 - N_2}{N_1}. \qquad (6.107)$$

Ignoring the wavelength dependence of the refractive index, and approximating the group index as $N_1 \simeq n_1$ and $N_2 \simeq n_2$, respectively, Equation (6.107) coincides with the propagation time difference Δt between the fundamental

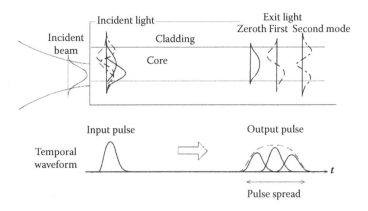

FIGURE 6.26 Concept of the intermode group delay difference of multimode fibers.

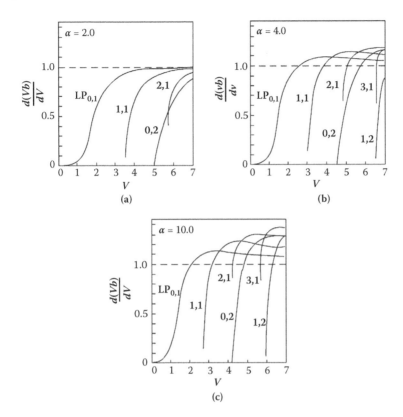

FIGURE 6.27 Intermode group delay difference $\frac{d(Vb)}{dV}$ of the α-power profile fiber: (a) $\alpha = 2$, (b) $\alpha = 4$, and (c) $\alpha = 10$.

mode and the highest order mode given by Equation (2.37) that was derived in Section 2.6.2 using $\Delta \simeq \frac{n_1 - n_2}{f} or n_1$.

On the other hand, for the intermode group delay difference of the parabolic index fiber with $\alpha = 2$, excluding the mode near the cutoff, the value of $\frac{d(Vb)}{dV}$ is concentrated near roughly 0, and it can be seen that this intermode group delay difference is much smaller than that for the step-index fiber. Next, let us derive the transmission characteristics in the case wherein α is close to 2.

6.3.2 Transmission Capacity of α-Power Profile Fibers

When Equation (6.52), which expresses the propagation constant of an α-power profile multimode fiber, is differentiated with respect to k to determine the group delay, the following equation is obtained [37]:

(Here, the β' of Equation (6.52) is β.)

$$
\begin{aligned}
\tau &= \frac{L}{c}\frac{d\beta}{dk} \\
&= \frac{L}{c}N_1\left[1+\frac{\alpha-2-\varepsilon}{\alpha+2}\Delta\left(\frac{\mu}{M}\right)^{\frac{\alpha}{\alpha+2}}\right. \\
&\quad \left.+\frac{3\alpha-2-2\varepsilon}{2(\alpha+2)}\Delta^2\left(\frac{\mu}{M}\right)^{\frac{2\alpha}{\alpha+2}}+O(\Delta^3)\right].
\end{aligned}
\tag{6.108}
$$

Here, ε is the parameter that represents the wavelength dependence of the relative index difference Δ and is defined by

$$
\varepsilon \equiv \frac{2n_1 k_0}{N_1 \Delta}\frac{d\Delta}{dk} = -\frac{2n_1\lambda}{N_1\Delta}\frac{d\Delta}{d\lambda}.
\tag{6.109}
$$

The terms showing the dependence of the group delay given in Equation (6.108) on the mode number μ are the second and the third terms inside the brackets [ ] on the right side of the equation. The second term is proportional to Δ, whereas the third term is proportional to Δ^2, so the impact of the second term is bigger. Consequently, the condition for the coefficient of the second term to be zero is the condition to obtain a broadband by making the intermode group delay difference to be zero. The exponent α_{opt} that gives the optimum refractive index distribution is

$$
\alpha_{opt} = 2+\varepsilon.
\tag{6.110}
$$

ε is the parameter that represents the wavelength dependence of Δ, but the parameter itself is dependent on the wavelength and the type of dopant,[*] so α_{opt} will change depending on the wavelength as shown in Figure 6.28 [38, 39]. The value of α_{opt} is slightly smaller than 2.0 in the wavelength range longer than 1.0 μm. Assuming a uniform mode distribution,[†] the transmission bandwidth of the α-power profile fiber is obtained as shown in Figure 6.29. If the exponent α deviates from the optimal value, the transmission bandwidth rapidly decreases. Thus, the precise control of the refractive index profile is essential to obtain a broadband.

---

[*] An impurity that is added to the SiO_2 of the core and the cladding to slightly change the refractive indices of the core and the cladding is called a dopant. Examples of dopants used to increase the refractive index of the core are GeO_2 and P_2O_5, whereas examples of dopants that decreases the refractive index of the cladding are F and B_2O_3.

[†] Approximation wherein the mode excitation distribution is assumed to be uniform from the fundamental mode to the highest order mode. In an actual fiber, the amplitude of the higher order mode is usually small.

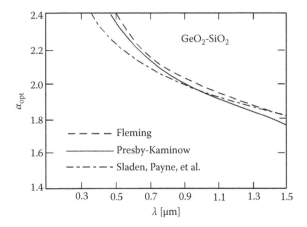

FIGURE 6.28 Wavelength dependence of the optimal exponent α_{opt} of the α-power profile fiber.

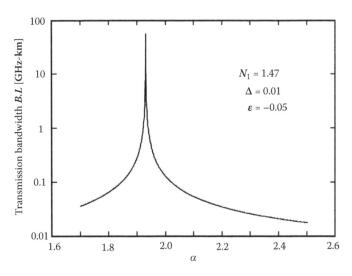

FIGURE 6.29 Exponent α dependence of the transmission bandwidth of the α-power profile fiber.

The product of the transmission bandwidth B and the distance L when $\alpha = \alpha_{opt}$ is given by

$$BL \simeq \frac{2c}{N_1 \Delta^2}. \tag{6.111}$$

On the other hand, in a step-index fiber ($\alpha = \infty$), it is given by

$$BL \simeq \frac{c}{N_1 \Delta}, \tag{6.112}$$

so the α-power profile fiber with the optimal coefficient has a transmission bandwidth that is $\frac{2}{\Delta}$ times compared to that of the step-index fiber. When $\Delta = 1\%$, the transmission bandwidth will be approximately 200 times.

Moreover, in an actual optical fiber, because of the scattering (intermode coupling) resulting from microscopic bending (called a microbend and occurs during cable assemblage), the bandwidth distance product is expressed in the form of

$$BL^{\gamma} = \frac{2c}{N_1 \Delta^2}. \tag{6.113}$$

The exponent γ has a value in the $0.5 \leq \gamma \leq 1.0$, and if the intermode coupling due to microbends is 0, then $\gamma = 1.0$. If the microbends are extremely strong, $\gamma = 0.5$, whereas in normal multimode fibers, it has a value of about $\gamma \simeq 0.8$–0.9.

6.4   OPTICAL FIBER COMMUNICATION

In Sections 6.2 and 6.3, we discussed how the pulse signal waveform in the optical fiber transforms (pulse width broadening) along with propagation. In communications, many of the signals initially generated are analog signals (continuous waveform signals), such as audio and video. The description of the process by which these are converted to digital signals will be left to other books; however, here, we will briefly explain the transmission and reception of digital signals.

In binary digital communication, the time axis of signals is divided by certain periods. If a pulse is sent out during a period, this will be designated as "1," whereas if a pulse is not sent out, this will be designated as "0." The pulse is propagated as a binary digital signal (only 0 or 1 are used).* The reciprocal of the time period is called the *clock frequency* (the unit is pulse per second, which is abbreviated as [pps]), and the "0" or "1" signal that is sent during the time period is counted as 1 bit. Consequently, for a binary signal, the unit of the clock frequency is [bit/s], which is referred to as the bit rate.[†] Moreover, the manner of sending of the pulse using "0" and "1" can be classified into

---

* Actually, multilevel coding, such as QAM (quadrature amplitude modulation), is also used in long-haul public communications.
[†] The bit rate for interoffice trunk lines as of 2011 is 40 Gbit/s per fiber in the fastest systems, with plans of making it faster. In a multilevel signal with a 2^n value, one pulse is n bits.

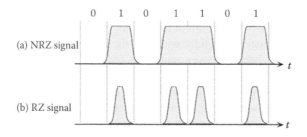

FIGURE 6.30 Two formats of binary digital signals: (a) NRZ signal and (b) RZ signal.

two kinds.[‡] When "1" is sent out, the signal that maintains the pulse amplitude throughout as "1" during one period, as shown in Figure 6.30a, is called
non-return-to-zero (NRZ) signal. In the NRZ signal, if "0" and "1" are sent out
alternately, it is easy to determine the boundary of the time period; however,
if the "1" signal is continuously sent, the amplitude will be the same throughout that period, whereas if the "0" signal is continuously sent, the amplitude
will be 0 throughout, so it will be difficult to determine the interval of one time
period. In the receiver circuit of an optical communication system, the clock frequency is determined from the incoming signal, and at a middle point, the time
interval that is synchronized with such a clock, the circuit operates to determine
whether the signal is "0" or "1." This operation is called the clock extraction.
However, for the NRZ signal, if the clock does not retain the time after the
clock has been extracted, the timing of the signal determination will deviate
when the "0" or "1" is continuous. On the other hand, in the RZ (return-to-zero)
signal, when "1" is sent out as shown in Figure 6.30b at an occupancy rate at
about half of one time period, the pulse amplitude is maintained at "1," and the
pulse amplitude returns to 0 at the remaining half of the time. For this signal,
the boundary of the time period is easy to determine even if the "1" signal is
continuous, but as the time period is further divided into half, the transmission
capacity required by the transmission line should be twice the transmission
capacity that was calculated from the clock frequency.

A digital signal is transmitted as described earlier. When optical pulses are
propagated in optical fibers, they will undergo changes that are roughly classified into two categories. One is attenuation, wherein the pulse amplitude
(intensity) is weakened by loss inside the optical fiber, and the other is pulse
width broadening caused by dispersion inside the optical fiber. First, let us
consider the case wherein there is no pulse broadening but the pulse intensity

---

[‡] In a slow link, CMI-modulated code and DMI-modulated code are used; however, these require
a band that is twice the original signal. Therefore, in high-speed transmission, a modulation code
that adds one bit for every eight bits or ten bits is used to facilitate code error detection and timing
extraction.

is attenuated because of optical fiber loss. The optical pulse that reaches the receiver is converted into a current signal (not voltage) in the photodetector.* When the pulse intensity rapidly weakens, the number of photons contained in one pulse will become fewer, and in an ordinary photodetector, one pair of carriers (electron–hole pair in semiconductors) is generated from one photon and the light intensity is converted into a current, so that finally, grains of photons will be seen in the receiving current and a noise corresponding to the number of photons will appear at the signal "1" amplitude. Such noise is called a shot noise (quantum noise). Because the ultimate receiver sensitivity is determined by this quantum noise, such is referred to as the quantum noise limit. On the other hand, in an ordinary photodetector (photodetectors used at room temperature), the thermal noise is greater than the shot noise, so the receiver sensitivity is determined roughly by the thermal noise. Then, as the thermal noise is superimposed onto the optical signal, which is converted to current signal, the signal current of the receiver will not have a clear "0" or "1" value as shown in Figure 6.30. Moreover, it will be difficult for a high-speed optical pulse to be generated as a perfect rectangular pulse, and so it will actually be a smooth pulse like that in Figure 6.30. The high-speed pulse waveform of the received signal involving noise cannot keep up with the speed of the electric circuit in depicting each of the pulses in the oscilloscope. Therefore, when the pulse period is set as T[s], instead of tracing one whole pulse during the time ΔT[s], the pulse train is sampled at the $nT + \Delta T$[s] interval (here, n is an integer greater than 1), wherein it is overlaid within the time interval for 1 period in the oscilloscope screen. This oscilloscope is called a sampling oscilloscope. For example, if an NRZ signal with noise is observed using this sampling oscilloscope, when the long-time observed waveform is overlaid, the observed waveform is not a single independent waveform, but the superimposed waveform of the following eight patterns, "0 → 0 → 0," "0 → 0 → 1," "0 → 1 → 0," "0 → 1 → 1," "1 → 0 → 0," "1 → 0 → 1," "1 → 1 → 0," and " 1 → 1 → 1," which denote the relationship with the signal before and after the observation. Moreover, as the sampled signal amplitude is represented as dots, the observed waveform will be a collection of dots as shown in Figure 6.31. Because this observed waveform resembles an eye, it is called an *eye pattern*. In the case shown in Figure 6.31a, as the signal points sampled at the middle of the period can be clearly distinguished into sets of "0" and "1," the eye is open. However, as in the case shown in Figure 6.31b, when "0" and "1" are indistinguishable, the eye is said to be closed.

As might be expected, when the eye is closed, the probability that the "0" and the "1" code will be incorrectly determined will increase. The probability

---

* If the current signal is placed in resistance, a voltage signal is generated; however, at higher frequencies, it is necessary to set the impedance while taking into consideration the input impedance of subsequent amplifier circuits and the characteristic impedance of the half-way wiring.

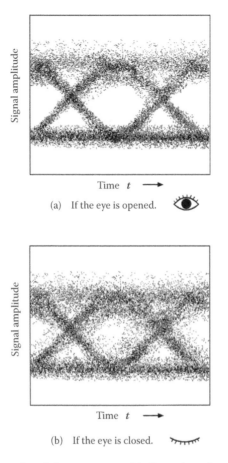

(a)   If the eye is opened.

(b)   If the eye is closed.

FIGURE 6.31 Examples of the eye pattern: (a) Case where the eye is opened and (b) case where the eye is closed.

that the receiver circuit will incorrectly determine the "0" and the "1" code is called the *bit-error-rate* (abbreviated as BER). The specifications of common public communication, such as telephones, ensure a BER of 10^{-9} (a probability of 1 error in the 10^9 pulses sent). The BER of 1 repeater (or 1 repeater span) is set to be about 10^{-11}; however, if a lower BER and reliable communication (in data communications and others) are required, an error correcting code is sometimes used.

Because thermal noise is considered to be roughly represented by a Gaussian distribution, its standard deviation is defined as σ_T. If the received signal is "1," as the shot noise and the signal intensity noise (noise due to amplitude fluctuation) will be added to the thermal noise, the standard deviation of the noise corresponding to the "1" signal will be bigger than σ_T. On the other hand, if the received signal is "0," noise due to the dark current that flows through the photodetector, even without input light, is added to the detected photocurrent (of course, the same will be added for a "1" signal). Here, let us designate the average of the received signal current for "1" as i_M, that for "0" as σ_S, the standard deviation of the additional noise when the code is "1" as σ_M (abbreviation for mark state), and the standard deviation of the noise when the code is "0" as σ_S (abbreviation for space state). Then, if each of these codes are received, the probability density function of the signal current i that flows to the receiver is expressed as

$$p_M(i) = \frac{1}{\sqrt{2\pi}\sigma_M} \exp\left[-\frac{(i-i_M)^2}{2\sigma_M^2}\right] \quad \text{if the code is "1",} \quad (6.114)$$

$$p_S(i) = \frac{1}{\sqrt{2\pi}\sigma_S} \exp\left[-\frac{(i-i_M)^2}{2\sigma_S^2}\right] \quad \text{if the code is "0". } (6.115)$$

Consequently, assuming that the "0" and the "1" code will be generated with a probability of 50% each, when the determination threshold of "0" and "1" at the midpoint of the time period is set as i_{th}, the BER is determined in terms of the probability density function by

$$\text{BER}(i_s) = \frac{1}{2}\int_{i_{th}}^{\infty} p_S(i)di + \frac{1}{2}\int_{-\infty}^{i_{th}} p_M(i)di$$
$$= \frac{1}{4}\left[\text{erfc}\left(\frac{i_{th}-i_S}{\sqrt{2}\sigma_S}\right) + \text{erfc}\left(\frac{i_M-i_{th}}{\sqrt{2}\sigma_M}\right)\right]. \quad (6.116)$$

Here, $\text{erf}(x) = \frac{2}{\sqrt{\pi}}\int_0^x e^{-t^2}dt$ is called the error function and the $\text{erfc}(x) = 1 - \text{erf}(x)$ is called the complimentary error function. The condition that would make the bit error probability to mistake "0" for "1," and conversely, the bit error probability to mistake "1" for "0" is given by

$$\frac{i_{th}-i_S}{\sqrt{2}\sigma_S} = \frac{i_M-i_{th}}{\sqrt{2}\sigma_M}. \quad (6.117)$$

To satisfy this condition, the determination threshold value i_{th} should be set as

$$i_{th} = \frac{\sigma_S i_M + \sigma_M i_S}{\sigma_M + \sigma_S}. \quad (6.118)$$

In such a case, Equation (6.116) will become the following equation:

$$\text{BER}(i_s) = \frac{1}{2}\text{erfc}\left[\frac{i_M-i_S}{\sqrt{2}(\sigma_M+\sigma_S)}\right]. \quad (6.119)$$

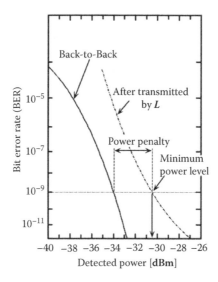

FIGURE 6.32 BER characteristics corresponding to the received signal level (a value which is one example of the received power value in the horizontal axis).

The bracket on the right side is called Q, where

$$Q = \frac{i_M - i_S}{\sigma_M + \sigma_S}. \tag{6.120}$$

For BER to be less than 10^{-9}, it is necessary that $Q > 6$ (7.8 dB).

From Equation (6.119), when the signal current i_s is greater than the noise, the BER will be small, but the bigger the BER, the smaller the signal current. This relation is depicted by the solid line in Figure 6.32, taking the received signal power (light intensity with a unit of dBm) in the horizontal axis and the logarithm of the BER in the vertical axis.* The characteristics of this solid line correspond to the signal without any waveform distortion (pulse broadening, etc.). In the actual measurement of an optical communications system, the optical signal from a transmitter will be transmitted on a very short optical fiber (about several meters), and the signal is sent to a receiver. Because the BER is measured and obtained while the received light intensity is being changed by a variable attenuator (a device that attenuates light intensity of any dB value) that is connected in front of the receiver, this is called the *back-to-back characteristic*. Then, the received power with a BER, which became 10^{-9}, is called the *receiver sensitivity*.

---

* Conversely, even if the received light intensity is too strong, a saturation phenomenon in the photodetector output takes place and the BER increases, so there is an optimum range of the received light intensity for the BER characteristic to be below the 10^{-9} level.

Next, let us consider the BER after a long-range transmission in an optical fiber. The received signal is not only attenuated, but also pulse broadening by dispersion and a phenomenon called *jitter*, wherein the pulse signal timing (position of the pulse center on the time axis) fluctuates irregularly, may occur. In such a case, a light intensity that is stronger than that needed to obtain the same BER is required. Then, the portion of the light intensity needed to obtain a 10^{-9} BER that increased from the back-to-back characteristic is called a *power penalty*. The power penalty is a measure that expresses the extent to which the pulse waveform deteriorated with transmission.

7 Propagation and Focusing of the Beam

The propagation of a beam can be classified as a propagation in free space (or in a homogeneous medium) and a propagation in a guided structure.

Free-space propagation can be described as propagation by diffraction of light, whereas propagation of light in a waveguide can be explained by the eigenmode expansion of the waveguide. In this chapter, we will discuss the propagation of light in a dielectric medium with a homogeneous, isotropic dielectic constant ε, including vacuum.

7.1 GAUSSIAN BEAM

The propagation of light as an electromagnetic wave exactly satisfies the following vector wave equation, which is derived from Maxwell's equations:

$$\nabla^2 E - \mu_0 \varepsilon \frac{\partial^2 E}{\partial t^2} = 0. \tag{7.1}$$

Here, with the propagation direction of the light as the z-axis, and considering a linearly polarized light beam of which electric field is in a particular direction of the xy plane, the electric field is expressed by

$$E(r,t) = e_p \Psi(r,t). \tag{7.2}$$

Here, e_p is the unit vector representing the polarization direction, and $\Psi(r,t)$ is the function expressing the electric field amplitude distribution of the beam. As described in Section 6.1.8, this approximation, wherein the electric field (and the magnetic field) is assumed to exist only in the plane, which is perpendicular to the direction of the propagation, is called the *transverse electromagnetic (TEM) wave approximation*. When this equation is substituted to Equation (7.1), the following *Helmholtz equation* can be obtained:

$$\nabla^2 \Psi - \frac{n^2}{c^2} \frac{\partial^2 \Psi}{\partial t^2} = 0. \tag{7.3}$$

Here, ε, ε_0, and the refractive index n can be related by the following equation:

$$n = \sqrt{\frac{\varepsilon}{\varepsilon_0}}. \tag{7.4}$$

Here, with the wavenumber in the vacuum set as $k_0(= \frac{2\pi}{\lambda})$, the time t and the propagation distance z dependence is assumed as

$$E(r,t) = e_p \psi(r) \exp[j(\omega t - k_0 n z)]. \tag{7.5}$$

When the change in the ψ and the $\frac{\partial \psi}{\partial z}$ is sufficiently small in the propagation distance of the wavelength λ, and so

$$\frac{\partial \psi}{\partial z} \ll k_0 n \psi, \tag{7.6}$$

$$\frac{\partial^2 \psi}{\partial z^2} \ll k_0^2 n^2 \psi, \tag{7.7}$$

can be approximated, and the Helmholtz equation with respect to the amplitude function $\psi(r)$ is rewritten as

$$\nabla_t^2 \psi - j2k_0 n \frac{\partial \psi}{\partial z} = 0, \tag{7.8}$$

(Here, $\nabla_t^2 = \frac{\partial^2}{\partial x^2} + \frac{\partial^2}{\partial y^2}$.)

Approximations such as those in Equations (7.6) and (7.7) are called *paraxial approximations*. Here, when the solution of the equation is assumed to be

$$\psi(r) = \frac{j z_0 A_0}{z + j z_0} \exp\left[-jk_0 n \frac{x^2 + y^2}{2(z + j z_0)}\right], \tag{7.9}$$

which is substituted to Equation (7.8), we can see that it is indeed the solution to the equation. Here, A_0 and z_0 are constants, and z_0 is called the Rayleigh range (depth of focus as confocal parameter). Moreover, when the parameters are defined as follows:

$$w(z) = w_0 \left[1 + \left(\frac{z}{z_0}\right)^2\right]^{1/2}, \tag{7.10}$$

$$R(z) = z \left[1 + \left(\frac{z_0}{z}\right)^2\right], \tag{7.11}$$

$$\zeta(z) = \tan^{-1}\left(\frac{z}{z_0}\right), \tag{7.12}$$

$$w_0 = \left(\frac{\lambda z_0}{\pi}\right)^{1/2} = \left(\frac{2z_0}{k_0 n}\right)^{1/2}, \tag{7.13}$$

$\psi(r)$ can be expressed by the following equation:

$$\psi(r) = A_0 \frac{w_0}{w(z)} \exp\left[-\frac{x^2 + y^2}{w^2(z)}\right] \cdot \exp\left[-jk_0 n \frac{x^2 + y^2}{2R(z)} + j\zeta(z)\right]. \tag{7.14}$$

As can be seen from Equation (7.14), the amplitude portion has the form of a Gaussian function, and as already explained in Section 3.3, this is a *Gaussian beam*. $w(z)$ is defined by the distance between the center axis and the point where the electric field amplitude of the beam falls down to $\frac{1}{e}$, and is called a spot size. Because the spot size $w(z)$ at $z = 0$ will have a minimum value w_0, and the diameter of the beam is smallest at this position, it is called a beam waist.

Equations (7.10) and (7.11) are often expressed using the spot size w_0 at the beam waist and not with z_0, so when Equation (7.13) is used, both equations can be rewritten as follows:

$$w(z) = w_0 \left[1 + \left(\frac{2z}{k_0 n w_0^2} \right)^2 \right]^{1/2}, \qquad (7.15)$$

$$R(z) = z \left[1 + \left(\frac{k_0 n w_0^2}{2z} \right)^2 \right]. \qquad (7.16)$$

7.2 PROPAGATION OF THE GAUSSIAN BEAM

The state of the propagation of the Gaussian beam has been described in Equation (7.14). Using the amplitude distribution, the intensity distribution can be expressed as $I(r) = \frac{1}{2}\sqrt{\frac{\varepsilon}{\mu_0}}|\psi(r)|^2 = \frac{1}{2}Y_0 n|\psi(r)|^2$ (here $Y_0 = \sqrt{\frac{\varepsilon_0}{\mu_0}}$ is the admittance of the vacuum), so

$$I(r, z) = \frac{A_0^2}{2} Y_0 n \left(\frac{w_0}{w(z)} \right)^2 \exp \left[-\frac{2r^2}{w^2(z)} \right] \qquad (7.17)$$

is obtained. Here, $r^2 = x^2 + y^2$. Because the total power P is determined by

$$P = \int_0^\infty I(r, z) 2\pi r \, dr = \frac{1}{4} A^2 Y_0 n (\pi w_0^2), \qquad (7.18)$$

the intensity distribution is normalized by the total power and expressed by the following equation:

$$I(r, z) = \frac{2P}{\pi w^2(z)} \exp \left[-\frac{2r^2}{w^2(z)} \right]. \qquad (7.19)$$

On the other hand, the *spot size* is given by Equations (7.10) and (7.15). If the beam is propagated far enough, that is, in the range of $z \gg z_0$ (Fraunhofer region), it can be approximated from Equation (7.15) as

$$w(z) \simeq \frac{2}{k_0 n w_0} z = \frac{\lambda}{\pi n w_0} z, \qquad (7.20)$$

so the *beam spread angle* $2\theta_D$ can be expressed as

$$2\theta_D = 2\tan^{-1}\left(\frac{w(z)}{z}\right) = 2\tan^{-1}\left(\frac{\lambda}{\pi n w_0}\right). \qquad (7.21)$$

Moreover, when $\theta_D \ll 1$, the $\tan\theta_D \simeq \theta_D$ (or $\tan^{-1} x \simeq x$) approximation can be used, so Equation (7.21) will be approximated as

$$2\theta_D \simeq \frac{2\lambda}{\pi n w_0} = 1.273\frac{\lambda}{2w_0 n}. \qquad (7.22)$$

When $n = 1.0$, we can see that it is identical to Equation (3.28). Moreover, as discussed in Section 3.3, since the NA of the beam is defined by $\sin\theta_D$, NA is expressed by

$$\text{NA} = \frac{\lambda}{\sqrt{\lambda^2 + (\pi n w_0)^2}}, \qquad (7.23)$$

and when $n = 1.0$, we can see that it is also identical to Equation (3.29). When $\theta_D \ll 1$, NA can be approximated as

$$\text{NA} \simeq \frac{\lambda}{\pi n w_0}. \qquad (7.24)$$

On the other hand, the phase term is given by the following equation:

$$\varphi(r, z) = k_0 n z - \zeta(z) + \frac{k_0 n r^2}{2R(z)}. \qquad (7.25)$$

Here, $\zeta(z)$ is a value of at most $\pm\frac{\pi}{2}$, so it is ignored. If the beam is propagated far enough so that $z \gg z_0$ holds true, then $R(z)$ can be approximated by the propagation distance L, and the point with the same phase as the $z = L$ position on the z-axis is on a curved surface that satisfies

$$\varphi(r, z) = k_0 n\left[z + \frac{r^2}{2z}\right] = k_0 n L. \qquad (7.26)$$

Equation (7.26) represents a parabolic surface; however, in the paraxial approximation range, the approximation $r \ll z$ can be used, so the equation will become an equation of a spherical surface such as the following:

$$\sqrt{z^2 + r^2} = L. \qquad (7.27)$$

That is, at a far enough distance, the amplitude distribution of the Gaussian beam is expressed by the Gaussian function, and the wave front forms the spherical surface, wherein the center is positioned at the beam waist. The propagation of the Gaussian beam is illustrated from these equations as shown in Figure 7.1. That is, the Gaussian beam diffraction in the Fraunhofer region was determined in Section 3.3, but it can be seen that even in the Fresnel region, the Guassian beam amplitude distribution can be expressed as a Gaussian function

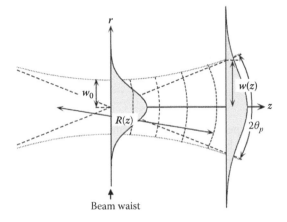

FIGURE 7.1   Propagation of a Gaussian beam.

and the wavefront as a spherical surface (however, the radius of curvature is longer than z, so the center of the curvature is farther than $z = 0$ to the negative side).

7.3   WAVE COEFFICIENT AND MATRIX FORMALISM

The changes in the amplitude distribution and phase distribution that a Gaussian beam passing through an optical component undergoes can be calculated easily using the following complex parameter called *wave coefficient*:

$$P = \frac{2}{w^2(z)} + j\frac{k_0 n}{R(z)}. \tag{7.28}$$

The real part of the wave coefficient represents the spot size, whereas the imaginary part expresses the radius of the curvature of the wave front. When this P is used, Equation (7.14) will be rewritten as follows (here, the $\zeta(z)$ term is ignored).

$$\psi(r) = A_0 \frac{w_0}{w(z)} \exp\left[-\frac{1}{2}P(x^2 + y^2)\right]. \tag{7.29}$$

This wave coefficient undergoes a certain transformation by passing through an optical component. First, when Equations (7.15) and (7.16) are substituted to Equation (7.28),

$$\frac{1}{P(z)} = \frac{w_0^2}{2} - j\frac{z}{k_0 n} = \frac{1}{P(0)} - j\frac{z}{k_0 n} \tag{7.30}$$

can be confirmed. Consequently, if the wave coefficients at the position z_1 and z_2 on the propagation axis are designed as P_1 and P_2, respectively, either of these can be related to the wave coefficient $P(0)$ at the beam waist by Equation (7.30). Therefore, by eliminating $P(0)$, the following equation is obtained:

$$P_1 = \frac{P_2}{1 + j\frac{1}{k_0 n}(z_2 - z_1)P_2}. \qquad (7.31)$$

This equation has a special form of linear transformation, and in general, when the wave coefficients before and after passing through the optical component are designated as P_1 and P_2, the transformation of the wave coefficient of the Gaussian beam is given by the following equation:

$$P_1 = \frac{AP_2 + B}{CP_2 + D}. \qquad (7.32)$$

Here, A, B, C, and D are the elements of the F matrix

$$F = \begin{bmatrix} A & B \\ C & D \end{bmatrix}. \qquad (7.33)$$

The F matrices of various optical components are shown in Table 7.1.

This F matrix is the same as the matrix used in calculating the impedance of the 2×2 terminal pair circuit of an electric circuit, and when the voltage and the current at the input end in the electric circuit is set as V_1, and I_1, and the voltage and the current at the output end is designated as V_2, and I_2, the relation among them can be expressed as

$$\begin{bmatrix} V_1 \\ I_1 \end{bmatrix} = \begin{bmatrix} A & B \\ C & D \end{bmatrix} \cdot \begin{bmatrix} V_2 \\ I_2 \end{bmatrix}. \qquad (7.34)$$

When the relation of the input end and the output end impedances $Z_1 = \frac{V_1}{I_1}$ and $Z_2 = \frac{V_2}{I_2}$ are determined from Equation (7.34), Z_1 can be written in terms of Z_2 as

$$Z_1 = \frac{AZ_2 + B}{CZ_2 + D}. \qquad (7.35)$$

Thus, we can see that the wave coefficient corresponds exactly to the impedance. This F matrix is called a cascade matrix, and when N of 2×2 terminal pair circuits is connected in a cascade, the F matrix of the whole circuit can be calculated easily by

$$F = F_1 \cdot F_2 \cdots F_N. \qquad (7.36)$$

Now, in an optical component, it is convenient to determine the wave coefficient at the output end from the wave coefficient at the input end. In such a case,

TABLE 7.1

F Matrices of Various Optical Components

Optical device	F matrix
Homogeneous medium	$\begin{bmatrix} 1 & 0 \\ j\dfrac{z}{k_0 n} & 1 \end{bmatrix}$
$n_1 \quad n_2$ Flat boundary	$\begin{bmatrix} 1 & 0 \\ 0 & 1 \end{bmatrix}$
R $n_1 \quad n_2$ Spherical boundary ($R>0$ when center is in the n_2 side)	$\begin{bmatrix} 1 & jk_0(n_2-n_1)\dfrac{1}{R} \\ 0 & \end{bmatrix}$
f $n_1 \;\; n_2 \;\; n_1$ Thin convex lens or concave mirror (f is the value in medium)	$\begin{bmatrix} 1 & jk_0\dfrac{1}{f} \\ 0 & 1 \end{bmatrix}$
r $0 \qquad z$ Distributed index rod lens ($n^2(r) = n^2(0)[1-(gr)^2]$)	$\begin{bmatrix} \cos gz & jk_0 n(0)g \sin gz \\ j\dfrac{1}{k_0 n(0)g}\sin gz & \cos gz \end{bmatrix}$

In a concave lens, $f<0$. When the focal length in vacuum is designated as f'; $\dfrac{1}{f} = \dfrac{n_2-n_1}{n_2-n_1}\dfrac{1}{f'}$

as the inverse matrix is easily determined using the property $|F| = 1$ of the
F matrix, P_2 is expressed in terms of P_1 and the equation is given as follows:*

$$P_2 = \frac{DP_1 - B}{-CP_1 + A}. \tag{7.37}$$

PROBLEM

1. A Gaussian beam is incident on the front side of the lens with a focal
   distance f and a diameter D. The incident plane of the lens is the
   beam waist, and the spot size is w_1 (here, $w_1 \simeq \frac{D}{4}$). Determine the
   focal position z and the spot size w_2 of the focal spot behind the lens.

   [Hint]: The F matrix of the lens and the free space should be multiplied. The beam waist will be formed again at the focusing position,
   so the imaginary part of the wave coefficient will become 0. The focal
   position is determined from this condition, but it will be slightly behind
   the focal distance f.

7.4   PROPAGATION OF NON-GAUSSIAN BEAM

In general, a light beam propagating in an optical system has an intensity distribution that is close to a Gaussian beam, but it is rare that they are exactly identical
to the Gaussian beam. Here, as the solution for the Helmholtz equation (7.8),
substituting not Equation (7.9) but the following approximated equation to
Equation (7.8)

$$\psi(r) = \mathcal{X}\left(\frac{\sqrt{2}x}{w(z)}\right) \mathcal{Y}\left(\frac{\sqrt{2}y}{w(z)}\right) \exp[j\mathcal{Z}(z)]\frac{jz_0 A_0}{z + jz_0}$$
$$\times \exp\left[-jk_0 n\frac{x^2 + y^2}{2(z + jz_0)}\right], \tag{7.38}$$

the following equation is obtained:

$$\frac{1}{\mathcal{X}}\left(\frac{\partial^2 \mathcal{X}}{\partial \xi^2} - 2\xi\frac{\partial \mathcal{X}}{\partial \xi}\right) + \frac{1}{\mathcal{Y}}\left(\frac{\partial^2 \mathcal{Y}}{\partial \eta^2} - 2\eta\frac{\partial \mathcal{Y}}{\partial \eta}\right) + k_0 n w^2(z)\frac{\partial \mathcal{Z}}{\partial z} = 0. \tag{7.39}$$

Here, the variables were converted by $\xi = \frac{\sqrt{2}x}{w(z)}$ and $\eta = \frac{\sqrt{2}y}{w(z)}$. Because the
variables of this equation are separated, it can be easily solved as follows:

---

* In several references [6], the wave coefficient is defined as $\frac{1}{q(z)} = \frac{1}{R(z)} - j\frac{\lambda}{\pi n w^2(z)}$, and there is
also another way of defining the transformation by the F matrix as $q_2 = \frac{A'q_1 + B'}{C'q_1 + D'}$. The transformation between the elements of both F matrices can be easily done by $A = A'$, $B = -jk_0 n_2 C'$, $C = -\frac{B'}{jk_0 n_1}$, and $D = \frac{n_2}{n_1}D'$.

$$\psi_{p,q}(x, y, z) = C_{p,q} N_p N_q \cdot H_p\left(\sqrt{2}\frac{x}{w(z)}\right) H_q\left(\sqrt{2}\frac{y}{w(z)}\right) \exp\left[-\frac{x^2 + y^2}{w^2(z)}\right]$$

$$\times \exp\left[-jk_0 n\frac{x^2 + y^2}{2R(z)} + j(p + q + 1)\zeta(z)\right]. \tag{7.40}$$

Here, $H_i(x)$ is an Hermite polynomial of order i, and some lower order functions are shown in Table 5.4 of Section 5.2.1(2). N_i is a normalization constant, which is given by

$$N_i = \frac{1}{\left[2^i i! w(z)\sqrt{\frac{\pi}{2}}\right]^{1/2}} \quad (i = p \text{ or } q). \tag{7.41}$$

In Equation (7.40), the mode order in the x-direction is p, and the mode order in the y-direction is q, and they are called the *Hermite–Gaussian mode*. $p = q = 0$ refers to a Gaussian beam. The Hermite–Gaussian function constitutes an orthonormal function system. That is, when $G_{p,q}(x, y)$ is defined as

$$G_{p,q}(x, y) = N_p N_q H_p\left(\sqrt{2}\frac{x}{w}\right) H_q\left(\sqrt{2}\frac{y}{w}\right) \exp\left[-\frac{x^2 + y^2}{w^2}\right], \tag{7.42}$$

the function $G_{p,q}(x, y)$ satisfies the following orthogonal relationship:

$$\iint_{-\infty}^{\infty} G_{p,q}(x, y) G_{\mu,\nu}(x, y) dx\, dy = \delta_{p\mu}\delta_{q\nu}. \tag{7.43}$$

The shape of the one-dimensional Hermite–Gaussian function $G_p(x)$ is shown in Figure 5.13.

Consequently, when an arbitrary amplitude function $U(x, y)$ (real function) is given at the position of the beam waist, this can be expressed by the orthogonal function expansion using the Hermite–Gaussian function as follows:

$$U(x, y) = \sum_{i=0}^{\infty}\sum_{j=0}^{\infty} C_{i,j} G_{i,j}(x, y). \tag{7.44}$$

The expansion coefficient in such a case is determined by

$$C_{\mu,\nu} = \frac{\displaystyle\iint_{-\infty}^{\infty} G_{\mu,\nu}(x, y) U(x, y) dx\, dy}{\displaystyle\iint_{-\infty}^{\infty} U^2(x, y) dx\, dy}. \tag{7.45}$$

Then, after the expansion coefficient of any amplitude distribution at the beam waist has been determined by Equation (7.45), each Hermite–Gaussian beam propagation is calculated from Equation (7.40), and finally, combining by multiplying with the expansion coefficient, the propagation of any beam can be calculated. For convenience in calculation, the function $U(x, y)$ is normalized as $\iint_{-\infty}^{\infty} U^2(x, y) dx\, dy = 1$.

7.5   CALCULATION FORMULA FOR SPOT SIZE

It was shown in Section 7.4 that a beam with an arbitrary amplitude distribution can be expanded in terms of an Hermite-Guassian function, but actually, the spot size of the function $G_{\mu,\nu}(x, y)$ that was used in Equation (7.45) for the expansion remains an arbitrary constant. In addition, the original shape of the beam is often close to a Gaussian function; however, if the expansion coefficient is determined by Equation (7.45) while setting the value of spot size far from the actual beam size, it will not converge up to a very high-order expansion coefficient, and the degree of approximation for the orthogonal function expansion will be poor. Thus, a calculation formula for an equivalent *spot size* of a beam, which has an amplitude distribution close to a Gaussian function, is necessary.

Several calculation formulas have been proposed so far for the simplest one-dimensional distribution (amplitude with a function $\psi(x)$ only in x and propagating to the z-axis). One of these is a calculation formula [10] based on the second moment of the intensity distribution given by

$$w_q = \left[\frac{4\int_{-\infty}^{\infty} x^2 \{\psi(x)\}^2 dx}{\int_{-\infty}^{\infty} \{\psi(x)\}^2 dx} \right]^{\frac{1}{2}}, \qquad (7.46)$$

and another one is the Villuendas calculation formula [11] 7.47. This formula is obtained by transforming the Petermann's formula [9] that is used in optical fibers into a one-dimensional coordinate system

$$w_V = \left[\frac{\int_{-\infty}^{\infty} \{\psi(x)\}^2 dx}{\int_{-\infty}^{\infty} \left\{ \frac{d\psi(x)}{dx} \right\}^2 dx} \right]^{\frac{1}{2}}. \qquad (7.47)$$

However, the accuracy of these formulas is not so good,* and originally, the form of the Gaussian beam could not be changed by Fourier transform along with diffraction, but in these calculation formulas, the form would change in the far-field pattern and the near-field pattern. That is, the Villuendas formula (7.47) and Equation (7.46) of the second moment of the intensity distribution are mutually

---

* The Gaussian beam spot size is defined as the distance from the center up to the point where the amplitude distribution becomes $\frac{1}{e}$, but one can come across definitions of the spot size for non-Gaussian beams, where approximations are done by simply expressing it as the distance from the center up to the point, when the amplitude distribution becomes $\frac{1}{e}$. This approximation is the most inaccurate, and the error in the spot size is determined by least squares approximation at the single-mode region, where $V < \frac{\pi}{2}$ ranges from 5% to 15%. Consequently, of late, the definition of the spot size as the distance up to the point where the amplitude becomes $\frac{1}{e}$ is not used anymore, even in international standards.

related by the Fourier transform and the inverse Fourier transform, so these are contradictory to the approximation formula for the Gaussian function.

Therefore, the authors have proposed the following calculation formula [12], based on a successive approximation, from a condition that any one-dimensional electric field distribution $\psi(x)$ is approximated by a least squares approximation* using a Gaussian function

$$
w_0 = \left[\frac{4 \int_{-\infty}^{\infty} x^2 \{\psi(x)\}^2 dx}{\int_{-\infty}^{\infty} \{\psi(x)\}^2 dx} \right]^{\frac{1}{2}},
\tag{7.48}
$$

$$
w_{\nu+1} = \left[\frac{4 \int_{-\infty}^{\infty} x^2 \exp\left[-\left(\frac{x}{w_\nu}\right)^2 \right] \psi(x) dx}{\int_{-\infty}^{\infty} \exp\left[-\left(\frac{x}{w_\nu}\right)^2 \right] \psi(x) dx} \right]^{\frac{1}{2}}.
\tag{7.49}
$$

Here, ν is the order of the successive approximation ($\nu = 0$ corresponds to Equation (7.48), and in the successive approximation for $\nu = 1$ or greater, Equation (7.49) is used), and $\psi(x)$ is the one-dimensional amplitude distribution function. In addition, Equation (7.48) for the zero order approximation has the same form as the equation for the second moment of the intensity distribution [10]. Now, Equation (7.49) does not change its form by Fourier transform, and the equation is also highly precise.

In the case of a two-dimensional distribution, the spot size w_x of the x-direction and the spot size w_y of the y-direction are determined separately and may be expressed by Equation (8.14).

In optical fibers, the following Petermann's formula is often used [9] (refer to Appendix K for the derivation)

$$
w_P = \left[\frac{2 \int_0^{\infty} \{\psi(r)\}^2 r\, dr}{\int_0^{\infty} \left\{ \frac{d\psi(r)}{dr} \right\}^2 r\, dr} \right]^{\frac{1}{2}}.
\tag{7.50}
$$

This formula is adopted in international standards; however, there is a problem in its precision. Therefore, for the least squares approximation using a Gaussian function, the following successive approximation formulas are appropriate:

---

* The least squares approximation using the Gaussian function is equivalent to the condition that the expansion coefficient of the zero-order Hermite–Gaussian function (i.e., the Gaussian function itself) is maximized, when the function is expanded in terms of the Hermite–Gaussian function. Equation (7.49) is derived from this condition.

$$w_0 = \left[\frac{2 \int_0^\infty \{\psi(r)\}^2 r^3 dr}{\int_0^\infty \{\psi(r)\}^2 r \, dr} \right]^{\frac{1}{2}}, \tag{7.51}$$

$$w_{\nu+1} = \left[\frac{2 \int_0^\infty \exp\left[-\left(\frac{r}{w_\nu}\right)^2 \right] \psi(r) r^3 dr}{\int_0^\infty \exp\left[-\left(\frac{r}{w_\nu}\right)^2 \right] \psi(r) r \, dr} \right]^{\frac{1}{2}}. \tag{7.52}$$

Now, Equations (7.48) and (7.49) for the approximation of the spot size of a two-dimensional waveguide, and Equations (7.51) and (7.52) for the approximation of the spot size of an optical fiber give the spot size in the near-field pattern. However, in the actual measurement, the spatial resolution of the light intensity in the far-field pattern measured by a photodetector and a fine manipulator with a motor is greater than that of the light intensity in the near-field pattern obtained from a microscope and an infrared imaging camera. Here, the calculation formula for the spot size in the far-field pattern is necessary, but fortunately, Equations (7.49) and (7.52) do not change their forms even with Fourier transform, so if the electromagnetic field distributions $\psi(x)$ and $\psi(r)$ in these equations are replaced with the electromagnetic field distributions $\Psi(\Omega_x)^*$ (Ω_x is the spatial angular frequency defined in Equation (3.21) of Section 3.2) and $\Psi(\Omega_r)$ in the far-field pattern, the spot size W in the far-field pattern can be obtained. Then, as the spot size of the near-field pattern and the spot size of the far-field pattern is related by $w = \frac{2}{W}$, the spot size of the near-field pattern (i.e., the guided mode) can be determined from this equation.

PROBLEM 7-1

Express the equation giving the spot size of the fundamental mode of the symmetrical three-layer slab waveguide in terms of the normalized parameters V, b, and $\frac{a}{\lambda}$.

SOLUTION 7-1

When the field distribution is expressed, the function in the core is expressed in the form of $\cos \kappa x$, whereas the function in the cladding

---

* Because the measured value of the far-field pattern is the light intensity distribution, then $I(x') \propto |\Psi(x')|^2$. So, the electromagnetic field distribution may be calculated by $\Psi(x') \propto \sqrt{I(x')}$ (An apostrophe was placed on the x coordinate because according to the coordinate system in Figure 3.1 of Section 3.1, the coordinate for the far-field pattern was set as x'). As can be seen from Equations (7.48), (7.49), (7.51), and (7.52), the electromagnetic field distribution function is included in both the denominator and the numerator, so it is not necessary to measure the absolute power, but it is enough if only the function of the distribution shape is determined to calculate the spot size.

is represented by $e^{-\gamma x}$, so κ and γ are included as parameters. Thus, these are expressed using Equations (6.13) and (6.14), and $\cos\{V\sqrt{1-b}(\frac{x}{a})\}$ and $\exp[-V\sqrt{b}(\frac{x}{a})]$, respectively, which are normalized representations. Substituting these field distribution equations to Equation (7.48),

$$\frac{w_0}{a} = 2\left[\frac{2b\sqrt{b}\,(1-b)\,V^3 + 6b\,(1-b)\,V^2 + 3\sqrt{b}\,(2-3b)\,V + 3\,(1-2b)}{6V^2\,(1-b)\,b\left(V\sqrt{b}+1\right)}\right]^{\frac{1}{2}}$$

(7.53)

is finally obtained.   **[End of Solution]**

From Equation (7.53), we can see that the spot size normalized by the core half-width a is dependent on the V parameter (as b is also dependent on V). Consequently, $\frac{w}{a}$ can be calculated with respect to the V parameter, as shown in Figure 7.2.

However, with the normalized spot size, it is difficult to understand the principle of a spot size converter that gradually changes the core thickness (this kind of structure is called a taper structure). Thus, when the refractive index difference between the core and the cladding is assumed to be constant, the dependence of the spot size on the core thickness is calculated as shown in Figure 7.3.

The spot size $\frac{wn_1}{\lambda}$ that was normalized by the λ/n_1 with respect to the core half-width $\frac{an_1}{\lambda}$ that was normalized by the wavelength λ/n_1 in the waveguide, when $\Delta = 1\%$, 5%, and 10% is shown in Figure 7.3. The straight dashed–dotted line that is rising diagonally up to the right shows the limit of the single-mode region, with the multimode region to the lower right of this line. From Figure 7.3, we can see that when the core thickness of the waveguide with constant refractive index difference between the core and the cladding decreases

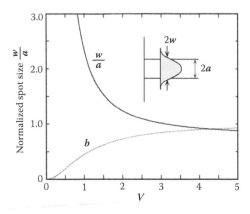

FIGURE 7.2 V parameter dependence of the normalized spot size w/a.

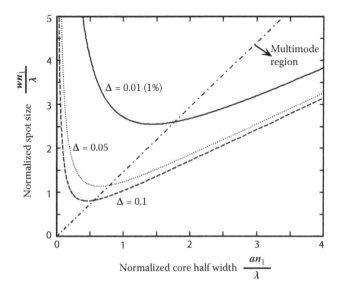

FIGURE 7.3 Normalized core half-width an_1/λ dependence of the normalized spot size wn_1/λ.

gradually, the spot size will decrease in proportion to the core thickness in the multimode region. In the single-mode region, however, the spot size will not decrease even if the core thickness is reduced. Consequently, with a waveguide device that converts the spot size (spot size converter), a high conversion rate is not achieved in a waveguide with the same refractive index difference by just changing the core thickness, so devises such as tapering and superimposition [13], [14] are necessary in a waveguide of different refractive index difference. On the other hand, we can see that even in the single-mode region, when the core thickness decreases below a certain limit, the spot size will increase. This phenomenon occurs because when the core thickness becomes thin, the optical confinement factor in the core decreases, and the electromagnetic field distribution will significantly penetrate into the cladding. Consequently, the field distribution will no longer be a rounded shape like a Gaussian function, but will be like that of Mt. Fuji, wherein the base spreads out like an exponential function. However, as long as the approximated spot size is close to the spot size of the Gaussian function, the overlap integral of the electromagnetic field with the Gaussian function will not be so small (coupling efficiency is more than 90%). Therefore, in the spot enlargement portion of the spot size converter, a structure [15] for thinning the core thickness is sometimes used for the purpose of high-efficiency coupling, specifically of semiconductor lasers and optical fibers.

Now, in Figure 7.3, the condition for each of the curved lines to have a minimum value is given by the same normalized value, that is, $V = 1.18$.

7.6 REPRESENTATION BY DIFFRACTION INTEGRAL

The propagation of an optical beam with an arbitrary complex amplitude distribution can be expressed by the *Fresnel-Kirchhoff diffraction integral* that was discussed in Section 3.1. When the complex amplitude distribution at $z = 0$ is designated as $\psi(x, y, 0)$ and the complex amplitude distribution after propagating at a distance z is designated as $\Psi(x', y', z)$, their relationship is given by

$$\Psi(x', y', z) = \frac{j}{\lambda} \exp(-jk_0 nz) \iint_{-\infty}^{\infty} \psi(x, y, 0) \frac{\exp(-jk_0 nr)}{r} \, dx \, dy. \quad (7.54)$$

Here,

$$r = \sqrt{z^2 + (x - x')^2 + (y - y')^2}. \quad (7.55)$$

Apart from the coefficient portion of this equation, the double integral portion is expressed by Huygen's principle, which is intuitively easy to understand. Moreover, using the paraxial approximation (i.e., in the Fresnel region), $\Psi(x', y', z)$ can be rewritten as

$$\Psi(x', y', z) = \frac{j}{\lambda z} \exp(-jk_0 nz) \iint_{-\infty}^{\infty} \psi(x, y, 0)$$
$$\times \exp\left[-jk_0 n \frac{(x - x')^2 + (y - y')^2}{2z} \right] dx \, dy. \quad (7.56)$$

In addition, when the medium is not homogeneous with refractive index of n but one of optical components as shown in Table 7.1, the diffraction integral can be calculated using the F matrix [40] and is expressed in the one-dimensional form for simplicity as

$$\Psi(x') = j \left(\frac{\lambda}{2\pi n_1 C} \right)^{\frac{1}{2}} \int_{-\infty}^{\infty} \psi(x) \exp\left[\frac{Ax^2 - 2xx' + Dx'^2}{2n_1 C} \right] dx. \quad (7.57)$$

On the other hand, although the Hermite–Gaussian function was derived in this section as the solution for the Helmholtz equation, the plane wave as another orthogonal function is also a solution. Therefore, expansion using a plane wave is also possible. In particular, when the paraxial approximation cannot be applied, a more accurate understanding of the optical beam propagation may be possible using the plane wave expansion.

8 Basic Optical Waveguide Circuit

8.1 COUPLING BY CASCADE CONNECTION OF OPTICAL WAVEGUIDES

8.1.1 GENERAL FORMULA FOR COUPLING EFFICIENCY

When a beam with an arbitrary electromagnetic field distribution is incident on an optical waveguide, the optical power is expanded into each eigenmode as shown in Figure 5.7 and the propagation starts. Then, the *coupling coefficient* for each mode is given by Equations (5.62a) and b. Using the same approach when two optical waveguides are connected, the electromagnetic field of each of the outgoing modes of the waveguide at the exit side may be expanded to each of the eigenmodes of the waveguide at the incident side. Then, the optical power coupled to the radiation mode will be the *coupling loss*.

In the case of both waveguides being single-mode waveguides, if the waveguide at the exit side is designated as #1 and the waveguide at the incident side is set as #2, as shown in Figure 8.1, the transverse electromagnetic field distribution of each of the fundamental modes is expressed as $E^{(1)}(x)$ and $E^{(2)}(x)$. When the characteristic polarization axis is assumed to be the y-direction (TE polarization), either of the components will be the E_y component. When the $E^{(1)}(x)$ is expanded in terms of the eigenmodes of waveguide #2, there is only one guided mode, so $E^{(1)}(x)$ will be expressed by

$$E^{(1)}(x) = C_0 E^{(2)}(x) + \int_0^\infty C_\sigma E_\sigma^{(2)}(x)d\sigma. \tag{8.1}$$

When the orthogonality of the eigenmode is used here, and $E^{(2)*}(x)$ (* represents the complex conjugate, which is necessary in negating the oscillation term $e^{j(\omega t - \beta z)}$) is multiplied to both sides of Equation (8.1), and then integrated, the following equation is obtained:

$$C_0 = \frac{\int_{-\infty}^\infty E^{(1)}(x)E^{(2)*}(x)dx}{\left[\int_{-\infty}^\infty |E^{(2)}(x)|^2 dx\right]^{\frac{1}{2}}}. \tag{8.2}$$

The optical power of each of the eigenmodes was normalized in Equations (5.62a) and b so that $P_z^{(\mu)}$ of Equation (5.58) is 1; however, here, this normalized

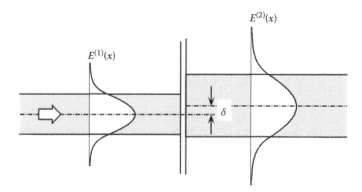

FIGURE 8.1   Butt coupling between two waveguides.

constant is expressed explicitly in this equation. Originally, the orthogonality relationship is obtained by multiplying both sides of Equation (8.1) by $H_x^{(2)}$ and then integrating them, but according to Equation (5.32), H_x is a constant multiple of E_y and so the calculation can be simplified this way. Moreover, when the optical power of mode $E^{(1)}(x)$ is also normalized based on Equation (5.63), and defined as

$$\eta = \frac{\int_{-\infty}^{\infty} |C_0 E^{(2)}(x)|^2 dx}{\left[\int_{-\infty}^{\infty} |E^{(1)}(x)|^2 dx \right]} = \frac{\left| \int_{-\infty}^{\infty} E^{(1)}(x) E^{(2)*}(x) dx \right|^2}{\left[\int_{-\infty}^{\infty} |E^{(1)}(x)|^2 dx \right] \left[\int_{-\infty}^{\infty} |E^{(2)}(x)|^2 dx \right]}, \quad (8.3)$$

Equation (8.3) expresses the *coupling efficiency* from waveguide #1 to #2. When $E^{(1)}(x)$ and $E^{(2)}(x)$ each satisfies

$$\int_{-\infty}^{\infty} |E^{(i)}(x)|^2 dx = 1 \qquad (i = 1 \text{ or } 2), \qquad (8.4)$$

the electromagnetic field distribution is referred to as normalized, and the coupling efficiency of such normalized mode electromagnetic field distribution will only be expressed using the numerator portion because the denominator of Equation (8.3) is 1.0.

As can be seen in Equation (8.3), when $E^{(1)}(x)$ and $E^{(2)}(x)$ are equal, η is 1.0, but if they are not equal, it will always be smaller than 1. Normally, the coupling efficiency that is expressed in the form of $-10 \log_{10} \eta$ in dB is called the *coupling loss*. Because this equation was derived by assuming the mode of a slab waveguide, the integral is expressed in the one-dimensional case, but in three-dimensional waveguides, such as optical fibers, the integral may be replaced with a double integral in the xy cross-section.

Now, if an optical beam is beamed from the incident end of the waveguide as shown in Figure 5.7, one has to be aware of the point where the situation is somewhat different from the eigenmode expansion that was discussed in Equations (5.59) and (5.60) in Section 5.1.4. In other words, it should be noted that the equation of eigenmode expansion represents an optical beam (not necessarily limited to a guided mode, but may be a transient state optical beam just after being beamed from the incident end) that was already propagated along the waveguide by the superposition of the eigenmode. Therefore, the electromagnetic field distribution that should be expanded is the "optical beam that was already propagated along the waveguide." However, in the situation where the optical beam is launched from the incident end of the waveguide, the determination of the expansion coefficient of the electromagnetic field of the optical beam in front of the incident end of the waveguide in terms of the guided modes is generally the problem that has to be solved. Thus, to apply the eigenmode expansion to this problem, the field distribution of the optical beam in front and behind the incident end of the waveguide is determined, and then, the eigenmode expansion needs to be applied with respect to the field distribution immediately following the incidence of the optical beam. Determining the optical beam in front and behind the incident end means the determination of the reflectance and transmittance of the optical beam at the boundary surface of the incident end. Even if the refractive index of space is uniform (if air, it is 1.0), the refractive indices of the core and the cladding are different for a waveguide, so the transmittance and the reflectance occurring at the boundary of the two media are problems that cannot be solved simply. That is, the plane wave is not incident, so the angle of incidence will have various components, and even at the boundary surface, the refractive indices of the core and the cladding differ. However, if the refractive index difference of the core and the cladding is small, the reflection and the transmission between the air and the core (or the cladding) may be taken into account. When the optical beam is focused using a lens and is incident on the optical waveguide at the beam waist, the wave front (equiphase surface) will be parallel to the incident end and will be closer to the incident status of the plane wave. Thus, a similar figure of the shape of the optical beam is maintained by the transmittance (as well as the reflection) occurring at the incident end. In such a case, the transmittance is approximated by the normal incidence transmittance

$$t_{eq} = \frac{2n_0}{n_0 + n_{eq}} \tag{8.5}$$

of the plane wave from a medium with a uniform refractive index n_0 to the medium with a uniform refractive index that is equal to the equivalent index n_{eq} of the guided mode. So, the coupling efficiency from the incident beam to the guided mode can be expressed by the product of the power transmittance and Equation (8.3), and represented by

$$\eta = \left(\frac{2n_0 n_{eg}}{n_0 + n_{eq}}\right)^2 \frac{\left|\int_{-\infty}^{\infty} E^{(1)}(x) E^{(2)*}(x) dx\right|^2}{\left[\int_{-\infty}^{\infty} |E^{(1)}(x)|^2 dx\right]\left[\int_{-\infty}^{\infty} |E^{(2)}(x)|^2 dx\right]}. \qquad (8.6)$$

8.1.2 MISALIGNMENT LOSS CHARACTERISTIC BY GAUSSIAN APPROXIMATION

As discussed in Sections 5.1.3 and 7.5, the field distribution of the fundamental mode of the waveguide can generally be approximated by a Gaussian function. Here, when two optical waveguides are connected with offset in the transverse direction as shown in Figure 8.1, Equation (5.56) may be calculated by shifting the central axis of one of the Gaussian functions with two different spot sizes. Here, one of the Gaussian functions with two different spot sizes w_1 and w_2 in Equation (5.56) is shifted from the center by δ in the x-axis. When these are substituted to Equation (8.3) to determine the coupling efficiency, η is obtained as

$$\eta = \frac{2w_1 w_2}{w_1^2 + w_2^2} \exp\left[-\frac{2\delta^2}{w_1^2 + w_2^2}\right]. \qquad (8.7)$$

(Refer to Appendix E for the derivation.) The characteristic of the *misalignment loss (or offset loss)* when two single-mode waveguides are butt-connected may be obtained by determining the spot sizes of each of the waveguides using Equations (7.48) and (7.49), and then substituting the results to Equation (8.7).

The offset loss characteristic is calculated using Equation (8.7), when the two spot sizes are equal, and the result is as shown in Figure 8.2.

On the other hand, the connection loss when the central axes of the two waveguides are in agreement but the spot sizes are different is shown in Figure 8.3. From this figure, we can see that *spot size matching* in the connection of two optical waveguides is extremely important. Here, several spot size converters that would convert the spot size of optical waveguides have been proposed and demonstrated [13]–[15].

Next, let us evaluate the connection loss if both axis deviation and angle deviation occur between the center axes of the optical waveguides, when both are connected as shown in Figure 8.4. Gaussian functions with different spot sizes w_1 and w_2 are considered, but now, one of the optical axes is inclined with respect to the other by an angle θ. Therefore, by setting the x'- and y'-axes as tilted around the y-axis for the coordinate axis of the waveguide at the light-receiving side with respect to the coordinate axis (x- and z-axes) of the exit waveguide, the exit Gaussian beam $\psi_1(x, z)$ (exit electric field is $E^{(1)}(x) = \psi_1(x, z)\, e^{j(\omega t - \beta^{(1)} z)}$) and the Gaussian beam $\psi_2(x', z')$ at the light-receiving side (the Gaussian beam approximating the fundamental mode of

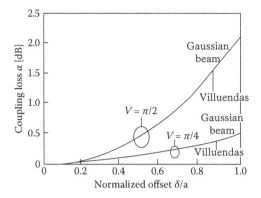

FIGURE 8.2 Characteristic of the offset loss when waveguides of equal spot sizes are butt-connected.

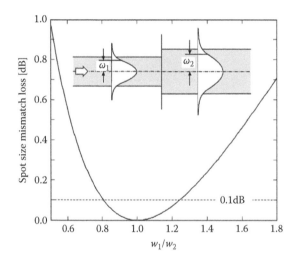

FIGURE 8.3 Characteristic of spot size mismatch loss in the butt-connection of waveguides.

the waveguide behind the connection) on the connecting surface $z = 0$ are expressed by the following equations:

$$\psi_1(x, z) = \exp\left[-\frac{x^2}{w_1^2}\right], \tag{8.8}$$

$$\psi_2(x', z') = \exp\left[-\frac{(x' - \delta)^2}{w_2^2}\right] e^{-jkz'}. \tag{8.9}$$

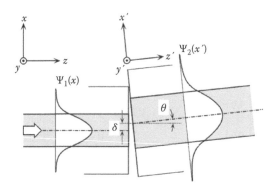

FIGURE 8.4 Butt-connection between two waveguides with axis deviation and angle deviation.

Here, the relationship of the coordinate (x', z') and the coordinate (x, z) can be expressed by

$$x' = x \cos \theta - z \sin \theta, \tag{8.10}$$

$$z' = x \sin \theta + z \cos \theta. \tag{8.11}$$

When these are used to rewrite Equation (8.9) with coordinate (x, z) (here, $z = 0$) and the overlap integral of the two Gaussian functions is calculated, it will be the following equation:

$$
\begin{aligned}
\int_{-\infty}^{\infty} \psi_1(x)\psi_2^*(x)dx &= \int_{-\infty}^{\infty} \exp\left[-\frac{x^2}{w_1^2} - \frac{(x\cos\theta - \delta)^2}{w_2^2} \right] e^{jkx\sin\theta} dx \\
&\simeq \exp\left[-\frac{\delta^2}{w_1^2 + w_2^2} - \frac{w_1^2 w_2^2 k^2 \theta^2}{4(w_1^2 + w_2^2)} \right] \\
&\quad \times \exp\left[jk\frac{w_2^2 \delta\theta}{w_1^2 + w_2^2} \right].
\end{aligned}
\tag{8.12}
$$

Here, the transformation from the first line to the second line of Equation (8.12) is obtained using Equations (A.19) and (A.7) of Appendix A because the integral has the form of the Fourier transform of the Gaussian function, and $\cos\theta \simeq 1.0$ and $\sin\theta \simeq \theta$ are used as approximations when the inclination angle is small. When this equation is substituted to Equation (8.3), the following formula is obtained:

$$\eta = \frac{2w_1 w_2}{w_1^2 + w_2^2} \exp\left[-\frac{2\delta^2}{w_1^2 + w_2^2} - \frac{w_1^2 w_2^2 k^2 \theta^2}{2(w_1^2 + w_2^2)} \right]. \tag{8.13}$$

From this formula, we can see that the coupling loss, when both axis deviation and angle deviation occur at the same time, will be the sum of the loss due to axis deviation and the loss due to angle deviation, which are each expressed in dB.

Now, Equation (8.7) is a formula for a slab waveguide, where light is confined only in the x-axis direction. However, when the spot size of the x-direction and that of the y-direction for a three-dimensional waveguide, where light is confined in the two-dimensional xy cross-section, are defined as w_x and w_y, respectively, the electromagnetic field distribution approximated using the Gaussian function is expressed by

$$ f(x, y) = \exp\left[-\left\{\left(\frac{x}{w_x}\right)^2 + \left(\frac{y}{w_y}\right)^2\right\}\right] \tag{8.14} $$

and will be of the separation of variables format. Consequently, the coupling efficiency based on the equation that expresses Equation (8.3) in two dimensions can be expressed as the product of the coupling efficiency of the x-direction and that of the y-direction, so the coupling loss expressed in dB will be expressed by the sum of the coupling loss of the x-direction and that of the y-direction.

PROBLEMS

1. Let us suppose the case where the central axes of two slab waveguides do not involve both the x-direction deviation and the angle deviation but the central axes deviation by a distance d in the z-direction. Prove that the coupling efficiency can be expressed by the following equation in terms of the respective spot sizes w_1, w_2, and d. Here, λ_g is the cavity wavelength $\frac{\lambda}{n}$.

$$ \eta = \frac{2}{\sqrt{\left(\frac{w_2}{w_1} + \frac{w_1}{w_2}\right)^2 + \frac{\lambda_g^2 d^2}{\pi^2 w_1^2 w_2^2}}}. \tag{8.15} $$

2. Determine the offset loss characteristic of an optical fiber with a circular cross-section using the Gaussian approximation.

8.1.3 Conditions for the Low Loss Connection of Optical Waveguides

Based on the calculation of the connection loss of optical waveguides that was discussed in Section 8.1.2, the following conditions must be satisfied to connect two optical waveguides (or from a spatial beam to optical waveguide) with low loss.

(a) Single-mode waveguide

*   **Alignment of the optical axis**: It is important to align the central axis and angle of the optical waveguides which are to be connected.
*   **Spot size matching**: For example, even if the sizes of the cores differ, the connection loss will be small if the spot sizes are matched as seen in Figure 8.3. Spot size matching is essentially equivalent to NA matching.
*   **Bringing the distance between the end surfaces closer**: If the end surfaces are far away, the optical beam will spread because of diffraction.
*   **Antireflection treatment at the end surfaces**: Fresnel reflection occurs when an air layer is formed between the end surfaces or when the refractive indices of the cores of the two waveguides are quite different from each other. As can be seen from Equation (8.6), this reflection may result in loss. In addition, it may bounce back to the light source causing instability in the oscillation condition of the semiconductor laser, or a Fabry–Perot resonator may be formed in between the reflection points, and cause a small wavelength dependence of the transmittance, and so it needs to be prevented. To prevent the reflection at the end surfaces, an antireflection coating is applied to the end surfaces (refer to Equation (4.69)), or a refractive index matching material of which the refractive index is close to that of the two cores is placed in between the end surfaces. In optical fiber connectors, to prevent an air layer from being formed in between the end surfaces, the end surfaces are polished into a spherical surface to induce physical contact (abbreviated as PC type).

(b) Multimode waveguide

*   **Alignment of the optical axis**: In the same way as in the single-mode waveguide, it is important to align the central axis and angle of the optical waveguides which are to be connected.
*   **Core size matching**: Unlike in the single-mode waveguide, it is necessary to match the core size and the cross-sectional shape.
*   **NA matching**: Unlike in the single-mode waveguide, in addition to the matching of core size and cross-sectional shape, it is also necessary to match the maximum acceptance angle and the NA. Strictly speaking, it is enough to make the maximum acceptance angle of the waveguide at the receiving side bigger than that of the exit waveguide; however, it may be better to match the NA from the view point of antireflective connection.
*   **Bringing the distance between the end surfaces closer**: As can be seen when taking into consideration the light rays, if the end

surfaces are far away, the number of light rays that can be beamed to the next waveguide will decrease.

- **Antireflection treatment at the end surfaces**: For the same reasons as that for the single-mode waveguide, it is important to form an antireflection coating to the end surfaces, or to prevent the formation of an air layer by physical contact or by a refractive-index-matching material.

8.2   OPTICAL COUPLING BETWEEN PARALLEL WAVEGUIDES

When two approximately equal cores (i.e., cores having approximately equal equivalent refractive indices of the fundamental modes) are brought closer up to a distance that is almost the same as that of the wavelength, the optical power between the two cores will be coupled. As shown in Figure 8.5, this kind of coupling means the exchange of the optical power in that the optical power in one of the waveguides will gradually move to the other waveguide and then bounce back gradually to the original waveguide.

The characteristics of the optical coupling between adjacent parallel waveguides can be understood precisely by analyzing the fundamental mode of the five-layer structured waveguide. That is, in the waveguide structure shown in Figure 8.5, there are modes with approximately equal (i.e., they almost degenerate) propagation constants called the even mode and the odd mode in the fundamental mode, as shown in Figure 8.6. The electric field profile of the even mode has two peaks in-phase of the two cores, whereas the odd mode has two peaks in reversed phase. Consequently, the optical power of the superimposed field profile that is dependent on this phase relationship (to be exact, in the case of the phase difference being 0 or π) will be localized to one of the cores. Then, because of the slight difference between the propagation constants of the even and the odd modes, the phase difference varies in the propagation direction, and so the optical power also periodically changes alternately in the propagation direction between the two cores, as shown in Figure 8.5.

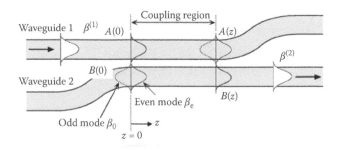

FIGURE 8.5   Optical coupling of two adjacent parallel waveguides.

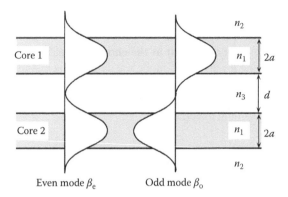

FIGURE 8.6 Two nearly degenerate eigenmodes of the coupled waveguide system.

Now, although the exact characteristics of this coupled waveguide should be analyzed by the superposition of the two quasi-degenerate modes of the five-layer waveguide as described earlier, if the spacing between the two cores is not so close (this state is called the weak coupling state), the periodic change in the optical power can be described in terms of the propagation constants of the fundamental modes of the cores when each core independently exists, using the coupled mode theory described below [41].

Let us define the electric field amplitude of the mode propagating in the cores as $A(z)$ and $B(z)$, and the propagation constants as $\beta^{(1)}$ and $\beta^{(2)}$. If these two cores exist independently, the electric field amplitudes will be expressed as $A(z)\overline{E}^{(1)}(x, y)$ and $B(z)\overline{E}^{(2)}(x, y)$ (where $\overline{E}^{(1)}(x, y)$ and $\overline{E}^{(2)}(x, y)$ are field distributions that do not include the $e^{j(\omega t - \beta z)}$ dependency), and the optical power in the core is constant. However, if the coupling occurs between the two cores, the change in the electric field amplitude is expressed by the following equation. (Refer to Appendix L and Equations (M.22) through (M.25) of Appendix M for the detailed derivation.)

$$A(z) = \left[\left(\cos \psi_w z + j\frac{\Delta\beta}{\psi_w} \sin \psi_w z\right) A(0) - j\frac{K_{12}}{\psi_w} \sin \psi_w z B(0)\right] e^{-j\beta_a z},$$

(8.16a)

$$B(z) = \left[-j\frac{K_{21}}{\psi_w} \sin \psi_w z A(0) + \left(\cos \psi_w z - j\frac{\Delta\beta}{\psi_w} \sin \psi_w z\right) B(0)\right] e^{-j\beta_a z}.$$

(8.16b)

Here,

$$\beta_a = \frac{\beta^{(1)} + \beta^{(2)}}{2}$$

(8.17)

is the average propagation constant of two modes, and

$$\Delta\beta = \frac{\beta^{(1)} - \beta^{(2)}}{2} \tag{8.18}$$

corresponds to the propagation constant difference of two modes. ψ_w is called the beat wave number and is given by the following equation in the case of codirectional coupling (if the guided modes are propagated in the same positive z-direction before the coupling):

$$\psi_w = \sqrt{\left(\frac{\beta^{(1)} - \beta^{(2)}}{2}\right)^2 + |K_{12}|^2}. \tag{8.19}$$

Here, K_{12} is the *coupling coefficient* from core 1 to core 2, and K_{21} is its reverse coupling coefficient, and in the case of codirecitonal coupling, they have the following relationship of power conservation condition:

$$K_{12} = K_{21}^*, \tag{8.20}$$

(refer to Appendix L, and as K_{12} is a real number for an optical waveguide with no absorption, it can also be written as $K_{12} = K_{21}$).

The coupling coefficient for two asymmetric three-layer slab waveguides shown in Figure 8.6 can be expressed by the following equation:

$$K_{12} = \frac{\kappa_0^2}{\beta_0} \frac{1}{\gamma_3 a} \frac{\exp(-\gamma_3 d)}{1 + \left(\frac{\kappa_0}{\gamma_3}\right)^2}. \tag{8.21}$$

Here, β_0 is the propagation constant of the mode when the core exists in isolation (the cladding is asymmetric), κ_0 is the transverse propagation constant defined by

$$\kappa_0 = \sqrt{k_0^2 n_1^2 - \beta_0^2} \tag{8.22}$$

and using the refractive index between the core and the intermediate cladding layer, γ_3 is defined as

$$\gamma_3 = \sqrt{\beta_0^2 - k_0^2 n_3^2}. \tag{8.23}$$

In general, the optical waveguide structure is not a slab but is often a complex structure like a rectangular cross-section, so it will be necessary to determine the coupling coefficient by numerical calculation using Equation (L.25) of Appendix L.

Now, as $A(z)$ and $B(z)$ are the electric field amplitudes of the modes, the $|A(z)|^2$ and $|B(z)|^2$ are proportional to the power (i.e., the optical power propagating in each of the cores) of each mode. Because this proportionality constant is omitted here as it is dependent on polarization and field distribution, let us just consider the normalized power P_0 that is incident on one core only. In this case, the initial conditions are

$$|A(0)|^2 = P_0, \tag{8.24a}$$
$$|B(0)|^2 = 0, \tag{8.24b}$$

and when these equations are substituted to Equations (8.16a) and b, the changes in the optical power that is propagating in each of the cores are expressed by the following equations:

$$P_1(z) = |A(z)|^2$$
$$= P_0[1 - F \sin^2 \psi_w z], \tag{8.25a}$$
$$P_2(z) = |B(z)|^2$$
$$= P_0 F \sin^2 \psi_w z. \tag{8.25b}$$

Here, F is

$$F = \frac{1}{1 + \left(\frac{\beta^{(1)} - \beta^{(2)}}{2|K_{12}|}\right)^2} \leq 1. \tag{8.26}$$

The dependence of the optical power on the propagation distance is shown in Figure 8.7. From Equation (8.26), we can see that if the propagation constants of the mode propagating in the two cores are equal ($\beta^{(1)} = \beta^{(2)}$), the optical power will fully migrate from core 1 to core 2 and will again bounce back in a periodic repeated manner. Such a case is called perfect coupling. On the other hand, when the system deviates from the perfect coupling, the optical power will not fully migrate from one core to the other. Here, the distance L_b of the migration of the optical power is called the *coupling length* (or the *beat length*) and from Equation (8.25b) it is given by

$$L_b = \frac{\pi}{2\psi_w}. \tag{8.27}$$

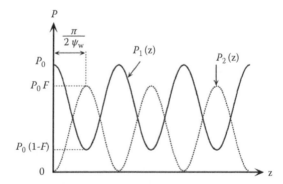

FIGURE 8.7 Variations in the optical power during the optical coupling between adjacent parallel waveguides.

8.3 MERGING AND BRANCHING OF OPTICAL WAVEGUIDES

When a three-dimensional optical waveguide branches into two, or two optical waveguides are merged into one, loss and mode conversion occurs according to certain laws. The analytical method for the merging and branching of optical waveguides differs for the multimode waveguides (in the case of very large number of modes) and the single-mode waveguides (including two and three modes).

8.3.1 MERGING AND BRANCHING OF MULTIMODE WAVEGUIDES

In multimode waveguides with a very large number of modes, the optical power distribution (or the modal distribution) in the cross-section that is perpendicular to the optical axis can be represented effectively by the ray density. In this case, as one ray of light can be represented by its position and direction at a certain time, the *mode distribution* cannot be expressed correctly as a distribution of the ray only on the spatial coordinate. To express the complete picture of the changes occurring in the mode distribution, it is also important to consider the direction of the ray. The authors have proposed a method for expressing such a mode distribution, which is the *phase space expression** of the ray [42].

First, as shown in Figure 8.8a, let us consider the case where the *ray trajectory* in the two-dimensional cross-section of an optical waveguide with an arbitrary refractive index distribution is given. In a phase space, the spatial coordinate x of this light ray at the propagation distance z is designated as the horizontal axis, and the slope $\dot{x}(= \frac{dx}{dz})$ with respect to the z-axis is set as the vertical axis, with the rays being represented by points. Thus, when the spatial

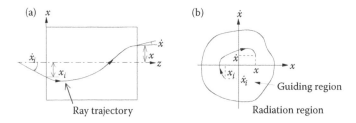

FIGURE 8.8 Ray trajectory in the real space of a multimode waveguide and ray trajectory in a phase space: (a) real space and (b) phase space.

---

* Phase space is a term used in analytical mechanics and refers to the space, wherein the position and its conjugate momentum are plotted onto coordinate axes. The area (or volume) occupying the phase space according to the natural movement of a system of particles is invariable even with a change of the state (Liouville's theorem).

coordinate of the light ray at $z = 0$ is set as x_i, the slope as $\dot{x}_i$, the position at the exit end of the light ray as x, and its slope as $\dot{x}$, the ray trajectory in real space as shown in Figure 8.8a is expressed in the phase space as shown in Figure 8.8b. The region in the phase space where the light ray can propagate is determined based on the waveguide structure and the refractive index distribution, and the rays that come out from this region during the propagation are radiated.

Only a single light ray was considered in Figure 8.8, but if many light rays are taken into account, a distribution of the points on the phase space will be obtained. For example, a group of light rays propagating all together almost parallel to the z-direction will be distributed only on the horizontal axis (x-axis) of the phase space, and the light ray that was radiated in all directions from the point light source is distributed only on the vertical axis ($\dot{x}$-axis). Moreover, the light ray that is radiated uniformly from any position on the radiating surface of the light source and uniformly in all directions in the angle (here, within the maximum acceptance angle in the optical waveguide), on the cross-section of the waveguide, is represented as a point group that is uniformly distributed in the region that is allowed for the guided mode on the phase space. In addition, in the parabolic distributed-index waveguide, the trajectory in the phase space of the light ray corresponding to the guided mode is a circle. Several features when the mode distribution is represented in this phase space for the parabolic distributed-index waveguide and the step-index waveguide is shown in Table 8.1. Here, in this table, the vertical axis and the horizontal axis are normalized by the core half-width a and the maximum value of the slope $\dot{x}_{max}(= \tan \theta_{max})$ of the light ray, respectively.

Using this phase space representation, let us consider the change in the mode distribution during the merging and branching of the multimode waveguide. In the case of the parabolic distributed-index waveguide, the vertical and the horizontal axes of the phase space are normalized in the same manner as in Table 8.1. The region that is allowed for the mode of the parabolic distributed-index waveguide is a circle, but in a merged waveguide, the optical axis of the waveguide is slightly inclined with respect to the z-axis,[*] and in the region allowed for the guided mode in the phase space of the incident waveguide, the center of the circle deviates in the vertical direction. Then, the two circles combine along with the gradual merging of the two waveguides (i.e., with the propagation of the light ray). If the merging angle is bigger than the maximum acceptance angle of the mode, the two circles completely deviate in the vertical direction, so the radiation loss and mode conversion because of the merging that will be discussed later will not occur. First, let us assume that the modes are incident on one of the merged waveguides with a uniform modal distribution

---

[*] This merging and branching angle should be sufficiently smaller than the propagation angle of the light ray in the waveguide and is usually less than one degree.

TABLE 8.1

Phase Space Representation of the Mode Distribution [43]

	Distribution index $\sqrt{X^2 + \dot{X}^2} \leq 1$	Step index $|X| \leq 1, |\dot{X}| \leq 1,$
Region of X and $\dot{X}$		
Expression of modal distribution in phase space (uniform mode excitation)		
Corresponding value to mode order	$\sqrt{X^2 + \dot{X}^2}$	$|\dot{X}|$
Lower mode	$\sqrt{X^2 + \dot{X}^2} \to 0$	$|\dot{X}| \to 0$
Higher mode	$\sqrt{X^2 + \dot{X}^2} \to 1$	$|\dot{X}| \to 1$
Propagation condition of ray		
Mode conversion (Lower mode →Higher mode)		

within the circle of half (normalized radius is $\frac{1}{\sqrt{2}}$), the region allowed for the guided mode in the phase space. The ray trajectories are expressed in real space and the phase space, as shown in Figures 8.9 and 8.10, respectively.

In Figure 8.9, the area occupied in the phase space of the modal distribution of the incident light before merging is $\frac{\pi}{2}$, in the same manner as the movement of point mass in analytical mechanics, this area does not change even after the merging. However, after the merging, the distribution spreads to the outer circumference of the circle that is allowed for the guided mode. This shows that a mode conversion from a lower order mode to a higher order mode has occurred. Consequently, if the incident mode distribution occupies a circle with

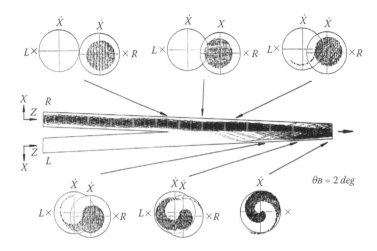

FIGURE 8.9 Change in the mode distribution with the merging of multimode waveguides [43].

a radius of 1, half of it will be radiated and lost. Thus, we can see that the mode distribution after the merging is not uniform.

On the other hand, when the mode distribution before the merging is incident on the branching waveguide, the ray trajectory is as shown in Figure 8.10. In this way, when a nonuniform modal distribution is incident on a branching

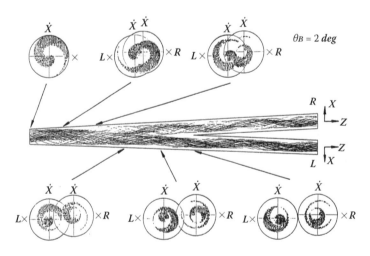

FIGURE 8.10 Change in the mode distribution with the branching of multimode waveguides [43].

waveguide, even if the branching is a fully symmetrical structure, the modal distribution differs at the two exit ends. In addition, although this figure may be hard to read clearly, as the occupied areas of the distribution are actually different, the branching ratio at the exit end is not 1:1. In this way, in the branching of the multimode waveguide, the branching ratio will change significantly depending on the modal distribution at the incident end, so care must be taken in the design of multimode waveguide circuit that includes branching. Therefore, to make the modal distribution at the incident end of the branching section almost uniform, mode scramblers that would bend the optical path to a zigzag shape and repeat the branching and merging a number of times have been proposed.

8.3.2 MERGING AND BRANCHING OF SINGLE-MODE WAVEGUIDES

The merging and branching of single-mode waveguides can be explained using the coupled mode theory in Section 8.3.1 [43], [45]. First, let us suppose the case that two light beams with equal electric field amplitudes are incident on both of the input ports of the symmetrical (the two incident ports are equal) merge circuit shown in Figure 8.11. The merging angle is assumed to be very small (i.e., the length of the merging section is significantly longer by several hundreds to several thousands than the wavelength and the waveguide width).

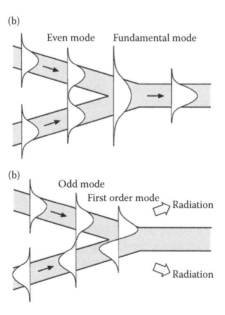

FIGURE 8.11 Merging of single-mode waveguides [43]–[45]: (a) If beamed in-phase and (b) if beamed in reversed phase.

While two merging cores are gradually getting closer to each other, these two waveguides can be considered as a coupled waveguide system of which the core interval gradually becomes smaller in the direction of propagation. An even mode is excited when coherent light with equal frequencies are beamed in the same phase to both waveguides of the coupled waveguide system, as shown in Figure 8.11a. Then, when the waveguide interval gradually decreases until it becomes zero, the valley at the center of the even mode in the coupled waveguide gradually decreases and becomes single peaked and is transformed into a fundamental mode (zero order mode) of a two-mode waveguide. Even if the core width of the two-mode waveguide decreases further in the tapered region and it becomes a single-mode waveguide, it will not be radiated as the fundamental mode does not have a cutoff. Consequently, in the in-phase input, there is hardly any loss due to merging (some loss occurs depending on the merging angle). In contrast, when coherent light is beamed in reversed phase, as shown in Figure 8.11b, both input ports are merged, resulting in a coupled waveguide, and an odd mode is excited. In addition, when the waveguide interval becomes zero, it will be transformed into a first-order mode of a two-mode waveguide. Furthermore, the two-mode waveguide is tapered, and in the process of becoming a single-mode waveguide, cutoff occurs, and it is radiated. Consequently, in the case of merging in reversed phase, all the optical power results in radiation loss.

Now, let us consider the case when light is incident on only one input port of the merged waveguides. In such a case, the even and the odd modes are excited in the coupled waveguide at the same ratio, so half of the incident power is coupled to the fundamental mode of the exit port, and the remaining half will be transformed into a first-order mode and radiated at the taper section. Consequently, a 3-dB radiation loss occurs with the merge.

Next, let us consider branching. This case corresponds to the reverse of that shown in Figure 8.11a, so light output of the same phase can be obtained from the exit port at a ratio of 1:1.

When the merging and branching above are combined, the operation of a Mach–Zehnder interferometer circuit shown in Figure 8.12 can be understood. Branched light can be separated at the same phase, so if the optical path lengths of the two arms are the same and the light is incident on the merging section, it will be inputted at the same phase as shown in Figure 8.12a. Consequently, if the merging angle is sufficiently smaller than the propagation angle of the guided mode, there is hardly any loss due to branching and merging. On the other hand, if the optical path length (in particular, the refractive index) of one of the two arms is changed using the electro-optic and the thermo-optic effects, and the phase of the light that passes through this arm is shifted relative to the other by π, a reverse phased input will occur at the merging section as shown in Figure 8.12b, so the light will be radiated. This

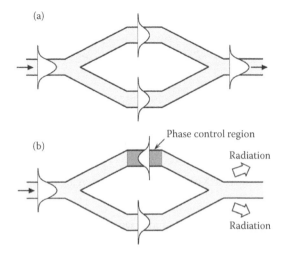

(a)

(b)

Phase control region

Radiation

Radiation

FIGURE 8.12 Waveguide-type Mach–Zehnder interferometer circuit by means of merging and branching. (a) If the two optical path lengths are the same and (b) if the phase of one of the optical paths deviates by π.

is the principle behind the Mach–Zehnder modulator that uses the electro-optic effect.

Conversely, how about in the case of branching after merging? First, if the light is incident from one of the two input ports, a 3-dB radiation loss occurs because of merging, and the remaining half of the optical power is outputted from the exit port at a ratio of 1:1. Moreover, when a coherent light of the same frequency is incident from the two input ports, merging loss will vary depending on the phase relationship. Consequently, as the operation of such merging and branching circuit (also considered as a 2×2 star coupler) is complex and unstable, it is not usually used. To avoid this merging loss, it may be better to start the branching right after the interval between the two input port waveguides becomes zero and the waveguide width becomes twice that of the single waveguide. However, this case corresponds to the case, wherein the waveguide interval of the coupled waveguide changes in the propagation direction, and the branching ratio changes because of the merging and branching angles and the wavelength, so designing the circuit is difficult. Consequently, this circuit is also not usually used.

The asymmetric X-shaped coupling and merging circuit [43] shown in Figure 8.13 is used as a 2×2 merging and branching circuit (star coupler). The action of this optical circuit is the same as that of the merging section up to just before the tapered section in Figure 8.11, so let us consider the asymmetric branching section using the asymmetrical branch that is branching from a

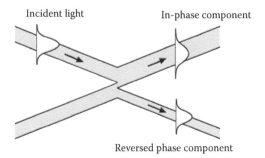

FIGURE 8.13 Asymmetric X-shaped coupling and merging circuit [43].

two-mode waveguide to two single-mode waveguides with different waveguide widths, as shown in Figure 8.14. In such a case, for the even and the odd modes that are generated in the coupled waveguide system in the branching, as shown in Figure 8.14, the even mode will have a large peak in the core with a wide waveguide width (the equivalent index is large), whereas the odd mode will have a large peak in the core with a narrow waveguide width. Consequently, the optical power of the fundamental mode of the two-mode waveguide will be outputted mostly at the waveguide with a wide waveguide width, whereas in the first-order mode, it will be outputted at the waveguide with a narrow waveguide width. When this action is connected to the merging circuit, and coherent light with equal frequencies is incident on two incident ports of the asymmetrical X-shaped merging and branching circuit as shown Figure 8.13, the in-phase component will be outputted at the exit port with a wide waveguide width, whereas the reversed phase component will be outputted at the narrow exit port. Moreover, when the light is incident on only one of the input ports, irrespective of the port, the light will be outputted at either of the two exit ports with a branching ratio of 1:1.

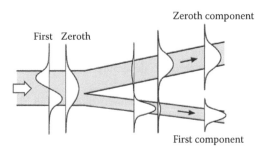

FIGURE 8.14 Asymmetrical Y-branching circuit [43], [45].

8.4 RESONATORS AND EFFECTIVE INDEX

In Section 4.2, the Fabry–Perot resonator of a free-space beam was discussed. Here, let us consider a Fabry–Perot resonator that is configured with an optical waveguide like that of an FP semiconductor laser. Even in a Fabry–Perot resonator of an optical waveguide, with the power reflectivity at both end surfaces of waveguide as R, the power transmittance T of the resonator is given by the following equation in the same manner as Equation (4.15):

$$T = \frac{1}{1 + \frac{4R}{(1-R)^2}\sin^2(\beta L)}. \tag{8.28}$$

The difference between Equation (8.28) and Equation (4.15) is that the propagation constant that was $k_0 n$ for the free-space beam propagating in a medium with refractive index n is designated as β in Equation (8.28). When the refractive index outside the waveguide (at the two exit ends in the FP laser) is lower (e.g., air) than that of the equivalent index of the waveguide, the standing wave in the resonator will look, as shown in Figure 8.15.

Unlike the Fabry–Perot resonator of the free-space beam that uses translucent mirrors, antinodes of the standing wave are formed at the end surfaces of the resonator, but in either case, the resonance condition (phase condition) is expressed by the following equation:

$$\beta L = N\pi \quad (N \text{ is a positive integer}). \tag{8.29}$$

The equation for FSR is derived in the same manner as Equation (4.19). When Equation (8.29) is transformed into an equation for N, and then subsequently differentiated for λ, the following equation is obtained:

$$d\lambda = -\frac{\lambda^2}{2L\frac{d\beta}{dk}}dN. \tag{8.30}$$

If dN in Equation (8.30) is set to be $dN = 1$, then the same equation for FSR as Equation (4.19) in Section 4.2 is obtained. When Equations (8.30) and (4.19)

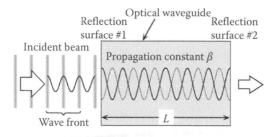

FIGURE 8.15 Standing wave in the waveguide-type Fabry–Perot resonator area.

are compared, we can see that the *effective index* of the *waveguide-type Fabry–Perot resonator* is given by

$$n_{\text{eff}} = \frac{d\beta}{dk}. \tag{8.31}$$

Because $\frac{d\beta}{dk}$ is given in Equations (6.87) and (5.81) as the formulas for a generalized waveguide, the effective index of a waveguide-type resonator is expressed by the following equation:

$$
\begin{aligned}
n_{\text{eff}} &= \frac{\left[n_2 N_2 + (n_1 N_1 - n_2 N_2)\left(b + \frac{1}{2}V\frac{db}{dV}\right)\right]}{[n_2^2 + (n_1^2 - n_2^2)b]^{\frac{1}{2}}} \\
&= \frac{1}{n_{\text{eq}}}[n_2 N_2 + \Gamma(n_1 N_1 - n_2 N_2)] + \frac{1}{2n_{\text{eq}}} \cdot \frac{\lambda^3}{4\pi^2} \cdot \frac{\partial}{\partial\lambda}\left(\frac{1}{w_x^2} + \frac{1}{w_y^2}\right).
\end{aligned}
\tag{8.32}
$$

The group index defined by Equation (6.88) in Section 6.2.2 is the same as the effective index defined by Equation (4.20) in Section 4.2, but it must be noted that the effective index of a waveguide-type resonator defined by Equation (8.32) involves the group index N_1 and N_2 of the core and the cladding. Either way, in a resonator or an interferometer that uses an optical waveguide, Equation (8.32) is necessary for the calculation of the resonant wavelength spacing of the resonator (or the free spectral range [FSR]), not just in Fabry–Perot resonators, but also in resonators that use diffraction grating (often used in semiconductor laser resonators) such as the DFB (distributed feedback) and the DBR (distributed Bragg reflector) resonators, Mach–Zehnder interferometers, and ring resonators.

Incidentally, as $\tau = \frac{L}{v_g}$ should hold true, comparing Equation (8.31) with (6.87), the following relationship is derived:

$$\frac{d\beta}{dk} = \frac{c}{v_g}. \tag{8.33}$$

When the above equation and Equation (8.32) are compared, and the relationship $n_{\text{eq}} = \frac{c}{v_p}$ is used, the following equation is obtained:

$$n_{\text{eq}}n_{\text{eff}} = \frac{c^2}{v_p v_g} = n_1 N_1 \Gamma + n_2 N_2(1 - \Gamma). \tag{8.34}$$

Here, the wavelength derivative terms of the inverse of the spot size are ignored because they are small when the spot size is much larger than the wavelength. If the spot size is not smaller than the wavelength, as in the case of silicon waveguides, they cannot be ignored. In this section, Equation (2.36) in Section 2.6.1 was generalized to the optical waveguide consisting of a dispersive medium (a medium wherein the refractive index is dependent on the wavelength).

8.5 WAVEGUIDE BENDS

If an optical waveguide bends in its propagation direction, *bending loss* and a change in the propagation constant will occur. Bending loss is further classified as *uniform bending loss*, *microbending loss*, and *bending loss* due to curvature change (there is still no accurate name for this type). This section will explain the concept of these phenomena; however, in recent years, the detailed quantitative calculations can often be assessed by numerical simulation methods such as the beam propagation method (see further) rather than an analysis using a mathematical formula.

Let us consider an optical waveguide when it bends in the z-direction as shown in Figure 8.16. The curvature radius of the propagation axis is R_0. The wave propagation in an optical waveguide is represented by the wave equation expressed by a Cartesian coordinate system (x, y, z coordinate system), but for a bend waveguide, it is convenient to describe the wave equation by the cylindrical coordinate system, as shown in Figure 8.16. The scalar wave equation (with respect to the electric field) using the cylindrical coordinate system of Figure 8.16 is given by the following equation:

$$\frac{1}{r}\frac{\partial E}{\partial r}\left(r\frac{\partial E}{\partial r}\right) + \frac{1}{r^2}\frac{\partial^2 E}{\partial \theta^2} + \frac{\partial^2 E}{\partial \zeta^2} + k_0^2 n^2(r,\theta,\zeta)E = 0. \tag{8.35}$$

Light propagates in the optical fiber toward the ζ-axis direction of the cylindrical coordinate system; however, here, light will propagate in the circumferential direction of radius R_0. Consequently, so as not to be confused with

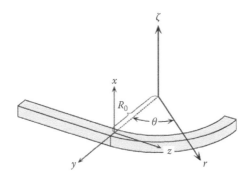

FIGURE 8.16 Coordinate system for the analysis of a bent waveguide.

the z coordinate of the Cartesian, the coordinates for the central axis of the cylindrical coordinate system is expressed as ζ. When the coordinate system is transformed as

$$x = \zeta, \tag{8.36}$$

$$y = R_0 \ln\left(\frac{r}{R_0}\right) \quad (y = 0 \text{ at } r = R_0), \tag{8.37}$$

$$z = R_0\theta. \tag{8.38}$$

Equation (8.35) is transformed into the following equation:

$$e^{-2\frac{y}{R_0}}\left(\frac{\partial^2 E}{\partial x^2} + \frac{\partial^2 E}{\partial y^2}\right) + e^{-2\frac{y}{R_0}}\left(\frac{\partial^2 E}{\partial z^2}\right) + k_0^2 n^2(x,y)E = 0. \tag{8.39}$$

Moreover, when the z dependence is assumed to be $e^{-j\beta z}$, the equation is transformed into

$$\frac{\partial^2 E}{\partial x^2} + \frac{\partial^2 E}{\partial y^2} + k_0^2 n^2(x,y,z)e^{2\frac{y}{R_0}}E = \beta^2 E. \tag{8.40}$$

Thus, the refractive index, which is $n(x,y,z)$ in a straight waveguide, is transformed to $n(x,y,z)e^{\frac{y}{R_0}}$ in a bent waveguide, and we can see that both are equivalent to each other. Moreover, as $e^{\frac{y}{R_0}} \simeq 1 + \frac{y}{R_0}$ from the first-order approximation of the Taylor series expansion, near the center $y = 0$ of the waveguide, the refractive index distribution is equivalent to the uniformly inclined distribution, as shown in Figure 8.17. In the refractive index distribution

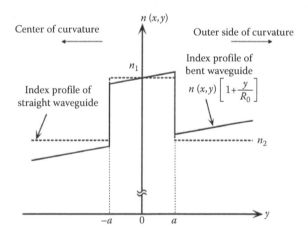

FIGURE 8.17 Transformation of the refractive index distribution in a bent waveguide.

that is transformed in this manner, the refractive index of the cladding will first become larger than that of the core when y increases, and then radiation loss will occur. This radiation loss is called the uniform bending loss. Here, if R_0 is large, the electromagnetic field that is confined to the core will be sufficiently attenuated inside the cladding, so the influence of the cladding is small even if the refractive index of the cladding is bigger than that in the core at a position farther away from the center $y = 0$. Consequently, the uniform bending loss is small in a bent waveguide with a large curvature radius. Thus, the change in the refractive index due to this uniform bending can be handled by perturbation (refer to Appendix N for details), and when the propagation constant of the vth-order mode of an unbent straight waveguide is designated as $\beta_0^{(v)}$, the propagation constant β of the vth-order mode of the bent waveguide is expressed as

$$
\begin{aligned}
\beta^{(v)2} &= \beta_0^{(v)2} + \frac{k_0^2 \iint_{-\infty}^{\infty} n^2(x, y) \left(e^{2\frac{y}{R_0}} - 1 \right) |E_0^{(v)}|^2 dx\, dy}{\iint_{-\infty}^{\infty} |E_0^{(v)}|^2 dx\, dy} \\
&\simeq \beta_0^{(v)2} + \frac{k_0^2 \iint_{-\infty}^{\infty} n^2(x, y) \left(2\frac{y}{R_0} \right) |E_0^{(v)}|^2 dx\, dy}{\iint_{-\infty}^{\infty} |E_0^{(v)}|^2 dx\, dy}.
\end{aligned}
\tag{8.41}
$$

The deformation of electromagnetic field profile rather than the propagation constant has a significant impact on the propagation characteristics of light. If the refractive index distribution is inclined along the y-axis, as shown in Figure 8.17, the electromagnetic field profile is shifted to the positive direction of y as if it gravitates toward the side of the larger refractive index. Consequently, when light is beamed from a straight waveguide to a bent waveguide, a significant loss (similar to an offset loss) will occur as the center axes of the field profiles are not aligned, as shown in Figure 8.18a. To reduce this loss, it is necessary to make the design such that the central axis of the bent waveguide is shifted so that the center of the field profile is aligned to that of the straight waveguide, as shown in Figure 8.18b. When two waveguides with a reverse direction of curvature are connected, the same scheme is also necessary.

On the other hand, although various formulas have been proposed (e.g., Reference [47]) as the evaluation formula for the uniform bending loss, numerical simulation technology has developed in recent years and softwares have since become commercially available. Therefore, numerical simulation methods such as the beam propagation method is often used in the layout design of waveguide devices to evaluate the bending loss by numerical calculation. On the other hand, when the optical fiber is assembled into cables, microscopic

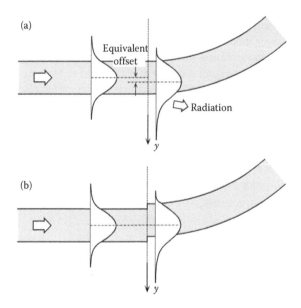

FIGURE 8.18 Electromagnetic field profile of the portion connecting the straight waveguide to the bent waveguide. (a) When the center of the waveguide is aligned; (b) when the center of the field profile is aligned.

bending occurs from various directions. Such bending is called a microbend, wherein the curvature is expressed by its statistical nature. It has been confirmed experimentally that the statistical nature of the curvature is expressed by a Gaussian distribution, and the following formula [48] has been obtained for the microbending loss:

Multimode fiber:

$$\alpha_{mb} = 2500N \overline{\left(\frac{1}{R}\right)^2} \overline{W}^2 \frac{1}{\Delta} \exp\left[-\left(\frac{\overline{W}}{2a}\right)^2 \Delta\right] \quad \text{[dB/km]}. \qquad (8.42)$$

Single-mode fiber:

$$\alpha_{mb} = xN \overline{\left(\frac{1}{R}\right)^2} \overline{W}^2 \frac{1}{\Delta} \exp\left[-y\left(\frac{\overline{W}n_1}{\lambda_c}\right)^2 \Delta^2\right] \quad \text{[dB/km]}. \qquad (8.43)$$

Here, N is the average number of bends per unit length (1 m) of the fiber, $\overline{\left(\frac{1}{R}\right)^2}$ is the square mean value of the curvature, $\overline{W}$ is the correlation length of the curvature distribution, and λ_c is the cutoff wavelength. The coefficients x and y for the single-mode fiber depends on the wavelength.

Here, the determination of the average number of bends N and the other parameters $\left(\left(\frac{1}{R}\right)^2, \overline{W}\right)$ is generally difficult, but essentially, if the cable assemblage is done without care to reduce the microbending loss, even if the loss in the fiber itself is small, the cable loss will increase. In addition, as the coefficient y decreases with the increase of $\frac{\lambda}{\lambda_c}$, when a wavelength that is farther than the cutoff wavelength is used, the microbending loss due to cable assemblage tends to increase. Consequently, it is necessary to carefully design the optical fiber configuration (a, Δ, λ_c) so that the microbending loss at the wavelength used after the cable assemblage does not increase.

PROBLEMS

1. In a parabolic distributed-index waveguide, the eigenmode is approximately expressed by an Hermite–Gaussian function, so the change in the propagation constant due to bending can be evaluated using Equation (8.41). Determine this change in the propagation constant using Equation (G.6) of Appendix G.

2. In the same manner, determine the deformation of the fundamental mode (Gaussian function) when the parabolic distributed index waveguide is bent using Equation (N.18) of Appendix N (only the center of the Gaussian profile is shifted).

8.6 POLARIZATION CHARACTERISTICS

In the guided mode of the optical waveguide except for optical fibers that have the axial symmetry, there are two polarization modes of which electric field is nearly parallel (TE polarization) and nearly perpendicular (TM polarization) to the substrate surface and their propagation constants are different. In channel waveguides, such as the rectangular waveguide, the electric field of the TE polarization does not only have the y component, which is completely parallel to the substrate, but it also has a little of the x and the z components (i.e., the line of electric force is slightly bent). Even in such a case, the polarization mode can be divided approximately into the TE and the TM polarizations. Consequently, these are sometimes referred to a bit accurately as the TE-like mode and the TM-like mode, but in this section, we will call them simply as the TE and the TM polarizations.

In the case of linear polarization, the direction of polarization indicates the direction of the oscillation of the electric field, but more accurately, if the oscillation of orthogonal electric field is maintained at a specific constant phase relationship, it would also be polarized. That is, the presence of constant phase relationship is closely related to coherence. However, here, we will deal with the polarization of perfectly coherent light for simplicity.

In an optical waveguide, the electromagnetic field distribution of the TE and the TM polarizations slightly differs, so for simplicity, the polarization in a plane wave will be discussed first. However, if the electromagnetic field distribution functions $E_{TE}(x, y)$ and $E_{TM}(x, y)$ are multiplied to the amplitude of each polarized light in the derivation below, the same manner of calculation can also be applied to a guided mode. If the direction of propagation of the plane wave is designated as the z-axis direction when the electric field oscillates in the x, y plane, as a vector, the electric field can be resolved into the x and y components and they are expressed by

$$E_x = A_x \exp[j(\omega t - \beta_x z + \phi_{x0})], \qquad (8.44a)$$

$$E_y = A_y \exp[j(\omega t - \beta_y z + \phi_{y0})]. \qquad (8.44b)$$

For the propagation in a birefringent medium that was derived in Section 1.5, $\beta_x = k_0 n_x$ and $\beta_y = k_0 n_y$; however, $\beta_x = \beta_y$ holds in an isotropic medium. Generally, in an optical waveguide, as the propagation constants of the TE and TM polarizations differ, the birefringence appears. Now, a vector with the x and y components of the electric field as elements is expressed as

$$[A] = \begin{bmatrix} A_x \exp(-j\phi_x) \\ A_y \exp(-j\phi_y) \end{bmatrix} \qquad (8.45)$$

and such a vector is called a *Jones vector*. Here, $\beta_x z - \phi_{x0} = \phi_x$ and $\beta_y z - \phi_{y0} = \phi_y$. When $|E|$ is normalized to be 1, the vector is expressed as

$$[A] = \begin{bmatrix} \dfrac{A_x}{\sqrt{A_x^2 + A_y^2}} \exp(-j\phi_x) \\ \dfrac{A_y}{\sqrt{A_x^2 + A_y^2}} \exp(-j\phi_y) \end{bmatrix} = \begin{bmatrix} \cos \psi \\ \sin \psi \exp(j\Gamma) \end{bmatrix}, \qquad (8.46)$$

and this vector is called the *normalized Jones vector*. It is a linear polarization, if the phase difference $\delta\phi = \phi_x - \phi_y$ of the x- and the y polarization components is 0; however, if there is a phase difference, it is an elliptical polarization. When the electric field vector of the light is observed from the negative z-direction to the positive z-direction (the opposite direction of Figure 1.21), the relationship of the trajectory and the normalized Jones vector is depicted as shown in Figure 8.19.

The polarization states of the incident light and the exit light can be described using this normalized Jones vector (hereafter referred to simply as the Jones vector), so the nature of the medium can be described by the matrix that relates the two vectors. For example, in the birefringent medium of Section 1.5, if the principal birefringent axis is at the x-axis and the y-axis, and if the x- and

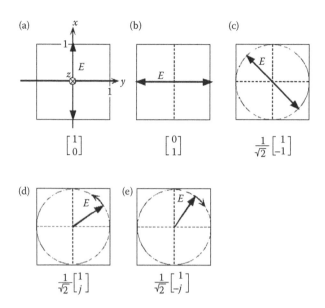

FIGURE 8.19 Trajectory of the electric field vector and the normalized Jones vector (take note that the observed direction is opposite to that of Figure 1.21): (a) $\begin{bmatrix} 1 \\ 0 \end{bmatrix}$. (b) $\begin{bmatrix} 0 \\ 1 \end{bmatrix}$. (c) $\frac{1}{\sqrt{2}}\begin{bmatrix} 1 \\ -1 \end{bmatrix}$. (d) $\frac{1}{\sqrt{2}}\begin{bmatrix} 1 \\ j \end{bmatrix}$. (e) $\frac{1}{\sqrt{2}}\begin{bmatrix} 1 \\ -j \end{bmatrix}$.

the y-direction refractive indices are denoted as n_x, n_y, respectively, the matrix that will relate the polarization of the incident light and the exit light is written as follows:

$$\begin{bmatrix} A_x^o \\ A_y^o \end{bmatrix} = \begin{bmatrix} e^{-jk_0 n_x z}, & 0 \\ 0 & e^{-jk_0 n_y z} \end{bmatrix} \begin{bmatrix} A_x^i \\ A_y^i \end{bmatrix} = [J] \begin{bmatrix} A_x^i \\ A_y^i \end{bmatrix}. \tag{8.47}$$

This matrix $[J]$ is called the *Jones matrix*. On the other hand, an optical component that rotates the polarization axis by θ is called an optical rotator, and in this case, the plane of polarization is rotated so it will be identical to a simple coordinate rotation. Thus, the Jones matrix is given by the following equation:

$$[T_\theta] = \begin{bmatrix} \cos\theta, & -\sin\theta \\ \sin\theta & \cos\theta \end{bmatrix}. \tag{8.48}$$

If $\theta > 0$, the x-axis will left rotate around the z-axis, so this is called levorotatory. The reverse is referred to as dextrorotatory. On the other hand, if the birefringent principal axis tilts by θ from the x-axis, the x, y coordinate system is rotated by an angle of θ, the Jones matrix of Equation (8.47) is operated, and returning to the original coordinate axis, the Jones matrix $[J_\theta]$ can be expressed by

$$[J_\theta] = [T_\theta][J][T_{-\theta}]. \tag{8.49}$$

Finally, the Jones matrix $[P_x]$ of the *polarizer* that transmits only the *x*-polarized light is

$$[P_x] = \begin{bmatrix} 1, & 0 \\ 0 & 0 \end{bmatrix}. \tag{8.50}$$

In other Jones matrices, the determinant of the matrix is 1, so they are unitary,* but the Jones matrix of a polarizer is not unitary.

8.7 DESCRIPTION OF THE OPTICAL CIRCUIT BY SCATTERING MATRIX AND TRANSMISSION MATRIX

Thus far, propagation has been described based on the wave equation, but the characteristics of an optical circuit as a black box is given by the relationship between the input light and the output light [46]. Thus, let us consider the N port device, as shown in Figure 8.20, and define the input light amplitude to each of the ports as a_i, and the output light amplitude as b_i (i is the port number). That is, a_i and b_i correspond to the coefficients in the eigenmode expansion in Section 5.1.4, wherein $|a_i|^2$ represents the input optical power to port i and $|b_i|^2$ expresses the output optical power from port i. In this case, the S matrix below that relates the vectors consisting of the input light amplitude a_i to all of

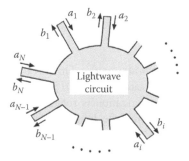

FIGURE 8.20 Definition of input and output to describe the scattering matrix of the N-port device.

---

* When the matrix is designated as A, and the complex conjugate of the transposed matrix A^t is set as $A^\dagger$, the matrix wherein $A = A^\dagger$ holds true is an Hermitian matrix, and the matrix wherein $AA^\dagger = 1$ is called a unitary matrix.

the ports, and the vectors consisting of the output light amplitude b_i from all of the ports is called the *scattering matrix*.

$$\begin{bmatrix} b_1 \\ b_2 \\ \vdots \\ b_N \end{bmatrix} = \begin{bmatrix} S_{11}, & S_{12}, & \cdots, & S_{1N} \\ S_{21}, & S_{22}, & \cdots, & S_{2N} \\ \vdots & \vdots & \ddots & \vdots \\ S_{N1}, & \cdots, & \cdots, & S_{NN} \end{bmatrix} \begin{bmatrix} a_1 \\ a_2 \\ \vdots \\ a_N \end{bmatrix} = [S] \begin{bmatrix} a_1 \\ a_2 \\ \vdots \\ a_N \end{bmatrix}. \tag{8.51}$$

In the case of an optical circuit that contains no loss or amplification, $S^\dagger S = S^{*t} S = 1$ holds. (That is, the scattering matrix is a unitary matrix.) (Refer to Problem **1**.)

The simplest scattering matrix is the two-port device. (Here, in the scattering matrix, all the ports are treated equally, so Equation (8.51) will not distinguish between the input port and the output port.) Such a scattering matrix is the following equation:

$$\begin{bmatrix} b_1 \\ b_2 \end{bmatrix} = \begin{bmatrix} S_{11}, & S_{12} \\ S_{21}, & S_{22} \end{bmatrix} \begin{bmatrix} a_1 \\ a_2 \end{bmatrix}. \tag{8.52}$$

Here, let us determine the sum of the optical power of the exit light from each of the ports. The optical power is given by $|b_i|^2 = b_i^* b_i$ ($i = 1$ or 2), so the matrix representation is given by

$$\begin{aligned} |b_1|^2 + |b_2|^2 &= [b_1{}^*, b_2{}^*] \begin{bmatrix} b_1 \\ b_2 \end{bmatrix} = [b^*]^t [b] \\ &= \{[S^*][a^*]\}^t \{[S][a]\} = [a^*]^t [S^*]^t [S][a] \\ &= [a_1{}^*, a_2{}^*] \begin{bmatrix} S_{11}{}^*, & S_{21}{}^* \\ S_{12}{}^*, & S_{22}{}^* \end{bmatrix} \begin{bmatrix} S_{11}, & S_{12} \\ S_{21}, & S_{22} \end{bmatrix} \begin{bmatrix} a_1 \\ a_2 \end{bmatrix} \\ &= \{|S_{11}|^2 + |S_{21}|^2\}|a_1|^2 + \{|S_{22}|^2 + |S_{12}|^2\}|a_2|^2 \\ &\quad + Re[\{S_{11} S_{12}{}^* + S_{22}{}^* S_{21}\} a_1 a_2{}^*]. \end{aligned} \tag{8.53}$$

Here, when an optical power of $P = 1$ is inputted from port 1 or port 2, $|b_1|^2 + |b_2|^2 = 1$ is derived from the power conservation condition, if there is no loss or gain in the optical circuit. Consequently, if $|a_1|^2 = 1$ and $|a_2|^2 = 0$, the following equation is derived from Equation (8.53):

$$|S_{11}|^2 + |S_{21}|^2 = 1, \tag{8.54}$$

whereas if $|a_1|^2 = 0$ and $|a_2|^2 = 1$

$$|S_{22}|^2 + |S_{12}|^2 = 1 \tag{8.55}$$

is obtained. In addition, if optical power is inputted from port 1 and port 2 at the same time, $|b_1|^2 + |b_2|^2 = |a_1|^2 + |a_2|^2$ holds, and so

$$S_{11} S_{12}{}^* + S_{22}{}^* S_{21} = 0 \tag{8.56}$$

can be derived. On the other hand, for the output amplitude b_2 from port 2, when an amplitude a_1 is beamed from port 1, when such output is time reversed, it will follow exactly the reverse of the same route (this is referred to as reversibility) in accordance to the nature of Maxwell's equations, so it will become input a_2 from port 2 and a_1 will become b_1 by time reversal. The following equation

$$S_{12} = S_{21} \tag{8.57}$$

is obtained from this reversibility. (Refer to Problem **2**.) $|S_{11}| = |S_{22}|$ can also be found in Equations (8.54), (8.55), and (8.57). S parameters $|S_{11}|$ to $|S_{22}|$ can be obtained, if one of the four components is known.

The scattering matrix is a matrix that expresses the sorted output of each of the ports in terms of the input to each of the ports, whereas the matrix that focuses on the input and output of one (or more) port and relates it to the input and output of another optical circuit is called the *transmission matrix*. Let us explain this using the simplest two-port device. In a transmission matrix, the primary direction of propagation of the optical signal is definitely defined and is different from the scattering matrix in the definition of the input port and the output port. Thus, when the input and output from each port is defined, as shown in Figure 8.21, the transmission matrix is given by the following equation:

$$\begin{bmatrix} A_1 \\ A_1' \end{bmatrix} = \begin{bmatrix} T_{11}, & T_{12} \\ T_{21}, & T_{22} \end{bmatrix} \begin{bmatrix} A_2 \\ A_2' \end{bmatrix}. \tag{8.58}$$

In the same way as the F matrix that was used in the representation of a 2×2 terminal pair circuit in the circuit theory, this matrix has a property, wherein the transmission matrix is given by the product of each of the transfer matrices when the optical circuits are cascaded. That is, in the case shown in Figure 8.21b, the transmission matrix of whole optical circuits is given as the product of the transfer matrices of each optical circuit, as shown in the following equation:

$$\begin{bmatrix} A_1 \\ A_1' \end{bmatrix} = \begin{bmatrix} T_{11}^{(1)}, & T_{12}^{(1)} \\ T_{21}^{(1)}, & T_{22}^{(1)} \end{bmatrix} \begin{bmatrix} A_2 \\ A_2' \end{bmatrix}$$

$$= \begin{bmatrix} T_{11}^{(1)}, & T_{12}^{(1)} \\ T_{21}^{(1)}, & T_{22}^{(1)} \end{bmatrix} \begin{bmatrix} T_{11}^{(2)}, & T_{12}^{(2)} \\ T_{21}^{(2)}, & T_{22}^{(2)} \end{bmatrix} \begin{bmatrix} A_3 \\ A_3' \end{bmatrix}. \tag{8.59}$$

In the case of a two-port device, transmission matrix is related to the scattering matrix by the following equations:

$$T_{11} = \frac{1}{S_{21}}, \qquad T_{12} = -\frac{S_{22}}{S_{21}},$$

$$T_{21} = \frac{S_{11}}{S_{21}}, \qquad T_{22} = -\frac{S_{11}S_{22} - S_{12}S_{21}}{S_{21}}. \tag{8.60}$$

Lightwave circuit

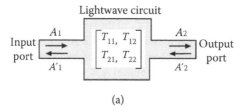

(a)

Lightwave circuit#1   Lightwave circuit#2

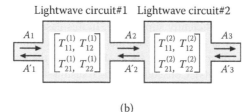

(b)

FIGURE 8.21 Definition of input and output to describe the transmission matrix of a two-port device: (a) single optical circuit and (b) cascaded optical circuits.

The transmission matrix introduced here expresses the incident light and the exit light (reflected light) A_1 and A_1' of the input port by the exit light and the input light (reflected light) A_2 and A_2' of the output port, but in the actual calculation, it is often more convenient to express in reverse the A_2 and A_2' of the output port by A_1 and A_1' of the input port.* The matrix that was used in the analysis of a Mach–Zehnder interferometer in Section 4.3 is also a 2×2 port device, so actually the format is also of the same form. Correcting the transmission matrix representation to be this kind of representation, the conversion is easy by the inverse matrix. That is, the following equation can be used to represent a two-port device:

$$\begin{bmatrix} A_2 \\ A_2' \end{bmatrix} = \begin{bmatrix} T_{11}, & T_{12} \\ T_{21}, & T_{22} \end{bmatrix}^{-1} \begin{bmatrix} A_1 \\ A_1' \end{bmatrix}. \tag{8.61}$$

Then, if matrix $[T^{-1}]$ is determined initially, the subsequent calculation will be easy. Here, note that the light amplitudes A_i and A_i' ($i = 1$ or 2) have two

---

* Representing the transmission matrix in the form of A_1 and A_1' for the input side by A_2 and A_2' for the output side is a convention derived from the transmission matrix of an electrical circuit. In an electrical circuit, the F matrix ($ABCD$ matrix) corresponds to the transmission matrix, so if the F matrix is used, the input impedance can be easily expressed by $Z_{in} = \frac{AZ_{out}+B}{CZ_{out}+D}$. In addition, the calculation of the impedance matching of a distributed constant transmission line will also be easy.

orthogonal polarization states, so they must be expressed by the Jones matrix. Consequently, each element S_{ij} and T_{ij} of the scattering matrix and transmission matrix should be further expressed by a 2×2 Jones matrix, so to accurately express the input and output characteristics of one input and one output port device (like a simple optical waveguide), a 4×4 column matrix is necessary. However, if coupling between orthogonal polarized lights can be practically ignored, the characteristics corresponding to the TE polarization and the TM polarization can be analyzed separately. Therefore, the scattering matrix and the transmission matrix for each of the polarization modes can be represented by a 2×2 column matrix.

Now, let us consider one polarization mode, when the reflected light of each of the ports is sufficiently suppressed. In this case, as there is no reflection in the circuit or at the output end, the amplitude of the light propagating toward the reverse direction can be set as $A_i' = 0$ (i is the number of cascade-connected optical circuits). Consequently, the transmission matrix of a two-port device can be given by the scalar number T_{11}, and its reverse matrix will be $\frac{1}{T_{11}}$. In assuming that there is no reflection, the number of ports and the matrix order will be the same. That is, the matrix that was used in the analysis of the Mach–Zehnder interferometer in Section 4.3 corresponds to this case.

In the Mach–Zehnder interferometer of Section 4.3, a translucent mirror was used to divide the free-space beam into two. In the Mach–Zehnder interferometer formed by an optical waveguide, a directional coupler takes on this role. Let us determine the transmission matrix (to be precise, the inverse matrix of the transmission matrix) for this directional coupler. First, when the input and the output amplitudes of the 2×2 port device (four-port device) are defined, as shown in Figure 8.22, the relationship between the input and the output amplitudes will have the same representation as Equation (8.52) of the scattering matrix of the two-port device. However, as the port corresponding to the input amplitude a_1 is different from that corresponding to the output amplitude b_1, to avoid confusion, the input is expressed as A_1 and B_1, whereas the output is denoted by A_2 and B_2,* and their relationship can be written as

$$\begin{bmatrix} A_2 \\ B_2 \end{bmatrix} = \begin{bmatrix} S_{11}^{(1)}, & S_{12}^{(1)} \\ S_{21}^{(1)}, & S_{22}^{(1)} \end{bmatrix} \begin{bmatrix} A_1 \\ B_1 \end{bmatrix}. \tag{8.62}$$

Thus, as the power conservation condition must be satisfied in the same manner as Equation (8.53), the relational expressions, Equations (8.54) through (8.56), hold in the same manner. However, even if the output A_2 is time reversed,

---

* That is, the amplitudes $A, B, \ldots$ expressed using the alphabets are the port numbers, and the subscript numbers represent the number of the connection stages when the circuits are cascaded in multiple stages. The superscript (k) that is attached to each of the elements $S_{ij}^{(k)}$ of the scattering matrix is also the number of the connection stages, when the circuits are cascaded in multiple stages.

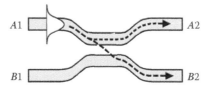

FIGURE 8.22 Transmission matrix representation of a directional matrix and definition of the input and the output amplitudes.

it will not become A_1, so Equation (8.57) does not necessarily hold. For Equation (8.57) to hold, the optical circuit must be the same even if the incident port and the output ports are switched. That is, this is possible only when the optical circuits are symmetrical at the intermediate point.

Now, in the relational Equations (8.16a) and b of the input and the output of the directional coupler, when two waveguide structures are equal such that $\beta^{(1)} = \beta^{(2)}$, then $\Delta\beta = 0$ and $\psi_w = |K_{12}|$. In addition, if the coupler length L is so designed that $|K_{12}|L = \frac{\pi}{4}$ (that is half of the coupling length L_b), the matrix in the same format of Equation (8.62) can be written as

$$\begin{bmatrix} A_2 \\ B_2 \end{bmatrix} = \frac{1}{\sqrt{2}} \begin{bmatrix} 1 & -j \\ -j & 1 \end{bmatrix} \cdot \begin{bmatrix} A_1 \\ B_1 \end{bmatrix}. \tag{8.63}$$

If the coupler length L is set to half of the coupling length L_b, and the light is beamed only from one of the incident ports so that $A_1 = 1$ and $B_1 = 0$, the output power is $|A_2|^2 = \frac{1}{2}$ and $|B_2|^2 = \frac{1}{2}$ and is outputted at both ports with a ratio of 1:1 (here, the phase deviates by $\frac{\pi}{2}$). This kind of directional coupler is called a *3-dB directional coupler*. This 3-dB directional coupler is used instead of the translucent mirror of the free-space beam Mach–Zehnder interferometer (MZ interferometer) in Figure 4.8. The MZ interferometer in which the optical path is all configured by optical waveguides is shown in Figure 8.23.

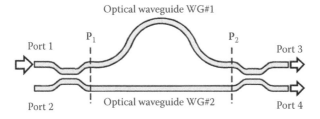

FIGURE 8.23 Optical waveguide-type Mach–Zehnder interferometer circuit using a directional coupler.

In this optical waveguide-type Mach–Zehnder interferometer, when the lengths of two optical waveguides WG#1 and WG#2, from point P_1, just after the coupling section, to point P_2, just before the merging section, are designated as L_1 and L_2, then the optical path lengths are $S_1 = n_{eq}L_1$ and $S_2 = n_{eq}L_2$, respectively. Consequently, after waveguide 1 and waveguide 2 undergo the respective phase changes $e^{-jk_0S_1}$ and $e^{-jk_0S_2}$, they will be merged at the directional coupler, and finally, in the same manner as Equation (4.29) of Section 4.3, the exit light amplitude A_{out} and B_{out} from the exit port of the MZ interferometer circuit are expressed by

$$
\begin{bmatrix} A_{out} \\ B_{out} \end{bmatrix} = \frac{1}{\sqrt{2}} \begin{bmatrix} 1 & -j \\ -j & 1 \end{bmatrix} \begin{bmatrix} e^{-jk_0S_1} & 0 \\ 0 & e^{-jk_0S_1} \end{bmatrix} \frac{1}{\sqrt{2}} \begin{bmatrix} 1 & -j \\ -j & 1 \end{bmatrix} \begin{bmatrix} A_1 \\ B_1 \end{bmatrix}
$$
$$
= -je^{-jk_0\frac{S_1+S_2}{2}} \begin{bmatrix} \sin\left(k_0\frac{S_1-S_2}{2}\right) & \cos\left(k_0\frac{S_1-S_2}{2}\right) \\ \cos\left(k_0\frac{S_1-S_2}{2}\right) & \sin\left(k_0\frac{S_1-S_2}{2}\right) \end{bmatrix} \cdot \begin{bmatrix} A_1 \\ B_1 \end{bmatrix}. \quad (8.64)
$$

Here, let us consider the case where the light is beamed from only one of the incident ports and set $B_1 = 0$. When the output light intensity P_3 from the exit port 3 is determined from $P_3 = |A_{out}|^2$

$$
P_3 = |A_1|^2 \sin^2\left[k_0\frac{S_1-S_2}{2}\right]
$$
$$
= |A_1|^2 \sin^2\left[\frac{\pi}{\lambda}\Delta S\right] \quad (8.65)
$$

is obtained. Here, $\Delta S = n_{eq}(L_1 - L_2)$ is the optical path length difference of waveguide 1 and waveguide 2.

When Equation (8.65) is illustrated, Figure 8.24 is obtained, and we can see that the optical output of a waveguide-type MZ interferometer with an

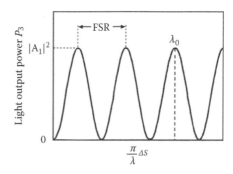

FIGURE 8.24 Wavelength filter characteristics of the optical waveguide-type MZ interferometer.

optical path length difference has a periodic dependence with respect to the wavelength in the same way as the beam-type MZ interferometer of Section 4.3. When we focus on the transmittance peak wavelength λ_0 with this wavelength filter, λ_0 can be expressed by the following equation:

$$\lambda_0 = \frac{2}{(2N+1)\pi} \Delta S \qquad (N = 0, 1, 2, \ldots). \qquad (8.66)$$

The $\sin^2[\frac{\pi}{\lambda}\Delta S]$ dependence of Equation (8.65), in contrast to the $\cos^2[\frac{\pi}{\lambda}\Delta S]$ dependence of Equation (4.30) of Section 4.3, is caused by the fact that the labeling of the corresponding output port is opposite.

PROBLEMS

1. Prove that the scattering matrix is a unitary matrix.

   [Hint]: The sum of the output light from all the ports is given by $\sum_{j=1}^{N} |b_j|^2 = \sum_{j=1}^{N} b_j{}^* b_j = [b^*]^t \cdot [b]$. Here, matrix $[b]$ is the vector at the left side of Equation (8.51). The sum of the input light is determined in the same manner, and if there is no loss or gain, $\sum_{j=1}^{N} |b_j|^2 = \sum_{j=1}^{N} |a_j|^2$.

2. Prove that from the reversibility of light, $S_{12} = S_{21}$ holds using the scattering matrix of a two-port device given by Equation (8.52).

   [Hint]: For example, the amplitude a_1 of the incident light of port 1 is expressed by $A_1 f_1(x, y) e^{-j\beta z}$, designating the electromagnetic field distribution function as $f_1(x, y)$ and the amplitude as A_1 when the amplitude involves a phase term except for the $e^{j\omega t}$ dependency. Consequently, when time reversal is done to propagate the light in the reverse direction, the z dependence may be reversed, so the light can be propagated in the reverse direction, if the amplitude is set to $a_1{}^*$. Then, it will become b_1. In the same manner, if a_2 is reversed, it will be set as $a_2{}^*$, and then it will become b_2. Consequently, the following equation is obtained:

$$\begin{bmatrix} a_1 \\ a_2 \end{bmatrix} = \begin{bmatrix} b_1{}^* \\ b_2{}^* \end{bmatrix} = \begin{bmatrix} S_{11}{}^*, & S_{12}{}^* \\ S_{21}{}^*, & S_{22}{}^* \end{bmatrix} \cdot \begin{bmatrix} a_1{}^* \\ a_2{}^* \end{bmatrix} = [S^*] \cdot \begin{bmatrix} b_1 \\ b_2 \end{bmatrix}$$

$$= \begin{bmatrix} S_{11}{}^*, & S_{12}{}^* \\ S_{21}{}^*, & S_{22}{}^* \end{bmatrix} \cdot \begin{bmatrix} S_{11}, & S_{12} \\ S_{21}, & S_{22} \end{bmatrix} \begin{bmatrix} a_1 \\ a_2 \end{bmatrix}. \qquad (8.67)$$

   $[S^*] \cdot [S] = I$ can be derived from this equation (here, I is a unit matrix in which all the diagonal elements are 1, and all other elements are 0). This relational expression and the $|S_{11}|^2 + |S_{21}|^2 = 1$, $|S_{22}|^2 + |S_{12}|^2 = 1$, and $S_{11} S_{12}{}^* + S_{22}{}^* S_{21} = 0$ which were derived from the unitary characteristics may be used to solve this problem.

3. From the nature of the scattering matrix of Problems **1** and **2**, prove that $T_{12} = T_{21}{}^*$ and $T_{22} = T_{11}{}^*$ in a lossless optical circuit.

4. In a 3-dB directional coupler, prove from the phase relationship of an even and an odd mode, shown in Figure 8.6, that when light is beamed only from one of the incident ports, the phase of the exit light from the two exit ports represented by Equation (8.63) deviates by $\frac{\pi}{2}$.

   [Hint]: The propagation constants of the even and the odd modes are expressed by β_e and β_o, respectively. The field distribution is expressed by the x-axis (vertical axis) and the amplitude axis (horizontal axis), as shown in Figure 8.6. In addition, if a three-dimensional figure is drawn to show $e^{-j\beta z}$ as a complex number and the x-axis of this figure is seen from the top (same representation as that of the sinusoidal steady state in terms of the complex voltage and complex current in the electrical circuit), the phase relationship between the even and the odd modes is easy to understand when the amplitudes of each of the exit ports are superimposed.

8.8 ANALYSIS OF AN OPTICAL WAVEGUIDE, INCLUDING STRUCTURE CHANGES IN PROPAGATION AXIS DIRECTION

As mentioned in Section 5.1.4, light propagation in the optical waveguide can be accurately described by the superposition of eigenmodes of the waveguide. However, in the analysis of light propagation for the tapered waveguide or the branching and merging waveguides of which the structure changes along the propagation direction, there is a limit to the eigenmode expansion method, and approximations such as the local normal mode [2] are necessary. However, even if it can be described analytically, the equation is complicated, so the outlook for such approximations is poor and the advantage of using them is small. Thus, the *beam propagation method* (BPM) or the propagating beam method [49]–[54], in which the light guided in the optical waveguide is treated as a propagation of a beam wave, was developed. In addition, the *finite difference time domain method* [55] (FDTD) has also been used recently to analyze optical waveguides and optical devices, in which the sequential changes with time of the electric and the magnetic fields are followed by tracing exactly Maxwell's equations and taking into consideration the boundary conditions at small intervals in space. However, this section will only describe a summary of BPM due to page volume constraints. Readers who want to learn more details may refer to [49] to [55].

First, with the refractive index n_0 as the basis, assume that the refractive index distribution of the waveguide is slightly higher in the core section and gradually changes in the z-direction, and designates $n(x, y, z)$ as

$$n(x, y, z) = n_0 + \delta n(x, y, z). \tag{8.68}$$

Assuming that $\delta n \ll n_0$, $n^2(x, y, z)$ can be written as

$$n^2(x, y, z) \simeq n_0^2 + 2n_0\delta n(x, y, z). \tag{8.69}$$

Here, using the form of the slightly modified field distribution for a plane wave in a medium of uniform refractive index n_0, the electric field is expressed by

$$E(r, t) = eU(r)\exp[j(\omega t - k_0 n_0 z)], \tag{8.70}$$

and when this is substituted to the wave equation (1.20),

$$j2k_0n_0\frac{\partial U}{\partial z} - \frac{\partial^2 U}{\partial z^2} = \nabla_t^2 U + [k_0^2 n^2 - k_0^2 n_0^2]U \tag{8.71}$$

is obtained. Here, as the amplitude function will increase gradually with respect to z, when it is approximated as

$$\frac{\partial^2 U}{\partial z^2} \ll k_0 n_0 \frac{\partial U}{\partial z}, \tag{8.72}$$

the following equation is obtained:

$$\frac{\partial U}{\partial z} = -j\left(\frac{\nabla_t^2}{2k_0 n_0} + k_0\delta n\right)U \tag{8.73}$$

This equation is called the *Fresnel equation* and is the same as Equation (7.8), which was derived in Section 7.1. Whereas the differentiation with respect to the transverse coordinates of space is a second-order differentiation, the z differentiation involves only the first-order derivative.* The solution for the propagation after infinitesimal distance Δz is given by

$$U(z + \Delta z) = \exp\left[-j\left(\frac{\nabla_t^2}{2k_0 n_0} + k_0\delta n\right)\Delta z\right]U(z). \tag{8.74}$$

When this equation is formally divided into three operators, it can be written as

$$U(z + \Delta z) = \exp\left[-j\frac{\nabla_t^2}{2k_0 n_0}\frac{\Delta z}{2}\right]\exp[-jk_0\delta n \Delta z]$$

$$\times \exp\left[-j\frac{\nabla_t^2}{2k_0 n_0}\frac{\Delta z}{2}\right]U(z). \tag{8.75}$$

---

* If z is replaced with time t, then this is similar to the Schrodinger equation of quantum mechanics. That is, if the propagation constant is assumed to be positive, only the solution for the wave (traveling wave) propagating in the positive z-direction is allowed. Therefore, the analysis based on this equation cannot be used to deal with the waveguide structure that accompanies a reflected wave.

The first and the third terms in the product on the right hand side are equivalent so as to solve Equation (7.8) using the initial conditions $U(x, y, z)$ and to determine the value at $z + \Delta z$, so this corresponds to the propagation in a medium of uniform refractive index n_0. On the other hand, the second term represents the phase change due to the refractive index distribution $\delta n(x, y)$ (approximated to be constant in the z-direction within the distance Δz). Consequently, this propagation is regarded as the first diffraction because of free-space propagation by a distance $\frac{\Delta z}{2}$ in a medium with a uniform refractive index n_0 and the repeated sequence of the phase change due to the distributed index $\delta n(x, y)$ of thickness Δz and the propagation at a distance Δz in the medium of refractive index n_0, as shown in Figure 8.25. Then, the change of the structure in the z-direction can be calculated by taking in $\delta n(x, y, z)$. The z-direction step size Δz is usually about the size of a wavelength.

In the calculation of Equation (8.75), the discrete Fourier series expansion was used previously, in which the amplitude distribution function $U(x, y, z)$ was decomposed into frequency components of various directions using a discrete Fourier series expansion. However, in recent years, methods that can obtain high-precision results even with larger step sizes in the z-direction and reduce computation time have been proposed, and some of them are the finite difference method [53] and the alternating direction implicit (ADI) method as direct solution methods of Equation (8.74), as well as the vector BPM [54]. In addition, as the normal BPM did not have high precision for the waveguide structure (Y-branch, etc.) that broadened with a large angle from the propagation axis because of the Fresnel approximation (equivalent to the paraxial approximation), the wide-angle BPM has been recently developed.

Finally, the state of the beam propagation of a simulated Y-merging waveguide that was determined using the semivectorial finite-difference BPM (SV-FDBPM) method is shown in Figure 8.26 (the waveguide parameters are

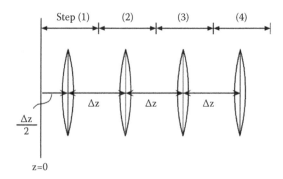

FIGURE 8.25 Equivalent lens in the calculation process of the BPM and arrangement in free space [49].

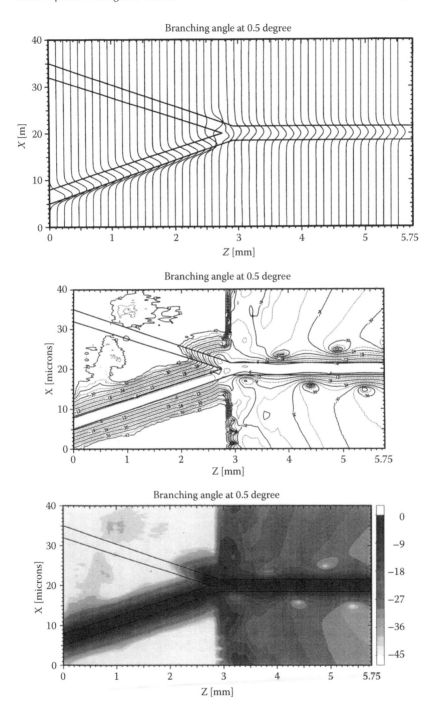

FIGURE 8.26 State of the beam propagation of the Y-merging waveguide that was calculated by SV-FDBPM. (a) View of the intensity distribution profile, (b) view of the intensity distribution contour, and (c) view according to the concentration of the intensity distribution.

$n_1 = 1.465$, $n_2 = 1.450$, waveguide width $W = 3$ μm, $\lambda = 1.3$ μm, branching angle $\theta = 0.5°$). As discussed in Section 8.3.2, a 3-dB radiation occurs in the merging circuit, and this is also seen in Figure 8.26c (it is not clearly seen in Figure 8.26a). As shown in this figure, the BPM program, by way of an interface that visualizes the calculation results, is an effective tool to easily understand the state of the propagation of light in a waveguide. However, in most cases, there is a significant difference in the vertical and the horizontal scales, so when looking at the figure, one must pay attention to the angle (as the change of the waveguide in the direction of propagation is very small, they often appear compressed in the direction of propagation).

A Fourier Transform Formulas

The Fourier transform $F(\omega)$ of function $f(t)$ is given by

$$F(\omega) = \int_{-\infty}^{\infty} f(t)e^{-j\omega t}\,dt. \tag{A.1}$$

Here, t is the time variable and ω is the angular frequency. On the other hand, the inverse Fourier transform of $F(\omega)$ is given by

$$f(t) = \frac{1}{2\pi} \int_{-\infty}^{\infty} F(\omega)e^{j\omega t}\,d\omega. \tag{A.2}$$

We can immediately see from the defining equation that these are linear transformations.

On the other hand, when the frequency $\nu\ (= \frac{\omega}{2\pi})$ is used instead of the angular frequency, the Fourier transform is expressed by

$$F(\nu) = \int_{-\infty}^{\infty} f(t)e^{-j2\pi\nu t}\,dt, \tag{A.3}$$

and the inverse Fourier transform is expressed by

$$f(t) = \int_{-\infty}^{\infty} F(\nu)e^{j2\pi\nu t}\,d\nu. \tag{A.4}$$

The Fourier transform has the following basic properties (as the proof is easy, it will be omitted). Here, $F(\omega)$ is expressed with the time function $f(t)$ being a Fourier-transformed function, the symbol $\longleftrightarrow$ expresses the relationship of the Fourier transform and the inverse Fourier transform, with the time variable being denoted to the left of the arrow and the angular frequency function being denoted to the right of the arrow.

(a) *Symmetry*

$$F(t) \longleftrightarrow 2\pi f(-\omega). \tag{A.5}$$

(b) *Time axis expansion/contraction*

$$f(at) \longleftrightarrow \frac{1}{|a|}F\left(\frac{\omega}{a}\right),^{*} \tag{A.6}$$

*(a is a real constant.)

293

(c) *Parallel translation of the time axis*

$$f(t - T_0) \longleftrightarrow F(\omega)e^{-jT_0\omega}. \tag{A.7}$$

(d) *Angular frequency axis parallel translation*

$$f(t)e^{j\omega_0 t} \longleftrightarrow F(\omega - \omega_0). \tag{A.8}$$

(e) *Time differentiation*

$$\frac{d^n f(t)}{dt^n} \longleftrightarrow (j\omega)^n F(\omega). \tag{A.9}$$

(f) *Frequency differentiation*

$$(-jt)^n f(t) \longleftrightarrow \frac{d^n F(\omega)}{d\omega^n}. \tag{A.10}$$

(g) *Complex conjugate*

$$f^*(t) \longleftrightarrow F^*(-\omega). \tag{A.11}$$

(h) *Integral*

$$\int_{-\infty}^{t} f(\tau)d\tau \longleftrightarrow F(\omega)\left[\frac{1}{j\omega} + \pi\delta(\omega)\right]. \tag{A.12}$$

Listed below are the Fourier transforms of some well-known functions.

$$\delta(t) \longleftrightarrow 1(\omega) \quad \text{(Impulse)} \tag{A.13}$$
$$1(t) \longleftrightarrow 2\pi\delta(\omega) \quad \text{(Direct current)} \tag{A.14}$$
$$\exp(j\omega_0 t) \longleftrightarrow 2\pi\delta(\omega - \omega_0) \tag{A.15}$$
$$\cos(\omega_0 t) \longleftrightarrow \pi[\delta(\omega - \omega_0) + \delta(\omega + \omega_0)] \tag{A.16}$$
$$\sin(\omega_0 t) \longleftrightarrow -j\pi[\delta(\omega - \omega_0) - \delta(\omega + \omega_0)] \tag{A.17}$$
$$U(t) \longleftrightarrow \frac{1}{j\omega} + \pi\delta(\omega) \quad \text{(Unit step function)} \tag{A.18}$$

$$U(t) - U(t-T) \longleftrightarrow \left[\frac{1}{j\omega} + \pi\delta(\omega)\right] \cdot (1 - e^{-jT\omega})$$
$$= \left[\frac{1}{j\omega} + \pi\delta(\omega)\right] \cdot 2j \exp\left(-j\frac{T\omega}{2}\right) \sin\frac{T\omega}{2}$$
$$= \exp\left(-j\frac{T\omega}{2}\right) \frac{\sin\frac{T\omega}{2}}{\frac{T\omega}{2}} \quad \text{(Window function)}$$
$$\tag{A.19}$$

$$\exp(-\alpha t^2) \longleftrightarrow \sqrt{\frac{\pi}{\alpha}} \exp\left[-\frac{\omega^2}{4\alpha}\right] \quad \text{(Gaussian function)} \quad \text{(A.20)}$$

$$\exp(j\alpha t^2) \longleftrightarrow \sqrt{\frac{\pi}{\alpha}} \exp\left[j\frac{\pi}{4} - j\frac{\omega^2}{4\alpha}\right] \quad \text{(Chirp function)} \quad \text{(A.21)}$$

$$\text{sech}(at) \longleftrightarrow \frac{\pi}{a}\text{sech}\left(\frac{\pi\omega}{2a}\right) \quad \text{(sech function)} \quad \text{(A.22)}$$

Here, the Fourier transform equation (A.21) of the chirp function where the exponent portion of the Gaussian function was made to be an imaginary number can actually be obtained when α is replaced with $-j\alpha$ in the Fourier transform equation (A.20) of the Gaussian function. That is, the α in the Fourier transform equation (A.20) of the Gaussian function need not be a real number but can be a complex number.

The convolution integral (also called a convolution) $f(x)$ of the two functions $h(x)$ and $g(x)$ is defined by the following equation:

$$f(x) = \int_{-\infty}^{\infty} h(y)g(x-y)\,dy = \int_{-\infty}^{\infty} h(x-y)g(y)\,dy \quad \text{(A.23)}$$

$$= h(x) * g(x). \quad \text{(A.24)}$$

In this way, instead of the integral, it can also be expressed as a convolution integral by the symbol $*$. The Fourier transform of the convolution integral is given by

$$F(\omega) = H(\omega) \cdot G(\omega) \quad \text{(A.25)}$$

as a product of the Fourier transform of $h(x)$ and $g(x)$. This is called the convolution theorem.

When this convolution theorem is used,

$$\int_{-\infty}^{\infty} f(t-\tau)f^*(\tau)d\tau \longleftrightarrow |F(\omega)|^2 \quad \text{(A.26)}$$

can be obtained from Equation (A.11), so by applying the inverse Fourier transform equation to the right side of Equation (A.26) and creating a equality with the left side

$$\int_{-\infty}^{\infty} f(t-\tau)f^*(\tau)\,d\tau = \frac{1}{2\pi}\int_{-\infty}^{\infty} |F(\omega)|^2 e^{j\omega t}\,d\omega \quad \text{(A.27)}$$

is obtained. Setting $t = 0$ in this equation and again replacing τ with t, the following equation is obtained:

$$\int_{-\infty}^{\infty} |f(t)|^2 dt = \frac{1}{2\pi}\int_{-\infty}^{\infty} |F(\omega)|^2 d\omega. \quad \text{(A.28)}$$

This identity is called the *Parseval's identity*. The left side of this equation expresses the energy of the waveform as the power of the time waveform is integrated with respect to the time. The right side expresses the energy on the angular frequency axis. That is, the energy of the waveform is equal whether we look at the time axis or the angular frequency axis. $|F(\omega)|^2$ is called the energy spectral density.

B Characteristics of the Delta Function

The δ *function (delta function)* is a kind of transcendental function and does not undergo differentiation and integration like that in normal functions (finite, differentiable functions). However, the function can be used conveniently by using the following characteristics:

$$\delta(x) = 0 \quad \text{at} \quad x \neq 0, \tag{B.1}$$

$$\int_{-\infty}^{+\infty} \delta(x) f(x)\, dx = \int_{0-\varepsilon}^{0+\varepsilon} \delta(x) f(x)\, dx = f(0). \tag{B.2}$$

Here, $f(x)$ is a continuous and finite function at $x = 0$, whereas ε is an infinitesimally small number. When Equation (B.2) is further modified, the following equation is obtained:

$$\int_{-\infty}^{+\infty} \delta(x - x_0) f(x)\, dx = \int_{-\infty}^{+\infty} \delta(u) f(u + x_0)\, du = f(x_0). \tag{B.3}$$

Here, x_0 is a constant. As can be seen from these equations, the delta function, in general, gives its value only after being integrated. Therefore, if a bare delta function is involved in an equation, one can calculate it by adding an integration from $-\infty$ to $+\infty$.

Now, Equation (B.3) expresses the characteristic of a delta function that gives the value $f(x_0)$ at x_0, when the argument for the real number function is swept. Then, this characteristic pairs with the following characteristic in Equation (B.5), which holds for the numerical sequence a_μ, when the argument is an integer for the integer function B.4 called the *Kronecker's delta* and this was introduced into Equation (2.22) of Section 2.4:

$$\delta_{\mu,\nu} = \begin{cases} 1 & : \text{if } \mu = \nu \\ 0 & : \text{if } \mu \neq \nu, \end{cases} \tag{B.4}$$

$$\sum_{\mu=i}^{j} \delta_{\mu,\nu} a_\mu = a_\nu \quad (i \leq \nu \leq j). \tag{B.5}$$

This kind of equation is frequently seen in the calculation of the expansion coefficient in the orthogonal function expansion of an arbitrary function.

C Derivation of Green's Theorem

Let us consider two mutually independent functions u and v. When the volume integral of both sides of the vector formula

$$\nabla \cdot (u\nabla v) = u\nabla^2 v + (\nabla u) \cdot (\nabla v) \tag{C.1}$$

is rewritten based on Gauss' theorem, it will become

$$\iiint_V u\nabla^2 v + \iiint_V (\nabla u) \cdot (\nabla v)\, dr^3 = \iiint_V \nabla \cdot (u\nabla v)\, dr^3$$

$$= \iint_S (u\nabla v)\, \boldsymbol{n} \cdot d\boldsymbol{a}. \tag{C.2}$$

This is called *Green's first theorem*. On the other hand, when u and v in Equation (C.2) are interchanged in a similar manner, the following equation is obtained:

$$\iiint_V v\nabla^2 u + \iiint_V (\nabla v) \cdot (\nabla u)\, dr^3 = \iint_S (v\nabla u)\, \boldsymbol{n} \cdot d\boldsymbol{a}. \tag{C.3}$$

When Equation (C.2) is subtracted from Equation (C.3),

$$\iiint_V [u\nabla^2 v - v\nabla^2 u]\, dr^3 = \iint_S [u\nabla v - v\nabla u]\, \boldsymbol{n} \cdot d\boldsymbol{a} \tag{C.4}$$

is obtained. This is called *Green's second theorem*. ∇u is expressed as $\frac{\partial u}{\partial n}$ in Equation (3.8), and both have the same meaning because they can be rewritten as grad u.

D Vector Analysis Formula

Triple vector product

$$A \times B \times C = (A \cdot C)B - (B \cdot C)A. \tag{D.1}$$

Pseudoscalar (triplet)

$$A \cdot (B \times C) = B \cdot (C \times A) = C \cdot (A \times B) = -A \cdot (C \times B). \tag{D.2}$$

(This is called pseudoscalar because even though it is a scalar quantity, when the sequence of the vector product section is switched, the sign will change.)

Vector differential operator (nabla) for the Cartesian coordinate system (x, y, z coordinate system)

$$\nabla = e_x \frac{\partial}{\partial x} + e_y \frac{\partial}{\partial y} + e_z \frac{\partial}{\partial z}. \tag{D.3}$$

Vector differential operator for the cylindrical coordinate system (r, θ, z coordinate system)

$$\nabla = e_r \frac{\partial}{\partial r} + e_\theta \frac{1}{r} \frac{\partial}{\partial \theta} + e_z \frac{\partial}{\partial z}. \tag{D.4}$$

Vector differential operator for the polar coordinate system (r, θ, ϕ coordinate system)

$$\nabla = e_r \frac{\partial}{\partial r} + e_\theta \frac{1}{r} \frac{\partial}{\partial \theta} + e_\phi \frac{1}{r \sin\theta} \frac{\partial}{\partial \phi}. \tag{D.5}$$

Principal identities (ψ is a scalar function, A and B are vector functions)

$$\nabla \cdot (\psi A) = \psi(\nabla \cdot A) + (\nabla\psi)A \tag{D.6}$$

$$\nabla \times (\psi A) = \psi(\nabla \times A) + (\nabla\psi) \times A \tag{D.7}$$

$$\nabla \cdot (A \cdot B) = (A \cdot \nabla)B + (B \cdot \nabla)A + A \times (\nabla \times B) + B \times (\nabla \times A) \tag{D.8}$$

$$\nabla \cdot (A \times B) = B \cdot (\nabla \times A) - A \cdot (\nabla \times B) \tag{D.9}$$

$$\nabla \times (A \times B) = (B \cdot \nabla)A - (A \cdot \nabla)B + A(\nabla \cdot B) - B(\nabla \cdot A) \tag{D.10}$$

$$\nabla \times (\nabla \times A) = \nabla(\nabla \cdot A) - \nabla^2 A \tag{D.11}$$

$$\nabla \cdot (\nabla \times A) = 0 \tag{D.12}$$

$$\nabla \times (\nabla\psi) = 0 \tag{D.13}$$

$$\nabla^2(\nabla \times A) = \nabla \times (\nabla^2 A) \tag{D.14}$$

$$\nabla^2(\nabla \psi) = \nabla(\nabla^2 \psi) \tag{D.15}$$

Laplacian (∇^2) for the Cartesian coordinate system (x, y, z coordinate system)

$$\nabla^2 = \frac{\partial^2}{\partial x^2} + \frac{\partial^2}{\partial y^2} + \frac{\partial^2}{\partial z^2}. \tag{D.16}$$

Laplacian for the cylindrical coordinate system (r, θ, z coordinate system)

$$\nabla^2 = \frac{1}{r}\frac{\partial}{\partial r}\left(r\frac{\partial}{\partial r}\right) + \frac{1}{r^2}\frac{\partial^2}{\partial \theta^2} + \frac{\partial^2}{\partial z^2}. \tag{D.17}$$

Laplacian for the polar coordinate system (r, θ, ϕ coordinate system)

$$\nabla^2 = \frac{1}{r^2}\frac{\partial}{\partial r}\left(r^2\frac{\partial}{\partial r}\right) + \frac{1}{r^2 \sin \theta}\frac{\partial}{\partial \theta}\left(\sin \theta \frac{\partial}{\partial \theta}\right) + \frac{1}{r^2 \sin \theta}\frac{\partial^2}{\partial \phi^2}. \tag{D.18}$$

Gauss' Theorem (Formula)

$$\iiint_V \nabla \cdot A dr^3 = \iint_S A \cdot da. \tag{D.19}$$

Here, V is a closed region of space and S is its surface (i.e., in a closed region of space that is like an egg, the whole three-dimensional space, including all the contents of the egg, is V, and the eggshell portion is S).

Stokes' Theorem (Formula)

$$\iint_S \nabla \times A da = \int_C A \cdot d\ell. \tag{D.20}$$

Here, S is a closed surface in space, and C is the integral region of its contour (i.e., S is the film when a bubble film is stretched inside a rubber band, and the rubber band is C).

E Infinite Integral of Gaussian Function

The infinite integral of the Gaussian function has the following value:

$$\int_{-\infty}^{\infty} \exp\left(-\alpha x^2\right) dx = \sqrt{\frac{\pi}{\alpha}}. \tag{E.1}$$

PROOF

$$\left(\int_{-\infty}^{\infty} \exp\left(-\alpha x^2\right) dx\right)^2 = \left(\int_{-\infty}^{\infty} \exp\left(-\alpha x^2\right) dx\right)\left(\int_{-\infty}^{\infty} \exp\left(-\alpha y^2\right) dy\right)$$

$$= \int_{-\infty}^{\infty}\int_{-\infty}^{\infty} \exp\left[-\alpha\left(x^2 + y^2\right)\right] dx\, dy$$

$$= \int_{0}^{\infty}\int_{0}^{2\pi} \exp\left(-\alpha r^2\right) r\, d\theta\, dr$$

$$= 2\pi \int_{0}^{\infty} \exp\left(-\alpha r^2\right) d\left(\frac{r^2}{2}\right)$$

$$= \frac{\pi}{\alpha} \tag{E.2}$$

[End of Proof]

If this formula is used, when two Gaussian beams with different spot sizes are connected with offset as in Equation (8.7) of Section 8.1.2, the coupling efficiency can be calculated. First, to express Equation (8.3) only in terms of the numerator portion, normalization is required for the mode electromagnetic field to satisfy Equation (8.4). Using Equation (E.1), when the amplitude normalization constant A_0 of the one-dimensional Gaussian function

$$\psi(x) = A_0 \exp\left[-\left(\frac{x}{w_0}\right)^2\right] \tag{E.3}$$

is determined, A_0 is obtained as

$$A_0 = \frac{1}{\left[\int_{-\infty}^{+\infty} \exp\left[-2\left(\frac{x}{w_0}\right)^2\right] dx\right]^{1/2}}$$

$$= \frac{1}{\left[w_0\sqrt{\frac{\pi}{2}}\right]^{1/2}}. \tag{E.4}$$

The same equation for the normalization constant is obtained, when i is set as $i = 0$ in Equation (7.41) of the normalization constant of the Hermite–Gaussian function.

Using the normalized Equation (E.3), with each of the spot sizes designated as w_1 and w_2, if one of the Gaussian functions with the offset δ from the central axis is substituted to the overlap integral of the numerator of Equation (8.3), the following equation is obtained:

$$A_1 A_2 \int_{-\infty}^{+\infty} \exp\left[-\left(\frac{x}{w_1}\right)^2\right] \exp\left[-\left(\frac{x-\delta}{w_2}\right)^2\right] dx, \qquad (E.5)$$

$$= A_1 A_2 \int_{-\infty}^{+\infty} \exp\left[-\left(\frac{1}{w_1^2}+\frac{1}{w_2^2}\right)\left\{x-\frac{w_1^2\delta}{w_1^2+w_2^2}\right\}^2 - \frac{\delta^2}{w_1^2+w_2^2}\right] dx$$

$$= A_1 A_2 \sqrt{\pi} \left(\frac{w_1^2 w_2^2}{w_1^2+w_2^2}\right)^{\frac{1}{2}} \exp\left[-\frac{\delta^2}{w_1^2+w_2^2}\right]. \qquad (E.6)$$

On the other hand, the normalization constant portion is given by

$$A_1 A_2 = \sqrt{\frac{2}{\pi w_1 w_2}}. \qquad (E.7)$$

Substituting these equations to Equation (8.3) (i.e., Equation (E.6) is squared), Equation (8.7) can be derived.

F Cylindrical Functions

When the wave equations (6.1) and (6.2) are made to undergo a separation of variables, for example, when E_z is separated into the r function and the θ function such that $E_z = R(r)\Theta(\theta)$, the θ function will become a trigonometric $\cos(\ell\theta + \phi_\ell)$ function, and $R(r)$ part will become the following Bessel differential equation:

$$\frac{\partial^2 R}{\partial r^2} + \frac{1}{r}\frac{\partial R}{\partial r} + \left[(k_0^2 n^2 - \beta^2) - \frac{\ell^2}{r^2}\right]R = 0. \tag{F.1}$$

When this equation is considered further by separating the core ($n = n_1 > \frac{\beta}{k_0}$) and the cladding ($n = n_2 < \frac{\beta}{k_0}$), the general solution for the wave equation inside the core can be expressed by the linear combination of the Bessel function $J_\ell(\kappa r)$ and the Neumann function $N_\ell(\kappa r)$, whereas the general solution in the cladding can be expressed as the linear combination of the modified Bessel functions of the first kind and the second kind $I_\ell(\kappa r)$ and $K_\ell(\kappa r)$, respectively. The functions corresponding to the $\cos \kappa x$ and $\sin \kappa x$ that appear in one-dimensional problems (a one-dimensional problem means that light is confined to the x-direction only) for slab waveguides are $J_\ell(\kappa r)$ and $N_\ell(\kappa r)$, whereas those corresponding to $e^{\gamma x}$ and $e^{-\gamma x}$ are $I_\ell(\gamma r)$ and $K_\ell(\gamma r)$. Here, the slight difference with the slab waveguide is the fact that $N_\ell(\kappa r)$ diverges to $-\infty$ at the center $r = 0$ of the core. Therefore, the function in the core is expressed with only $J_\ell(\kappa r)$. In addition, $I_\ell(\gamma r)$ diverges to ∞ at infinity $r = \infty$ in the cladding, which is a physically meaningless solution. Consequently, the function inside the cladding is only expressed by $K_\ell(\gamma r)$.

The principal formulas for these cylindrical functions are listed below. Here, $J_\ell(x)'$ and $K_\ell(x)'$ are expressed as the derivatives $\frac{dJ_\ell}{dx}$ and $\frac{dK_\ell}{dx}$ with respect to the argument x.

(a) *Bessel function*

$$\frac{\ell J_\ell(x)}{x} = \frac{1}{2}[J_{\ell-1}(x) + J_{\ell+1}(x)], \tag{F.2}$$

$$J_\ell(x)' = \frac{1}{2}[J_{\ell-1}(x) - J_{\ell+1}(x)], \tag{F.3}$$

$$J_{-\ell}(x) = (-1)^\ell J_\ell(x), \tag{F.4}$$

From Equations (F.2) and (F.3), the following equation can be obtained:

$$\frac{J_\ell(x)'}{x J_\ell(x)} = \frac{J_{\ell-1}(x)}{x J_\ell(x)} - \frac{\ell}{x^2}, \tag{F.5}$$

$$= -\frac{J_{\ell+1}(x)}{x J_\ell(x)} + \frac{\ell}{x^2}. \tag{F.6}$$

(b) *Modified Bessel function of the second kind*

$$\frac{\ell K_\ell(x)}{x} = \frac{1}{2}[K_{\ell+1}(x) - K_{\ell-1}(x)], \tag{F.7}$$

$$K_\ell(x)' = -\frac{1}{2}[K_{\ell+1}(x) + K_{\ell-1}(x)], \tag{F.8}$$

$$K_{-\ell}(x) = K_\ell(x). \tag{F.9}$$

From Equations (F.7) and (F.8), the following equation can be obtained:

$$\frac{K_\ell(x)'}{x K_\ell(x)} = -\frac{K_{\ell-1}(x)}{x K_\ell(x)} - \frac{\ell}{x^2}, \tag{F.10}$$

$$= -\frac{K_{\ell+1}(x)}{x K_\ell(x)} + \frac{\ell}{x^2}. \tag{F.11}$$

When x is small, the following asymptotic approximation can be used:

$$K_0(x) = \ln\left(\frac{2}{\gamma x}\right), \tag{F.12}$$

$$K_\ell(x) = (\ell - 1)! 2^{\ell-1} x^{-\ell} \quad (\ell \geq 1). \tag{F.13}$$

Here, γ is an Euler constant, which is about 1.781.

G Hermite–Gaussian Functions

The differential equation

$$\frac{d^2y}{dx^2} - 2x\frac{dy}{dx} + \lambda y = 0 \tag{G.1}$$

has a solution that is expressed by a finite polynomial, only if $\lambda = 2p$ (p is 0 or a positive integer). That is, as the polynomial has a finite number of terms, when $e^{-x^2/2}$ is multiplied to Equation (G.1) at $x = \pm\infty$, it will converge to 0. This solution is a polynomial of degree p and is called an Hermite polynomial. It is written as $H_p(x)$. The function of the product of the Hermite polynomial and the Gaussian function is called the Hermite–Gaussian function. An equation that has a solution only for a specific value of the parameter λ, like that in Equation (G.1), is called an *eigenvalue equation*. λ is called the *eigenvalue*, whereas the solution y is called the *eigenfunction*. The representative formulas for the Hermite–Gaussian functions are listed below.

(1) Representation based on the differential (Rodrigues' formula):

$$H_p(x) = (-1)^p e^{x^2} \frac{d^p e^{-x^2}}{dx^p}. \tag{G.2}$$

(2) Representation based on the generating function:

$$e^{-t^2 + 2tx} = e^{x^2} e^{-(t-x)^2} = \sum_{p=0}^{\infty} H_p(x)\frac{t^p}{p!}. \tag{G.3}$$

(3) Differential equation of the Hermite–Gaussian function $y = H_p(x)e^{-x^2/2}$:

$$\frac{d^2y}{dx^2} + (2p + 1 - x^2)y = 0. \tag{G.4}$$

(4) Differential:

$$\frac{dH_p(x)}{dx} = 2pH_{p-1}(x). \tag{G.5}$$

(5) Recurrence formula:

$$H_{p+1}(x) - 2xH_p(x) + 2pH_{p-1}(x) = 0. \tag{G.6}$$

(6) Orthogonal relation equation:

$$\int_{-\infty}^{\infty} e^{-x^2} H_p(x)H_q(x)dx = \begin{cases} 0 & (p \neq q) \\ 2^p p!\sqrt{\pi} & (p = q). \end{cases} \tag{G.7}$$

Here, as the Gaussian function portion $\exp[-\frac{x^2}{2}]$ of the Hermite–Gaussian function keeps its form as $\exp[-\frac{\omega^2}{2}]$ (the coefficient is different) with a Fourier transform, it is mathematically beautiful.* There are textbooks that describe the Guassian beam that follow this format as

$$f(x) = A \exp\left[-\frac{1}{2}\left(\frac{x}{w'}\right)^2\right].$$

However, as noted in the footnote in Section 3.3, spot size is defined by the JIS Standards and international standards, such as ISO, as "the distance between the center axis and the point where the electric field amplitude is $\frac{1}{e}$ of that of the center of the beam" or "the distance between the center of the beam and the point where the light intensity is $\frac{1}{e^2}$ of that of the center." Therefore, based on this definition, the description of the Gaussian beam is

$$f(x) = A \exp\left[-\left(\frac{x}{w}\right)^2\right].$$

In this book, the Gaussian beam is also described based on this definition. Consequently, one must be careful with variable transformation when applying the above equation.

NOTE

There are Hermite polynomials that are expressed as

$$\mathrm{He}_p(x) = (-1)^p e^{\frac{x^2}{2}} \frac{d^p}{dx^p}\left(e^{-\frac{x^2}{2}}\right), \tag{G.8}$$

and the series of equations for this polynomial are different from the equations mentioned here, so one should be careful not to get confused when reading textbooks on special functions and textbooks on quantum mechanics.

---

* In addition, the condition for a function that does not change forms even after undergoing a Fourier transform is the solution for the following differential equation:

$$\frac{d^2 f(t)}{dt^2} - t^2 f(t) + (2n+1)f(t) = 0, \quad \text{where } n \text{ is a positive integer.}$$

This can be confirmed by the fact that the form of the differential equation does not change in the Fourier transform. The Hermite–Gaussian function satisfies this differential equation.

H Derivation of the Orthogonality of the Eigenmode

Let us consider the electromagnetic fields $E^{(\mu)}$, $H^{(\mu)}$, $E^{(\nu)}$, and $H^{(\nu)}$ of two modes with different mode numbers. (Here, these fields involve only the part that is dependent on the spatial coordinates, excluding $e^{j\omega t}$.) First, let us consider the following vector operations:

$$\nabla \cdot (E^{(\mu)} \times H^{(\nu)*} + E^{(\nu)*} \times H^{(\mu)}). \tag{H.1}$$

When the vector formula (D.9) is used here, Equation (H.1) is rewritten by the following equation:

$$\begin{aligned}
\nabla \cdot (E^{(\mu)} \times H^{(\nu)*} &+ E^{(\nu)*} \times H^{(\mu)}) \\
&= H^{(\nu)*} \cdot \nabla \times E^{(\mu)} - E^{(\mu)} \cdot \nabla \times H^{(\nu)*} \\
&+ H^{(\mu)} \cdot \nabla \times E^{(\nu)*} - E^{(\nu)*} \cdot \nabla \times H^{(\mu)}.
\end{aligned} \tag{H.2}$$

Moreover, when Equations (5.7) and (5.8) are used in $\nabla \times E$ and $\nabla \times H$, and the refractive index is expressed as a real number (i.e., a medium with no gain and no absorption loss),

$$\nabla \cdot (E^{(\mu)} \times H^{(\nu)*} + E^{(\nu)*} \times H^{(\mu)}) = 0 \tag{H.3}$$

is derived.

Here, the z dependence is expressed as

$$E(x, y, z) = \overline{E}(x, y)e^{-j\beta z}, \tag{H.4}$$
$$H(x, y, z) = \overline{H}(x, y)e^{-j\beta z}. \tag{H.5}$$

When Equations (H.4) and (H.5) are substituted to Equation (H.3) and then a surface integration is executed at an arbitrary region S in the xy cross-section of the optical waveguide, Equation H.3 is rewritten by the following equation:

$$\iint_S \nabla \cdot [(\overline{E}^{(\mu)} \times \overline{H}^{(\nu)*} + \overline{E}^{(\nu)*} \times \overline{H}^{(\mu)})e^{j(\beta^{(\nu)} - \beta^{(\mu)})z}]dx\,dy = 0. \tag{H.6}$$

Here, the divergence (div) of the vector function and the scalar function are determined from the vector formula (D.7), and the vector part of Equation (H.6) is dependent only on x and y, whereas the $e^{j(\beta^{(v)}-\beta^{(\mu)})z}$ part is dependent only on z. Thus, when the vector part of Equation (H.6) is designated as A, the following equation is obtained:

$$\nabla \cdot (A e^{j(\beta^{(v)}-\beta^{(\mu)})z}) = e^{j(\beta^{(v)}-\beta^{(\mu)})z} \nabla \cdot A + A(\nabla e^{j(\beta^{(v)}-\beta^{(\mu)})z})$$

$$= e^{j(\beta^{(v)}-\beta^{(\mu)})} [\nabla_t \cdot A + j(\beta^{(v)} - \beta^{(\mu)}) e_z \cdot A]. \quad \text{(H.7)}$$

Here, $\nabla_t = \frac{\partial^2}{\partial x^2} + \frac{\partial^2}{\partial y^2}$. When this equation is used, Equation (H.6) is transformed into the following equation:

$$\iint_S \nabla_t \cdot [\overline{E}^{(\mu)} \times \overline{H}^{(v)*} + \overline{E}^{(v)*} \times \overline{H}^{(\mu)}] dx\, dy$$

$$+ j(\beta^{(v)} - \beta^{(\mu)}) \iint_S e_z \cdot [\overline{E}^{(\mu)} \times \overline{H}^{(v)*} + \overline{E}^{(v)*} \times \overline{H}^{(\mu)}] dx\, dy = 0. \quad \text{(H.8)}$$

Furthermore, using Gauss' theorem in a two-dimensional plane (the equation where the volume integral and the surface integral in Equation (D.19) fall, one dimension at a time, on each of the surface and line integrals), the first term on the left side of Equation (H.8) will be

$$\iint_S \nabla_t \cdot [\overline{E}^{(\mu)} \times \overline{H}^{(v)*} + \overline{E}^{(v)*} \times \overline{H}^{(\mu)}] dx\, dy$$

$$= \int_C [\overline{E}^{(\mu)} \times \overline{H}^{(v)*} + \overline{E}^{(v)*} \times \overline{H}^{(\mu)}] d\ell. \quad \text{(H.9)}$$

If the integral region C is set to an infinite distance from the core of the waveguide, and if at least one of the modes μ and v is a guided mode, the field distribution of guided mode reaches zero at infinite distance, so Equation (H.9) is zero. (If both modes μ and v are guided modes, it will become a function that includes $\delta(\beta^{(v)} - \beta^{(\mu)})$; however, the proof of this is omitted.) Consequently, this will finally become the following equation:

$$j(\beta^{(v)} - \beta^{(\mu)}) \iint_S e_z \cdot [\overline{E}^{(\mu)} \times \overline{H}^{(v)*} + \overline{E}^{(v)*} \times \overline{H}^{(\mu)}] dx\, dy = 0. \quad \text{(H.10)}$$

Now, if modes μ and v differ, in general $\beta^{(v)} \neq \beta^{(\mu)}$, so

$$\iint_S e_z \cdot [\overline{E}^{(\mu)} \times \overline{H}^{(v)*} + \overline{E}^{(v)*} \times \overline{H}^{(\mu)}] dx\, dy = 0 \quad (\mu \neq v) \quad \text{(H.11)}$$

is obtained. Here, the z component of $[\overline{E}^{(\mu)} \times \overline{H}^{(v)*} + \overline{E}^{(v)*} \times \overline{H}^{(\mu)}]$ in Equation (H.11) is surface integrated; however, the z-direction component (longitudinal direction component) of the cross product of the two vectors is equal to the cross product of each of the xy plane components of the vector (transverse direction component), so Equation (H.11) can be written as

$$\iint_S [\overline{E}_t^{(\mu)} \times \overline{H}_t^{(\nu)*} + \overline{E}_t^{(\nu)*} \times \overline{H}_t^{(\mu)}] dx\, dy = 0 \quad (\mu \neq \nu). \tag{H.12}$$

Here, the subscript t of $\overline{E}_t^{(\mu)}$ expresses the transverse component (xy components). If $\beta^{(\nu)} = \beta^{(\mu)}$ (this case is referred to as "modes μ and ν are degenerate"), even though the modes μ and ν differ, another linear combination can be created from the mode electromagnetic fields $E^{(\mu)}, H^{(\mu)}$ and $E^{(\nu)}, H^{(\nu)}$, using the similar orthogonalization method as the Schmidt orthogonalization in the vector algebra in which an orthogonal electromagnetic field is produced.

Now, Equation (H.12) should hold for all the eigenmodes of waveguides, and the eigenmode should include not just the forward traveling wave guided toward the positive z-direction, but also the backward traveling wave (or the reflected wave) that is guided toward the negative z-direction. Then, as the waveguide structure is symmetrical with respect to $z = 0$ (plane symmetry), the electromagnetic field distribution $\overline{E}^{(-)}(x, y)$ (the part excluding $e^{j(\omega t + \beta z)}$) of the backward traveling wave that replaced the $e^{j(\omega t - \beta z)}$ of the forward traveling wave with $e^{j(\omega t + \beta z)}$ should satisfy the following relationship with the $\overline{E}^{(+)}(x, y)$ of the forward traveling wave:

$$\overline{E}_t^{(-)}(x, y) = \overline{E}_t^{(+)}(x, y), \quad \overline{E}_z^{(-)}(x, y) = -\overline{E}_z^{(+)}(x, y), \tag{H.13}$$

$$\overline{H}_t^{(-)}(x, y) = -\overline{H}_t^{(+)}(x, y), \quad \overline{H}_z^{(-)}(x, y) = -\overline{H}_z^{(+)}(x, y). \tag{H.14}$$

Actually, only the relationship in the transverse direction is necessary for Equation (H.12), and the reason why the signs for the relationship between $\overline{H}_t^{(-)}(x, y)$ and $\overline{H}_t^{(+)}(x, y)$ is inverted in contrast to the signs that do not invert in the relationship between $\overline{E}_t^{(-)}(x, y)$ and $\overline{E}_t^{(+)}(x, y)$ is that the Poynting vector faces the opposite direction (this relationship also appears in Figure 1.16 for the reflection on the metal surface). Now, applying Equations (H.13) and (H.14) to mode μ and expressing the backward wave of the μth-order mode as $-\mu$, the following equations are obtained:

$$\overline{E}_t^{(-\mu)}(x, y) = \overline{E}_t^{(\mu)}(x, y), \quad \overline{H}_t^{(-\mu)}(x, y) = -\overline{H}_t^{(\mu)}(x, y). \tag{H.15}$$

Here, when the Equation (H.12) is applied to the $-\mu$th-order mode (backward wave of the μth-order mode) and the νth-order mode, and Equation (H.15), which expresses the relationship between the $-\mu$th-order mode and the μth-order mode, is used, the following equation is obtained:

$$\iint_S [\overline{E}_t^{(\mu)} \times \overline{H}_t^{(\nu)*} - \overline{E}_t^{(\nu)*} \times \overline{H}_t^{(\mu)}] dx\, dy = 0 \quad (\mu \neq \nu). \tag{H.16}$$

Consequently, the sum of Equations (H.12) and (H.16) gives the following equation:

$$\iint_S [\overline{\boldsymbol{E}}_t^{(\mu)} \times \overline{\boldsymbol{H}}_t^{(\nu)*}] = \iint_S \boldsymbol{e}_z [\overline{\boldsymbol{E}}^{(\mu)} \times \overline{\boldsymbol{H}}^{(\nu)*}] dx\, dy = 0 \quad (\mu \neq \nu), \quad (H.17)$$

and this completes the proof of Equation (5.58), when $\mu \neq \nu$.

On the other hand, when $\nu = \mu$,

$$\iint_S \boldsymbol{e}_z \cdot [\boldsymbol{E}^{(\mu)} \times \boldsymbol{H}^{(\mu)*} + \boldsymbol{E}^{(\mu)*} \times \boldsymbol{H}^{(\mu)}] dx\, dy$$

$$= 2 \iint_S \boldsymbol{e}_z \cdot Re[\boldsymbol{E}^{(\mu)} \times \boldsymbol{H}^{(\mu)*}] dx\, dy = 4P^{(\mu)} \qquad (H.18)$$

is obtained from Equation (1.51), so Equation (5.58) can also be proven even when $\mu = \nu$. Now, the complex conjugate of the electric field is used in Equation (H.18). Since the refractive index was already assumed to be a real number in Equation (H.3), if the z dependence of both the electric field and the magnetic field are expressed by Equations (H.4) and (H.5), $\overline{\boldsymbol{E}}^{(\mu)}$ and $\overline{\boldsymbol{E}}^{(\nu)}$ are real numbers. Consequently, in such a case, the symbol $Re[\ ]$ of Equation (H.18) can be omitted.

Lorentz Reciprocity Theorem

The Lorentz reciprocity theorem is the relationship that generally holds between the eigenmodes guided in each of the waveguides, when two different waveguide structures exist. This relational expression can be derived from a calculation that is similar to that for the mode orthogonality in the Appendix H.

First, the refractive index distributions of two different waveguides are designated as $n^{(1)}(x, y)$ and $n^{(2)}(x, y)$, and the electromagnetic fields of the eigenmodes of these waveguides are designated as $E^{(1)}, H^{(1)}$ and $E^{(2)}, H^{(2)}$. Excluding the complex conjugate in Equation (H.1), the sign of the second term is inverted, and the divergence (div) of the following vector is considered.

$$\nabla \cdot (E^{(1)} \times H^{(2)} - E^{(2)} \times H^{(1)}). \tag{I.1}$$

When the vector formula (D.9) is used here, Equation (I.1) is rewritten as the following equation:

$$\begin{aligned}
\nabla \cdot (E^{(1)} & \times H^{(2)} - E^{(2)} \times H^{(1)}) \\
&= H^{(2)} \cdot \nabla \times E^{(1)} - E^{(1)} \cdot \nabla \times H^{(2)} \\
&\quad - H^{(1)} \cdot \nabla \times E^{(2)} + E^{(2)} \cdot \nabla \times H^{(1)}.
\end{aligned} \tag{I.2}$$

Furthermore, if Equations (5.7) and (5.8) are used in $\nabla \times E$ and $\nabla \times H$,

$$\text{Right side of Equation (I.2)} = j\omega\varepsilon_0(n^{(1)^2} - n^{(2)^2})E^{(1)} \cdot E^{(2)} \tag{I.3}$$

is derived. Consequently, setting Equation (I.1) to be equal to the right side of Equation (I.3),

$$\begin{aligned}
\nabla \cdot (E^{(1)} & \times H^{(2)} - E^{(2)} \times H^{(1)}) \\
&= j\omega\varepsilon_0(n^{(1)^2} - n^{(2)^2})E^{(1)} \cdot E^{(2)}
\end{aligned} \tag{I.4}$$

is obtained. Moreover, when the waveguide structures are the same (when the electric fields $E^{(1)}, H^{(1)}$ and $E^{(2)}, H^{(2)}$ belong to the same waveguide structure), $n^{(1)}(x, y) = n^{(2)}(x, y)$, and so

$$\nabla \cdot (E^{(1)} \times H^{(2)} - E^{(2)} \times H^{(1)}) = 0 \tag{I.5}$$

is obtained. This equation is called the Lorentz reciprocity theorem.

J WKB Method

Let us solve Equation (6.43) using the WKB method. The WKB method is an approximate solution method developed by Wentzel, Kramers, and Brillouin in the field of quantum mechanics. First, assuming

$$F(r) = Ae^{jk_0 S(r)} \tag{J.1}$$

and substituting it to Equation (6.43), the following equation for $S(r)$ is obtained:

$$jk_0 \frac{d^2 S(r)}{dr^2} - k_0^2 \left(\frac{dS(r)}{dR} \right)^2 + U^2(r) = 0. \tag{J.2}$$

In addition, when $S(r)$ is expanded to the power of $\frac{1}{k_0} (= \frac{\lambda}{2\pi})$ (this is the key-point of the WKB method, where the higher order terms of wavelength is ignored as its impact is small),

$$S(r) = S_0(r) + \frac{1}{k_0} S_1(r) + \frac{1}{k_0^2} S_2(r) + \cdots . \tag{J.3}$$

When Equation (J.3) is substituted to Equation (J.2), and then arranged in the order starting from the k_0^2 order term, the following equation is obtained:

$$\left[-k_0^2 \left(\frac{dS_0}{dr} \right)^2 + U^2(r) \right] + \left(jk_0 \frac{d^2 S_0}{dr^2} - 2k_0 \frac{dS_0}{dr} \frac{dS_1}{dr} \right)$$

$$+ [\text{the terms in the order of } k_0^0, k_0^{-1}, k_0^{-2} \cdots] = 0. \tag{J.4}$$

Here, the terms for each of the orders of k_0 (k_0^2, k_0, $\cdots$) are independently set to 0. Here, the term that is proportional to the higher order λ can be ignored, if the λ is small (in quantum mechanics, this is the case when energy is large, that is, in the case of particle-like behavior), so considering only the k_0^2 and k_0 terms, the following solution is obtained:

$$k_0^2 \text{ term}: \quad S_0 = \pm \frac{1}{k_0} \int U(r) dr + C_1, \tag{J.5}$$

$$k_0 \text{ term}: \quad \frac{S_1(r)}{k_0} = \pm \frac{j}{2k_0} \ln \left[\frac{1}{k_0^2} U(r) \right] + C_2. \tag{J.6}$$

Here, C_1 and C_2 are constants. Substituting Equations (J.5) and (J.6) to Equation (J.1),

$$F(r) = A \frac{1}{\sqrt{U(r)}} \sin \left[\int_{r_1}^{r} U(\rho)d\rho + \varphi \right] \qquad (J.7)$$

is obtained. Here, A and φ are constants, and r_1 is the smaller solution of the two solutions for $U(r) = 0$.

Equation (J.7) is the solution of the oscillation region, and Equation (6.45) can be derived from the condition for forming a standing wave in this region. (If the $F(r_1) = 0$ condition is imposed, $\varphi = 0$ is obtained.)

K Derivation of the Petermann's Formula for the Optical Fiber Spot Size

The variational representation (or the stationary representation) Equation (5.80) of Section 5.1.6 is expressed by the surface integral in the x, y plane; however, as there is an axial symmetry in the fundamental mode ($LP_{0,1}$ mode) of a cylindrical optical fiber, the refractive index distribution and the field distribution are expressed as a function of r only. Then, the integration can be transformed to that with respect to r and θ. Because each of the 2π resulting from the integral for the θ in the denominator and numerator disappears, the integral for r only remains, and the following equation is obtained:

$$\beta^2 = \frac{\int_0^\infty \left[k_0^2 n^2(r) \psi(r)^2 - \left(\frac{\partial \psi(r)}{\partial r} \right)^2 \right] r \, dr}{\int_0^\infty \psi^2(r) r \, dr}. \tag{K.1}$$

Because this equation is stationary, even if a function, which is somewhat similar to the eigenfunction in form, is substituted to Equation (K.1), the error in the eigenvalue β^2 would be small, which is the characteristic of the stationary expression. Here, the objective is to approximate the original field distribution $\psi(r)$ of the fundamental mode by the Gaussian function

$$F_G(r) = \exp\left[-\frac{r^2}{w^2} \right]. \tag{K.2}$$

So, when F_G is substituted for the $\psi(r)$ at the right side of Equation (K.1), it can be confirmed that the second term of Equation (K.1) is equal to $\frac{2}{w^2}$

$$\frac{\int_0^\infty \left(\frac{d F_G(r)}{dr} \right)^2 r \, dr}{\int_0^\infty \left[F_G(r) \right]^2 r \, dr} = \frac{2}{w^2}. \tag{K.3}$$

Equation (7.50) was derived based on mere prediction that $F_G(r)$ can be substituted for $\psi(r)$, but this equation is based only on the fact that Equation (K.1) is stationary. Thus, $\psi(r)$ does not give the least squares approximation equation using the Gaussian function $F_G(r)$. The equations that give the least squares approximation are Equations (7.51) and (7.52).

L Derivation of the Coupling Mode Equation

Here, let us derive the equation that describes the coupling between two wave-guides, which are in close proximity by a distance of about a wavelength, as shown in Figure L.1c.

First, if the electromagnetic field distributions of the mode propagating in waveguide I are designated as $E^{(1)}$ and $H^{(1)}$ in the case of Figure L.1a where waveguide I exists alone, and the electromagnetic field distributions are designated as $E^{(2)}$ and $H^{(2)}$ in the case of Figure L.1b where waveguide II exists alone, then they will satisfy the following Maxwell's equations:

$$\nabla \times E^{(\sigma)} = -j\omega\mu_0 H^{(\sigma)}, \tag{L.1}$$

$$\nabla \times H^{(\sigma)} = j\omega\varepsilon_0 n^{(\sigma)^2}(x, y)E^{(\sigma)}. \tag{L.2}$$

Here, $\sigma = 1$ or 2, and $n^{(1)}(x, y)$ and $n^{(2)}(x, y)$ represent the refractive index distributions, when waveguide I exists alone and when waveguide II exists alone, respectively. Moreover, if both waveguides I and II are present at the same time, the refractive index distribution is defined as $n(x, y)$.

Now, the electromagnetic field distributions E and H in the case of Figure L.1c, when both waveguides I and II are present, are assumed to be expressed approximately by the linear combination of the electromagnetic field distributions of waveguides I and II as

$$E = a(z)E^{(1)} + b(z)E^{(2)}, \tag{L.3}$$

$$H = a(z)H^{(1)} + b(z)H^{(2)}. \tag{L.4}$$

These electric field distributions must also satisfy the Maxwell's equations

$$\nabla \times E = -j\omega\mu_0 H, \tag{L.5}$$

$$\nabla \times H = j\omega\varepsilon_0 n^2(x, y)E. \tag{L.6}$$

Here, unlike that in Equations (L.1) and (L.2), note that the refractive index distribution is $n^2(x, y)$ in the case, when both waveguides I and II are present at the same time.

Here, when terms such as $\nabla \times (a(z)E^{(1)})$ that appear when Equations (L.3) and (L.4) are substituted to Maxwell's equations (L.5) and (L.6) are calculated

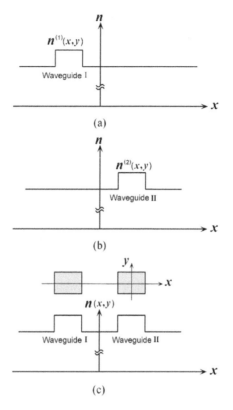

FIGURE L.1 Refractive index distribution of coupled waveguide systems: (a) when waveguide I exists alone, (b) when waveguide II exists alone, and (c) when both waveguide I and waveguide II are present.

using Equation (D.7), $a(z)$ and $b(z)$ are dependent on only z, so the following four equations are obtained:

$$\nabla \times (a(z)\boldsymbol{E}^{(1)}) = a(z)(\nabla \times \boldsymbol{E}^{(1)}) + (\nabla a(z)) \times \boldsymbol{E}^{(1)}$$

$$= a(z)(-j\omega\mu_0 \boldsymbol{H}^{(1)}) + \frac{da(z)}{dz}(\boldsymbol{e}_z \times \boldsymbol{E}^{(1)}), \qquad (\text{L.7})$$

$$\nabla \times (b(z)\boldsymbol{E}^{(2)}) = b(z)(-j\omega\mu_0 \boldsymbol{H}^{(2)}) + \frac{db(z)}{dz}(\boldsymbol{e}_z \times \boldsymbol{E}^{(2)}), \qquad (\text{L.8})$$

$$\nabla \times (a(z)\boldsymbol{H}^{(1)}) = a(z)(j\omega\varepsilon_0 n^{(1)2} \boldsymbol{E}^{(1)}) + \frac{da(z)}{dz}(\boldsymbol{e}_z \times \boldsymbol{H}^{(1)}), \qquad (\text{L.9})$$

$$\nabla \times (b(z)\boldsymbol{H}^{(2)}) = b(z)(j\omega\varepsilon_0 n^{(2)2} \boldsymbol{E}^{(2)}) + \frac{db(z)}{dz}(\boldsymbol{e}_z \times \boldsymbol{H}^{(2)}). \qquad (\text{L.10})$$

Here, when Equation (L.3) is substituted to Maxwell's equation (L.5), and then Equations (L.7) and (L.8) are used,

$$\frac{da(z)}{dz}(e_z \times E^{(1)}) + \frac{db(z)}{dz}(e_z \times E^{(2)}) = 0 \qquad (L.11)$$

is obtained. In the same way, when Equation (L.4) is substituted to Maxwell's equation (L.6), and then Equations (L.9) and (L.10) are used,

$$\frac{da(z)}{dz}(e_z \times H^{(1)}) + \frac{db(z)}{dz}(e_z \times H^{(2)})$$
$$= j\omega\varepsilon_0(n^2 - n^{(1)^2})a(z)E^{(1)} + j\omega\varepsilon_0(n^2 - n^{(2)^2})b(z)E^{(2)} \qquad (L.12)$$

is obtained.

Now, the objective is to derive a simultaneous differential equation for $a(z)$ and $b(z)$ from Equations (L.11) and (L.12), but these equations still include the electromagnetic field distributions $E^{(1)}$, $H^{(1)}$, $E^{(2)}$, and $H^{(2)}$, so the coefficients are dependent on x and y. Here, let us normalize either of the coefficients $\frac{da(z)}{dz}$ or $\frac{db(z)}{dz}$ of Equations (L.11) and (L.12) by the optical power carried by the mode (modal power). That is,

$$\frac{1}{4} \iint_{-\infty}^{\infty} (E^{(\mu)} \times H^{(\mu)*} + E^{(\mu)*} \times H^{(\mu)})dx\,dy$$
$$= \frac{1}{2}Re\left[\iint_{-\infty}^{\infty} E^{(\mu)} \times H^{(\mu)*}dx\,dy\right] = P^{(\mu)} \qquad (L.13)$$

holds, when $\mu = \nu$ for Equation (5.58), so this form can be produced from Equations (L.11) and (L.12).

First, when the vector formula (D.2) is applied to e_z, $E^{(1)}$, and $H^{(1)*}$, and the following resulting equations

$$H^{(1)*} \cdot (e_z \times E^{(1)}) = e_z \cdot (E^{(1)} \times H^{(1)*}), \qquad (L.14)$$
$$E^{(1)*} \cdot (e_z \times H^{(1)}) = -e_z \cdot (E^{(1)*} \times H^{(1)}), \qquad (L.15)$$

are used, and when $H^{(1)*}$ is multiplied to Equation (L.11)

$$[e_z \cdot (E^{(1)} \times H^{(1)*})]\frac{da(z)}{dz} + [e_z \cdot (E^{(2)} \times H^{(1)*})]\frac{db(z)}{dz} = 0 \qquad (L.16)$$

is obtained. If $H^{(2)*}$ is multiplied to Equation (L.11) in the same manner, then

$$[e_z \cdot (E^{(1)} \times H^{(2)*})]\frac{da(z)}{dz} + [e_z \cdot (E^{(2)} \times H^{(2)*})]\frac{db(z)}{dz} = 0 \qquad (L.17)$$

is obtained. On the other hand, when $E^{(1)*}$ and $E^{(2)*}$ are multiplied to Equation (L.12), the following two equations are obtained:

$$[e_z \cdot (E^{(1)*} \times H^{(1)})]\frac{da(z)}{dz} + [e_z \cdot (E^{(1)*} \times H^{(2)})]\frac{db(z)}{dz}$$

$$+ j\omega\varepsilon_0(n^2 - n^{(1)^2})[E^{(1)} \cdot E^{(1)*}]a(z)$$

$$+ j\omega\varepsilon_0(n^2 - n^{(2)^2})[E^{(2)} \cdot E^{(1)*}]b(z) = 0, \quad \text{(L.18)}$$

$$[e_z \cdot (E^{(2)*} \times H^{(1)})]\frac{da(z)}{dz} + [e_z \cdot (E^{(2)*} \times H^{(2)})]\frac{db(z)}{dz}$$

$$+ j\omega\varepsilon_0(n^2 - n^{(1)^2})[E^{(1)} \cdot E^{(2)*}]a(z)$$

$$+ j\omega\varepsilon_0(n^2 - n^{(2)^2})[E^{(2)} \cdot E^{(2)*}]b(z) = 0. \quad \text{(L.19)}$$

Here, when the propagation constants of the individual waveguides are designated as $\beta^{(1)}$ and $\beta^{(2)}$, $E^{(i)}$ and $H^{(i)}$ can be expressed as

$$E^{(i)} = \overline{E}^{(i)} e^{j(\omega t - \beta^{(i)} z)} \quad (i = 1 \text{ or } 2), \quad \text{(L.20)}$$

$$H^{(i)} = \overline{H}^{(i)} e^{j(\omega t - \beta^{(i)} z)} \quad (i = 1 \text{ or } 2). \quad \text{(L.21)}$$

After Equations (L.16) through (L.19) are integrated over the whole infinite cross-section of the waveguide, the sum of Equations (L.16) and (L.18) and the sum of Equations (L.17) and (L.19) give the following two equations:

$$\frac{da(z)}{dz} + c_{12}\frac{db(z)}{dz}e^{-j(\beta^{(2)} - \beta^{(1)})z} + jK_{11}a(z) + jK_{12}b(z)e^{-j(\beta^{(2)} - \beta^{(1)})z} = 0,$$

$$\text{(L.22)}$$

$$\frac{db(z)}{dz} + c_{21}\frac{da(z)}{dz}e^{-j(\beta^{(1)} - \beta^{(2)})z} + jK_{21}a(z)e^{-j(\beta^{(1)} - \beta^{(2)})z} + jK_{22}b(z) = 0.$$

$$\text{(L.23)}$$

Here, the coefficients c_{ij} and K_{ij} (i and j are 1 and 2, respectively) are given by the following equations:

$$c_{ij} = \frac{\int_{-\infty}^{\infty}\int_{-\infty}^{\infty}\left[e_z \cdot (\overline{E}^{(i)*} \times \overline{H}^{(j)} + \overline{E}^{(j)} \times \overline{H}^{(i)*})\right]dx\,dy}{\int_{-\infty}^{\infty}\int_{-\infty}^{\infty}\left[e_z \cdot (\overline{E}^{(i)} \times \overline{H}^{(i)*} + \overline{E}^{(i)*} \times \overline{H}^{(i)})\right]dx\,dy}, \quad \text{(L.24)}$$

$$K_{ij} = \frac{\omega\varepsilon_0 \int_{-\infty}^{\infty}\int_{-\infty}^{\infty}\left[(n^2 - n^{(j)^2})\overline{E}^{(j)} \cdot \overline{E}^{(i)*}\right]dx\,dy}{\int_{-\infty}^{\infty}\int_{-\infty}^{\infty}\left[e_z \cdot (\overline{E}^{(i)} \times \overline{H}^{(i)*} + \overline{E}^{(i)*} \times \overline{H}^{(i)})\right]dx\,dy}. \quad \text{(L.25)}$$

The denominator in Equation (L.24) and Equation (L.25) expresses the mode power of Equation (L.13), so when the power is normalized such that $P^{(1)} = 1$

and $P^{(2)} = 1$, it is easy to see from Equation (L.24) that $c_{12} = c_{21}*$ holds and from Equation (L.25) that $K_{ii} = K_{ii}*$ ($i = 1$ or 2) holds. Here, the electric field $\overline{E}^{(i)}$ and the magnetic field $\overline{H}^{(i)}$ are expressed as complex numbers so that generality will not be lost in these equations, but in waveguides without absorption, the electromagnetic field that does not include the $e^{j(\omega t - \beta z)}$ dependency is a real number, and so $c_{12} = c_{21}$ holds (i.e., c_{12} is a real number).

Here, let us consider the coefficients given by Equations (L.24) and (L.25). When $i \neq j$, c_{ij} is expressed by the combination of the inner product of the electric field of waveguide I and the magnetic field of waveguide II (or vice versa). Because the magnitude of the magnetic field is about $\sqrt{\frac{\mu_0}{\varepsilon}}$ times that of the magnitude of the electric field (impedance of space in the case of plane waves), c_{ij} is in the order of about $\sqrt{\frac{\mu_0}{\varepsilon}} |\overline{E}^{(i)} \cdot \overline{E}^{(j)}|$. Consequently, when waveguides I and II are somewhat far apart, the overlap of the electric fields $\overline{E}^{(i)}$ and $\overline{E}^{(j)}$ is small, so c_{ij} is also small. On the other hand, for K_{ij}, when $i \neq j$, the overlap integral of the electric fields $\overline{E}^{(i)}$ and $\overline{E}^{(j)}$ is calculated in the core portion of only the waveguide I or II. However, when $i = j$, the electric field $|\overline{E}^{(i)}|^2$ is integrated in the region where $n^2 - n^{(i)2}$ is not 0, that is, the core region of the waveguide j, so K_{ii} is smaller compared to K_{ij}. Moreover, when $i \neq j$, K_{ij} is in the order of $\omega \varepsilon_0 |\overline{E}^{(i)} \cdot \overline{E}^{(j)}|$ and c_{ij} is $\frac{Z}{\omega}$ ($Z = \sqrt{\frac{\mu_0}{\varepsilon}}$ is the impedance of the medium and is in the order of several hundred of magnitude at most, but ω is in the order of 10^{15}) times smaller than this. Consequently, if waveguides I and II are far enough, the approximations $c_{ij} \simeq 0$ and $K_{ii} \simeq 0$ may be used. Such an approximation is called a weakly coupled approximation.

Up to this point in the derivation, the z dependency $e^{-j\beta z}$ was included in the electromagnetic field, and the electric and magnetic fields were expressed by Equations (L.20) and (L.21). Consequently, the change of $a(z)$ and $b(z)$ along the z-direction is gradual enough compared to that of the wavelength. However, here, $a(z)$ and $b(z)$ including the $e^{-j\beta z}$ dependence are expressed as

$$A(z) = a(z)e^{-j\beta^{(1)}z}, \tag{L.26}$$

$$B(z) = b(z)e^{-j\beta^{(2)}z}. \tag{L.27}$$

That is, the electromagnetic fields $E^{(1)}$, $E^{(2)}$, $H^{(1)}$, and $H^{(2)}$ in Equations (L.3) and (L.4) include the $e^{-j\beta z}$ dependency, and the coefficients $a(z)$ and $b(z)$ show a sufficiently gradual change compared to that of the wavelength. On the other hand, using $A(z)$ and $B(z)$, the electromagnetic field of the coupled mode can be expressed as

$$E(x, y, z) = A(z)\overline{E}^{(1)}(x, y) + B(z)\overline{E}^{(2)}(x, y), \tag{L.28}$$

$$H(x, y, z) = A(z)\overline{H}^{(1)}(x, y) + B(z)\overline{H}^{(2)}(x, y). \tag{L.29}$$

Here the electromagnetic fields $\overline{E}^{(1)}$, $\overline{E}^{(2)}$, $\overline{H}^{(1)}$, and $\overline{H}^{(2)}$ are dependent only on the transverse coordinates that appear in Equations (L.20) and (L.21). Thus, when Equations (L.26) and (L.27) are used, and Equations (L.22) and (L.23) are transformed into equations for $A(z)$ and $B(z)$, and then expressed in a matrix formalism, they can be expressed as follows:

$$
\begin{bmatrix} 1, & c_{12} \\ c_{21}, & 1 \end{bmatrix} \cdot \begin{bmatrix} \dfrac{dA(z)}{dz} \\ \dfrac{dB(z)}{dz} \end{bmatrix}
$$

$$
= -j \left\{ \begin{bmatrix} \beta^{(1)}, & c_{12}\beta^{(2)} \\ c_{21}\beta^{(1)}, & \beta^{(2)} \end{bmatrix} + \begin{bmatrix} K_{11}, & K_{12} \\ K_{21}, & K_{22} \end{bmatrix} \right\} \cdot \begin{bmatrix} A(z) \\ B(z) \end{bmatrix}
$$

$$
= -j \{[C][B] + [K]\} \cdot \begin{bmatrix} A(z) \\ B(z) \end{bmatrix}. \tag{L.30}
$$

Here, $[C]$, $[B]$, and $[K]$ are the following matrices:

$$
[C] = \begin{bmatrix} 1, & c_{12} \\ c_{21}, & 1 \end{bmatrix}, \tag{L.31}
$$

$$
[B] = \begin{bmatrix} \beta^{(1)}, & 0 \\ 0, & \beta^{(2)} \end{bmatrix}, \tag{L.32}
$$

$$
[K] = \begin{bmatrix} K_{11}, & K_{12} \\ K_{21}, & K_{22} \end{bmatrix}. \tag{L.33}
$$

Consequently, the equation containing only one of the differentials of A and B can be determined as follows using the inverse matrix:

$$
\begin{bmatrix} \dfrac{dA(z)}{dz} \\ \dfrac{dB(z)}{dz} \end{bmatrix} = -j \{[C]^{-1}[C][B] + [C]^{-1}[K]\} \cdot \begin{bmatrix} A(z) \\ B(z) \end{bmatrix}
$$

$$
= -j \begin{bmatrix} \xi_1, & k_{12} \\ k_{21}, & \xi_2 \end{bmatrix} \cdot \begin{bmatrix} A(z) \\ B(z) \end{bmatrix}. \tag{L.34}
$$

Here, the elements of the coefficient matrix are expressed by

$$
\xi_1 = \beta^{(1)} + \frac{K_{11} - c_{12}K_{21}}{1 - |c_{12}|^2}, \tag{L.35}
$$

$$
\xi_2 = \beta^{(2)} + \frac{K_{22} - c_{21}K_{12}}{1 - |c_{12}|^2}, \tag{L.36}
$$

$$
k_{12} = \frac{K_{12} - c_{12}K_{22}}{1 - |c_{12}|^2}, \tag{L.37}
$$

$$
k_{21} = \frac{K_{21} - c_{21}K_{11}}{1 - |c_{12}|^2}. \tag{L.38}
$$

Equation (L.34) is called the *coupled mode equation*.

Now, let us consider the power

$$P = \frac{1}{2}Re\left[\iint_{-\infty}^{\infty} E \times H^* dx\, dy\right]$$ (L.39)

of the coupled mode. First, when the power of each of the modes is normalized as $P^{(1)} = 1$ and $P^{(2)} = 1$ (i.e., if both the modes are propagating in the positive z-direction), from Equation (L.13), the denominators of Equations (L.24) and (L.25) that represent c_{ij} and K_{ij} can be normalized to four. In a waveguide with no absorption, the power of the entire coupled system is conserved, and so $\frac{dP}{dz} = 0$ must hold. Thus, when Equations (L.28) and (L.29) are substituted to Equation (L.39) and differentiated, since $c_{12} = c_{21}^*$ holds,

$$\frac{dP}{dz} = \frac{d}{dz}[AA^* + BB^* + AB^*c_{21} + A^*Bc_{12}]$$

$$= \frac{d}{dz}[AA^* + BB^* + 2Re[AB^*c_{21}]] = 0$$ (L.40)

is obtained. When Equation (L.34) is used to the z differential of A and B of this equation and the result is sorted out, the coefficients of AA^* and BB^* will be zero and the following is obtained:

$$jAB^*[\xi_2^*c_{21} - \xi_1 c_{21} + k_{12}^* - k_{21}]$$
$$- jA^*B[\xi_2 c_{21}^* - \xi_1^*c_{21}^* + k_{12} - k_{21}^*] = 0.$$ (L.41)

For this equation to hold at any position of z (i.e., it holds with respect to any arbitrary A and B),

$$(\xi_2^* - \xi_1)c_{21} + k_{12}^* - k_{21} = 0$$ (L.42)

has to be satisfied. On the other hand, when Equations (L.35) to (L.38) are substituted to Equation (L.42)

$$(\beta^{(2)} - \beta^{(1)})c_{21} + K_{12}^* - K_{21} = 0$$ (L.43)

is derived. If weak coupling is applied for this equation such that $c_{12} = c_{21}^* = 0$, or if $\beta^{(1)} = \beta^{(2)}$ holds,

$$K_{12} = K_{21}^*$$ (L.44)

is obtained. K_{12} and K_{21} are called the *coupling coefficients*. Because $c_{12} = c_{21}^* = 0$ and $K_{11} = K_{22} = 0$ hold in the case of weak coupling, we can easily see from Equations (L.35) to (L.38) that for $\xi_1 = \beta^{(1)}$, $\xi_2 = \beta^{(2)}$, $k_{12} = K_{12}$, $k_{21} = K_{21}$ holds.

On the other hand, in the case where one of the guided modes of the waveguides before coupling is propagating in the positive z-direction, whereas the other one is propagating in the negative z-direction, coupling will not occur by just bringing both waveguides close together. However, by compensating for the difference in the propagation constants by a diffraction grating, and setting

$$\beta^{(1)} = \beta^{(2)} + \frac{2\pi}{\Lambda} \tag{L.45}$$

coupling can be made to occur. Here, Λ is the period of the diffraction grating. The condition for the power to be conserved in the case of a reverse coupling is

$$\frac{dP}{dz} = \frac{d}{dz}[AA^* - BB^*] = 0 \tag{L.46}$$

and when Equation (L.34) is substituted to this equation, and weak coupling is assumed,

$$K_{12} = -K_{21}{}^* \tag{L.47}$$

is obtained.

M General Solution of the Coupled Mode Equation

Let us determine the general solution for the coupled mode equation (L.34). Because Equation (L.34) can be broken down into a set of simultaneous first-order differential equations of $A(z)$ and $B(z)$, it can be transformed into a constant coefficient second-order differential equation of $A(z)$ and $B(z)$, when one of the variables is eliminated. The general solution of a constant coefficient second-order differential equation is expressed by the linear combination of two particular solutions, and these particular solutions are determined by substituting $A_0 e^{pz}$, which is assumed to be the solution (conventional means of determining the response of linear electrical circuits). As the same approach can be used for simultaneous differential equations, the solution can be assumed to be

$$\begin{bmatrix} A(z) \\ B(z) \end{bmatrix} = \begin{bmatrix} A_0 \\ B_0 \end{bmatrix} e^{-j\beta_c z}. \tag{M.1}$$

Here, A_0 and B_0 are constants. Substituting Equations (M.1) to (L.34), the following simultaneous equation can be obtained for A_0 and B_0:

$$\begin{bmatrix} \xi_1 - \beta_c, & k_{12} \\ k_{21}, & \xi_2 - \beta_c \end{bmatrix} \begin{bmatrix} A_0 \\ B_0 \end{bmatrix} = 0. \tag{M.2}$$

The condition for this equation to have a solution other than $A_0 = B_0 = 0$ is for the determinant of these coefficients to be 0. Therefore, the following equation is obtained for β_c (eigenvalue equation or characteristic equation for electrical circuits):

$$(\xi_1 - \beta_c)(\xi_2 - \beta_c) - k_{12}k_{21} = 0. \tag{M.3}$$

The solution for this equation is given by

$$\beta_c = \frac{\xi_1 + \xi_2}{2} \pm \sqrt{\left(\frac{\xi_1 - \xi_2}{2} \right)^2 + k_{12}k_{21}}, \tag{M.4}$$

so defining ψ and $\Delta\xi$ as

$$\psi = \sqrt{\Delta\xi^2 + k_{12}k_{21}}, \tag{M.5}$$

$$\Delta\xi = \frac{\xi_1 - \xi_2}{2}. \tag{M.6}$$

Equation (M.4) can be expressed as

$$\beta_c = \frac{\xi_1 + \xi_2}{2} \pm \psi. \tag{M.7}$$

Here, β_c that corresponds to the double sign of Equation (M.7) is designated as β_c^+ and β_c^-, and the eigenvector that corresponds to them are designated as A_0^+, B_0^+ and A_0^-, B_0^-, respectively.

When β_c satisfies Equation (M.3), the solution of Equation (M.2) includes the arbitrary coefficients that are common to A_0 and B_0, so eventually, it is determined by the ratio of A_0 and B_0

$$\frac{A_0}{B_0} = \frac{k_{12}}{\beta_c - \xi_1} = \frac{\beta_c - \xi_2}{k_{21}}. \tag{M.8}$$

When $\beta_c^{\pm}$ is substituted to this solution, as $\Delta\xi < \psi$, (assuming $k_{12} > 0$ and $k_{21} > 0$)

$$\frac{A_0^+}{B_0^+} = \frac{k_{12}}{-\Delta\xi + \psi} = \frac{\Delta\xi + \psi}{k_{21}} > 0, \tag{M.9}$$

$$\frac{A_0^-}{B_0^-} = \frac{k_{12}}{-\Delta\xi - \psi} = \frac{\Delta\xi - \psi}{k_{21}} < 0 \tag{M.10}$$

will hold. That is, the coupled mode that propagates with β_c^+ is an in-phase mode with a peak at the same direction of the amplitude A_0^+ and B_0^+. Conversely, the coupled mode that propagates with β_c^- is an antiphase mode with a peak at the reverse direction of the amplitude A_0^- and B_0^-. When the propagation constants for each of the modes before coupling have different ω dependency (dispersion characteristic), such as $\beta^{(1)}(\omega)$ and $\beta^{(2)}(\omega)$, and the two propagation constants are assumed to be equal at $\omega = \omega_0$, the dispersion characteristics of propagation constant β_c^+ of the in-phase mode and the propagation constant β_c^- of the antiphase mode are as shown in Figure M.1. On the other hand, if one of the individual modes before the coupling propagates in the reverse direction, and the sign of the propagation constants $\beta^{(1)}(\omega)$ and $\beta^{(2)}(\omega)$ are opposite, the propagation constants will not be equal, even if $\omega = \omega_0$ (let us assume that $\beta^{(2)}(\omega) = -\beta^{(1)}(\omega) < 0$ at $\omega = \omega_0$). However, by compensating for the difference in the propagation constants using a diffraction grating like that in Equation (L.45), when the period of the diffraction grating is set as

$$\beta^{(1)}(\omega_0) = \beta^{(2)}(\omega_0) + \frac{2\pi}{\Lambda} = -\beta^{(1)}(\omega_0) + \frac{2\pi}{\Lambda} \tag{M.11}$$

at $\omega = \omega_0$, coupling can be made to occur. Then, by assuming a weak coupling, Equation (M.4) is written as

$$\beta_c = \frac{\beta^{(1)} + \beta^{(2)}}{2} \pm \sqrt{\left(\frac{\beta^{(1)} - \beta^{(2)}}{2}\right)^2 + K_{12}K_{21}}. \tag{M.12}$$

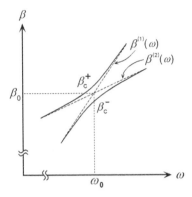

FIGURE M.1 Dispersion relationship for the codirectional optical coupling.

However, in a reverse direction coupling, $K_{12} = -K_{21}{}^*$ will hold from Equation (L.44), so $K_{12}K_{21} = -|K_{12}|^2$, and in the vicinity of $\omega = \omega_0$, a region that is

$$\left| \frac{\beta^{(1)}(\omega) - \beta^{(2)}(\omega)}{2} \right| < |K_{12}| \tag{M.13}$$

will appear. Consequently, in that region, $\beta_c^{\pm}$ is a complex number. The real parts are equal at $\frac{1}{2}[|\beta^{(1)}(\omega)| - |\beta^{(2)}(\omega)| + \frac{2\pi}{\Lambda}]$, but the presence of the imaginary part implies the attenuation of the electromagnetic field along with propagation, and the coupled mode cannot substantially exist in this region. Thus, this region is called the *stop band*. The dispersion relationship of the reverse direction coupling is shown in Figure M.2.

To simplify the problem further, a weak coupling is assumed, and setting $c_{12} = c_{21} = 0$ and $K_{11} = K_{22} = 0$, let us further assume that $\beta^{(1)} = \beta^{(2)}$. In the case of a weak coupling, as discussed in the last part of Appendix L, $\xi_1 = \beta^{(1)}$, $\xi_2 = \beta^{(2)}$, $k_{12} = K_{12}$, $k_{21} = K_{21}$, and $K_{12} = K_{21}{}^*$ hold. Therefore, the assumption of $\beta^{(1)} = \beta^{(2)}$ will result in $\Delta\xi = 0$ and $\psi = |K_{12}|$, and (assuming that $K_{12} > 0$) Equations (M.9) and (M.10) will each become

$$\frac{A_0^+}{B_0^+} = \frac{K_{12}}{|K_{12}|} = 1, \tag{M.14}$$

$$\frac{A_0^-}{B_0^-} = \frac{K_{12}}{-|K_{12}|} = -1. \tag{M.15}$$

That is, the coupled mode that propagates with β_c^+ is an even mode, as shown in Figure 8.6 of Section 8.2, and in the same manner, the coupled mode that

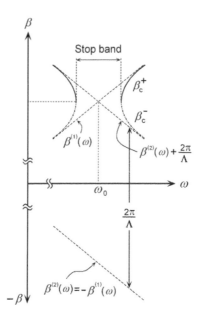

FIGURE M.2 Dispersion relationship for the contradirectional optical coupling.

propagates with β_c^- is an odd mode. In addition, in the dispersion relationship of Figure M.1 for a weak coupling, as $\beta^{(1)} = \beta^{(2)}$ holds at $\omega = \omega_0$,

$$\beta_c^+ - \beta_c^- = 2|K_{12}| \tag{M.16}$$

is obtained.

Now, let us return to the discussion on the general solution of Equation (L.34). Because the coefficients of the particular solution are given by Equations (M.9) and (M.10), the general solution as the linear combination of these particular solutions is expressed as

$$
\begin{aligned}
\begin{bmatrix} A(z) \\ B(z) \end{bmatrix} &= C_1 \begin{bmatrix} k_{12} \\ \Delta\xi + \psi \end{bmatrix} e^{-j\beta_c^+ z} + C_2 \begin{bmatrix} k_{12} \\ \Delta\xi - \psi \end{bmatrix} e^{-j\beta_c^- z} \\
&= \begin{bmatrix} k_{12}, & k_{12} \\ \Delta\xi + \psi, & \Delta\xi - \psi \end{bmatrix} \begin{bmatrix} C_1 e^{-j\beta_c^+ z} \\ C_2 e^{-j\beta_c^- z} \end{bmatrix} \\
&= [V] \begin{bmatrix} e^{-j\beta_c^+ z}, & 0 \\ 0, & e^{-j\beta_c^- z} \end{bmatrix} \begin{bmatrix} C_1 \\ C_2 \end{bmatrix}.
\end{aligned}
\tag{M.17}
$$

Here,

$$[V] = \begin{bmatrix} k_{12}, & k_{12} \\ \Delta\xi + \psi, & \Delta\xi - \psi \end{bmatrix}. \tag{M.18}$$

To determine C_1 and C_2 from the initial conditions, when $z = 0$ is substituted to Equation (M.17), the following equation is obtained:

$$\begin{bmatrix} A(0) \\ B(0) \end{bmatrix} = [V] \begin{bmatrix} 1, & 0 \\ 0, & 1 \end{bmatrix} \begin{bmatrix} C_1 \\ C_2 \end{bmatrix}. \tag{M.19}$$

Then, conversely, the equation that expresses C_1 and C_2 by $A(0)$ and $B(0)$ is determined as follows:

$$\begin{bmatrix} C_1 \\ C_2 \end{bmatrix} = [V]^{-1} \begin{bmatrix} A(0) \\ B(0) \end{bmatrix}. \tag{M.20}$$

When Equation (M.20) is substituted to Equation (M.17) and then sorted out, the following equation will be obtained as the general solution of the coupled mode equation:

$$\begin{bmatrix} A(z) \\ B(z) \end{bmatrix} = [V] \begin{bmatrix} e^{-j\beta_c^+ z}, & 0 \\ 0, & e^{-j\beta_c^- z} \end{bmatrix} [V]^{-1} \begin{bmatrix} A(0) \\ B(0) \end{bmatrix}$$

$$= e^{-j\frac{\xi_1+\xi_2}{2}z} \begin{bmatrix} \cos \psi z + j\frac{\Delta\xi}{\psi} \sin \psi z, & -j\frac{k_{12}}{\psi} \sin \psi z \\ -j\frac{k_{21}}{\psi} \sin \psi z, & \cos \psi z - j\frac{\Delta\xi}{\psi} \sin \psi z \end{bmatrix} \begin{bmatrix} A(0) \\ B(0) \end{bmatrix}. \tag{M.21}$$

This equation is a general solution that includes even the solution for strong coupling. In the case of weak coupling, designating $\xi_1 = \beta^{(1)}, \xi_2 = \beta^{(2)}, k_{12} = K_{12}$, and $k_{21} = K_{21}$, and substituting to Equation (M.21),

$$\begin{bmatrix} A(z) \\ B(z) \end{bmatrix} = e^{-j\beta_a z} \begin{bmatrix} \cos \psi_w z + j\frac{\Delta\beta}{\psi_w} \sin \psi_w z, & -j\frac{K_{12}}{\psi} \sin \psi_w z \\ -j\frac{K_{21}}{\psi_w} \sin \psi z, & \cos \psi_w z - j\frac{\Delta\beta}{\psi_w} \sin \psi_w z \end{bmatrix}$$

$$\times \begin{bmatrix} A(0) \\ B(0) \end{bmatrix} \tag{M.22}$$

is obtained. Here,

$$\beta_a = \frac{\beta^{(1)} + \beta^{(2)}}{2}, \tag{M.23}$$

$$\Delta\beta = \frac{\beta^{(1)} - \beta^{(2)}}{2}, \tag{M.24}$$

$$\psi_w = \sqrt{\Delta\beta^2 + K_{12}K_{21}} = \begin{cases} \sqrt{\Delta\beta^2 + |K_{12}|^2} & : \text{codirectional coupling} \\ \sqrt{\Delta\beta^2 - |K_{12}|^2} & : \text{contradirectional coupling.} \end{cases} \tag{M.25}$$

N Perturbation Theory

In the problem of determining eigenvalues as in a wave equation, the perturbation theory is an approach for determining the impact of infinitesimal changes in the potential function to the eigenvalue.

When the time and z dependence is assumed to be $e^{j(\omega t - \beta z)}$, the wave equation for optical waveguides can be written using the following equation in Cartesian coordinates (x, y, z coordinates):

$$\frac{\partial^2 E^{(\nu)}}{\partial x^2} + \frac{\partial^2 E^{(\nu)}}{\partial y^2} + k_0^2 n^2(x, y) E^{(\nu)} = \beta^{(\nu)2} E. \tag{N.1}$$

Here, ν is the mode order. When the operator $\mathcal{H}$ is defined as

$$\mathcal{H} = \frac{\partial^2}{\partial x^2} + \frac{\partial^2}{\partial y^2} + k_0^2 n^2(x, y), \tag{N.2}$$

Equation (N.1) can be written as

$$\mathcal{H} E^{(\nu)} = \beta^{(\nu)2} E^{(\nu)}. \tag{N.3}$$

In physical mathematics, $E(x, y)$ is the eigenfunction, with $n^2(x, y)$ as the potential, $\mathcal{H}$ as the operator, and β^2 as the eigenvalue.

Now, let us change the potential $n^2(x, y)$ by an infinitesimal quantity to result in $n^2(x, y) + \lambda \delta n^2(x, y)$ [the infinitesimal quantity of change may be dependent on z as in $\delta n^2(x, y, z)$, but it will be omitted here]. Here, λ is not a wavelength but is a coefficient that expresses the infinitesimal quantity. A similar coefficient may appear in an actual problem, and after using this in the derivation of an equation, it may be designated as $\lambda = 1$. Now, the operator is divided into the operator part $\lambda \mathcal{H}'$ that corresponds to the infinitesimal changes and the operator part $\mathcal{H}_0$ that does not include the infinitesimal changes, and Equation (N.3) is written as

$$(\mathcal{H}_0 + \lambda \mathcal{H}') E^{(\nu)}(x, y) = \beta^{(\nu)2} E^{(\nu)}(x, y), \tag{N.4}$$

$$\text{where} \quad \mathcal{H}_0 = \frac{\partial^2}{\partial x^2} + \frac{\partial^2}{\partial y^2} + k_0^2 n^2(x, y), \tag{N.5}$$

$$\mathcal{H}' = k_0^2 \delta n^2(x, y). \tag{N.6}$$

Here, the operator $\lambda \mathcal{H}'$ is called the perturbation. On the other hand, the eigenvalue and the eigenfunction is expressed by the power series of the infinitesimal coefficient, and the eigenvalue in the case where the infinitesimal change is not included (no perturbation) is defined as β_0^2, and the eigenfunction with no perturbations is defined as E_0. Then, by placing a subscript according to the power of λ, the eigenvalue and eigenfunction are expressed as follows:

$$\beta^{(v)^2} = \beta_0^{(v)^2} + \lambda\beta_1^{(v)^2} + \lambda^2\beta_2^{(v)^2} + \lambda^3\beta_3^{(v)^2} + \cdots, \tag{N.7}$$

$$E^{(v)} = E_0^{(v)} + \lambda E_1^{(v)} + \lambda^2 E_2^{(v)} + \lambda^3 E_3^{(v)} + \cdots. \tag{N.8}$$

If Equations (N.7) and (N.8) are substituted to Equation (N.4) and the coefficients of the same power of λ are equated on both sides, the following equations are obtained (the order of the power of λ is called the perturbation order):

$$\mathcal{H}_0 E_0^{(v)} = \beta_0^{(v)^2} E_0^{(v)}, \tag{N.9}$$

$$\mathcal{H}_0 E_1^{(v)} + \mathcal{H}' E_0^{(v)} = \beta_0^{(v)^2} E_1^{(v)} + \beta_1^{(v)^2} E_0^{(v)}, \tag{N.10}$$

$$\mathcal{H}_0 E_2^{(v)} + \mathcal{H}' E_1^{(v)} = \beta_0^{(v)^2} E_2^{(v)} + \beta_1^{(v)^2} E_1^{(v)} + \beta_2^{(v)^2} E_0^{(v)}. \tag{N.11}$$

The first equation corresponds to the case of no perturbation, and its eigenvalue $\beta_0^{(v)^2}$ and the eigenfunction $E_0^{(v)}(x, y)$ are assumed to be known. Here, as the eigenfunction E_0 (eigenmode in the optical waveguide) of the operator $\mathcal{H}_0$ is a complete orthonormal system, the first-order perturbation eigenfunction E_1 is expanded in terms of the non-perturbed eigenfunction $E_0^{(v)}$ and is expressed as

$$E_1^{(v)}(x, y) = \sum_i c_i^{(v)} E_0^{(i)}(x, y). \tag{N.12}$$

When this equation is substituted to the first-order perturbation Equation (N.10), the following equation is obtained:

$$\sum_i c_i^{(v)}\beta_0^{(i)^2} E_0^{(i)} + \mathcal{H}' E_0^{(v)} = \beta_0^{(v)^2} \sum_i c_i^{(v)} E_0^{(i)} + \beta_1^{(v)^2} E_0^{(v)}. \tag{N.13}$$

Multiplying this equation by $E_0^{(\mu)*}$, then integrating over the cross-section of the waveguide, and using the complete orthonormal condition (to be accurate, this should be normalized to the transmitted power in a waveguide, but mathematically, unity can be used as the complete orthonormal condition)

$$\iint_{-\infty}^{\infty} E_0^{(v)} E_0^{(\mu)*} dx\, dy = \delta_{v,\mu}, \tag{N.14}$$

the following equation can be derived.

$$c_\mu^{(v)}(\beta_0^{(\mu)^2} - \beta_0^{(v)^2}) + \iint_{-\infty}^{\infty} E_0^{(\mu)*} \mathcal{H}' E_0^{(v)} dx\, dy = \beta_1^{(\mu)^2}\delta_{\mu,v}. \tag{N.15}$$

First, if we designate $\mu = \nu$ in this equation, the eigenvalue $\beta_1^{(\nu)^2}$ based on the first-order perturbation is determined as

$$\beta_1^{(\nu)^2} = \iint_{-\infty}^{\infty} E_0^{(\nu)*} \mathcal{H}' E_0^{(\nu)} dx \, dy. \tag{N.16}$$

Consequently, when taking into consideration the first-order perturbation, and assuming also that the mode power is not normalized (λ is just a parameter that is incorporated for calculation purposes, so here it is designated as 1), the propagation constant is expressed as

$$\beta^{(\nu)^2} = \beta_0^{(\nu)^2} + \frac{k_0^2 \iint_{-\infty}^{\infty} \delta n^2(x, y) |E_0^{(\nu)}|^2 dx \, dy}{\iint_{-\infty}^{\infty} |E_0^{(\nu)}|^2 dx \, dy}. \tag{N.17}$$

On the other hand, designating $\mu \neq \nu$ in Equation (N.15), and then again setting as $\mu = i$, the expansion coefficient $c_i^{(\nu)}$ of the eigenfunction based on the first-order perturbation is obtained as follows:

$$c_i^{(\nu)} = \frac{k_0^2}{\beta_0^{(\nu)^2} - \beta_0^{(i)^2}} \frac{\iint_{-\infty}^{\infty} \delta n^2(x, y) E_0^{(\nu)} E_0^{(i)*} dx \, dy}{\iint_{-\infty}^{\infty} |E_0^{(\nu)}|^2 dx \, dy}. \tag{N.18}$$

Bibliography

1. H. Kogelnik and R. V. Ramaswamy. "Scaling rules for thin-film optical waveguides," *Appl. Opt.*, 13, no. 8, pp. 1857–1862, 1974.
2. D. Marcuse. *Theory of Dielectric Optical Waveguides* (2nd ed.), Academic Press, Boston, 1991.
3. H. Kogelnik. *Guided-Wave Optoelectronics*, Edited by T. Tamir, Springer-Verlag, Berlin, 1988, Chap. 2.
4. M. J. Adams. *An Introduction to Optical Waveguides*, John Wiley & Sons, Chichester, 1981.
5. A. R. Mickelson. *Guided Wave Optics*, Thomson International, London, 1993.
6. A. Yariv and P. Yeh. *Optical Waves in Crystals*, John Wiley & Sons, New York, 1984.
7. B. E. A. Saleh and M. C. Teich. *Fundamentals of Photonics*, 2nd Ed., John Wiley & Sons, New York, 2007.
8. M. Born and E. Wulf. *Principles of Optics* (7th (expanded) ed.), Cambridge University Press, Cambridge, 1999.
9. K. Petermann. "Constraints for fundamental-mode spot size for broadband dispersion-compensated single-mode fibres," *Electron. Lett.*, 19, no. 18, pp. 712–714, 1983.
10. K. Hayata, M. Koshiba, and M. Suzuki. "Modal spot size of axially nonsymmetrical fibers," *Electron. Lett.*, 22, no. 3, pp. 127–129, 1986.
11. F. Villuendas, F. Calvo, and J. B. Marqués. "Measurement of mode field radius in axially nonsymmetrical single-mode fibers with arbitrary power distribution," *Opt. Lett.*, 12, no. 11, pp. 941–943, 1987.
12. Y. Kokubun and S. Tamura. "Precise recursive formula for calculating spot size in optical waveguides and accurate evaluation of splice loss," *Appl. Opt.*, 34, no. 30, pp. 6862–6873, 1995.
13. N. Yamaguchi, Y. Kokubun, and K. Sato. "Low loss spot size transformer by dual tapered waveguides (DTW-SST)," *J. Lightwave Technol.*, 8, no. 4, pp. 587–594, 1990.
14. Y. Kokubun, S. Tamura, and T. Kondo. "Spot size transformer with a type-B antiresonant reflecting optical waveguide," *Opt. Lett.*, 17, no. 24, pp. 1746–1748, 1992.
15. O. Mitomi, K. Kasaya, Y. Tohmori, Y. Suzaki, H. Fukano, Y. Sakai, M. Okamoto, and S. Matsumoto. "Optical spot-size converters for low-loss coupling between fibers and otpoelectronic semiconductor devices," *J. Lightwave Technol.*, 14, no. 7, pp. 1714–1720, 1989.
16. Y. Kokubun and K. Iga. "Formulas for TE_{01} cutoff in optial fibers with arbitrary index profile," *J. Opt. Soc. Am.*, 70, no. 1, pp. 36–40, 1980.
17. L. M. Brekhovskikh. *Waves in Layered Media*, Academic Press, New York, 1960.
18. Y. Suematsu and K. Furuya. "Propagation mode and scattering loss of a two-dimensional dielectric waveguide with gradual distribution of refractive index," *IEEE Trans. Microwave Theory Tech.*, MTT-20, pp. 524–531, 1972.

19. P. Yeh. *Optical Waves in Layered Media*, John Wiley & Sons, New York, 1988.

20. M. A. Duguay, Y. Kokubun, T. L. Koch, and L. Pfeiffer. "Anti-resonant reflecting optical waveguides in SiO_2-Si multi-layer structures," *Appl. Phys. Lett.*, 49, no. 1, pp. 13–15, 1986.

21. T. Baba and Y. Kokubun. "Dispersion and radiation loss characteristics of antiresonant reflecting optical waveguides—Numerical results and analytical expressions," *IEEE J. Quantum Electron.*, 28, no. 7, pp. 1689–1700, 1992.

22. E. A. J. Marcatili. "Dielectric rectangular waveguide and directional coupler for integrated optics," *Bell Sys. Tech. J.*, 48, pp. 2071–2102, 1969.

23. R. Ulrich and R. J. Martin. "Geometrical optics in thin film light guides," *Appl. Opt.*, 10, no. 9, pp. 2077–2085, 1971.

24. E. Snitzer. "Cylindrical dielectric waveguide modes," *J. Opt. Soc. Am.*, 51, no. 5, pp. 491–498, 1961.

25. D. Gloge. "Weakly guiding fibers," *Appl. Opt.*, 10, no. 10, pp. 2252–2258, 1971.

26. Y. Kokubun and K. Iga. "Single-mode condition of optical fibers with axially symmetric refractive index distribution," *Radio Sci.*, 17, no. 1, pp. 43–49, 1982.

27. T. Tanaka and Y. Suematsu. "An exact analysis of cylindrical fiber with index distribution by matrix method and its application to focusing fiber," *Trans. IECEJ*, E59, no. 11, pp. 1–8, 1976.

28. K. Okamoto and T. Okoshi. "Vectorial wave analysis of inhomogeneous optical fibers using finite element method," *IEEE Trans. Microwave Theory Tech.*, MTT-26, no. 2, pp. 109–114, 1978.

29. K. Okamoto and T. Okoshi. "Computer-aided synthesis of the optimum refractive index profile for a multimode fiber," *IEEE Trans. Microwave Theory Tech.*, MTT-25, no. 3, p. 213, 1977.

30. Y. Kokubun and K. Iga. "Mode analysis of graded-index optical fibers using a scalar wave equation including gradient-index terms and direct numerical integration," *J. Opt. Soc. Am.*, 70, no. 4, pp. 388–394, 1980.

31. I. H. Malitson. "Interspecimen comparison of the refractive index of fused silica," *J. Opt. Soc. Am.*, 55, pp. 1205–1209, 1965.

32. J. W. Fleming. "Material and mode dispersion in $GeO_2 \cdot B_2O_3 \cdot SiO_2$ glasses," *J. Am. Ceram. Soc.*, 59, pp. 503–507, 1976.

33. N. Shibata and T. Edahiro. "Refractive-index dispersion for GeO_2-, P_2O_5-, and B_2O_3-doped silica glasses in optical fibers," *Trans. IECE Japan*, 65-E, no. 3, pp. 166–172, 1982.

34. J. W. Fleming. "Dispersion in GeO_2-SiO_2 glasses," *Appl. Opt.*, 23, no. 24, pp. 4486–4493, 1984.

35. D. Gloge. "Dispersion in weakly guiding fibers," *Appl. Opt.*, 10, no. 11, pp. 2442–2445, 1971.

36. K. Furuya, M. Miyamoto, and Y. Suematsu. "Bandwidth of single mode optical fibers," *Trans. Inst. Electron. & Commun. Eng. Japan*, 62-E, no. 5, pp. 305–310, 1979.

37. R. Olshansky and D. B. Keck. "Pulse broadening in graded-index optical fibers," *Appl. Opt.*, 15, no. 2, pp. 483–491, 1976.

38. D. Marcuse and H. M. Presby. "Fiber bandwidth-spectrum studies," *Appl. Opt.*, 18, no. 19, pp. 3242–3248, 1979.

39. D. Marcuse. "Calculation of bandwidth from index profiles of optical fibers, 1: Theory," *Appl. Opt.,* 18, no. 12, pp. 2073–2080, 1979.
40. P. A. Bélanger. "Beam propagation and the ABCD ray matrices," *Opt. Lett.,* 16, no. 4, pp. 196–198, 1991.
41. S. E. Miller. "Coupled wave theory and waveguide applications," *Bell Syst. Tech. J.,* 33, no. 3, pp. 661–719, 1954.
42. Y. Kokubun, S. Suzuki, and K. Iga. "Phase space evaluation of distributed-index branching waveguides," *IEEE J. Lightwave Technol.,* LT-4, no. 10, pp. 1534–1541, 1986.
43. M. Izutsu, Y. Nakai, and T. Sueta. "Operation mechanism of the single-mode optical-waveguide Y junction," *Opt. Lett.,* 7, no. 3, pp. 136–138, 1982.
44. H. Yajima. "Coupled mode analysis of dielectric planar branching waveguides," *IEEE J. Quantum Electron.,* QE-14, no. 10, pp. 749–755, 1978.
45. S. K. Burns and A. F. Milton. "An analytic solution for mode coupling in optical waveguide branches," *IEEE J. Quntum Electron.,* QE-16, no. 4, pp. 446–454, 1980.
46. Y. Weissman. *Optical Network Theory,* Artech House, Norwood, MA, 1992.
47. D. Gloge. "Bending loss in multimode fibers with graded and ungraded core index," *Appl. Opt.,* 11, no. 11, pp. 2506–2513, 1972.
48. K. Furuya and Y. Suematsu. "Random-bend loss in single-mode and parabolic-index multimode optical fiber cables," *Appl. Opt.,* 19, no. 9, pp. 1493–1500, 1980.
49. M. D. Feit and J. A. Fleck, Jr. "Light propagation in graded-index optical fibers," *Appl. Opt.,* 17, no. 24, pp. 3990–3998, 1978.
50. M. D. Feit and J. A. Fleck, Jr. "Computation of mode properties in optical fiber waveguides by propagating beam method," *Appl. Opt.,* 19, no. 7, pp. 1154–1164, 1980.
51. L. Thylen and D. Yevick. "Beam propagation method in anisotropic media," *Appl. Opt.,* 21, no. 15, pp. 2751–2754, 1982.
52. R. Baets and P. E. Lagasse. "Calculation of radiation loss in integrated-optic tapers and Y-junction," *Appl. Opt.,* 21, no. 11, pp. 1972–1978, 1982.
53. Y. Chung and N. Dagli. "An assesment of finite difference beam propagation method," *J. Quantum Electron.,* 26, no. 8, pp. 1335–1339, 1990.
54. W. Huang, C. Xu, S. T. Chu, and S. K. Chaudhuri. "The finite-difference vector beam propagation method: Analysis and assessment," *J. Lightwave Technol.,* 10, no. 3, pp. 295–305, 1992.
55. S. T. Chu and S. K. Chaudhuri. "A finite-difference time-domain method for the design and analysis of guided-wave optical structures," *J. Lightwave Technol.,* LT-7, pp. 2033–2038, 1989.

Index